Roman Woodworking

ROMAN Woodworking

Roger B. Ulrich

YALE UNIVERSITY PRESS
New Haven and London

Published with assistance from the office of the Dean
of the Faculty, Dartmouth College.

Set in Quadraat by Tseng Information Systems, Inc., Durham, North Carolina.
Designed by Gregg Chase.

Library of Congress Cataloging-in-Publication Data
Ulrich, Roger Bradley.
Roman woodworking / Roger B. Ulrich
p. cm.
Includes bibliographical references and index.
ISBN-13: 978-0-300-19984-0 (alk. paper)

1. Building, Wooden—Rome. 2. Carpentry—History—To 1500.
3. Woodwork—Rome. 4. Rome—Antiquities. I. Title.
TH1111.U47 2006
684'.080937—dc22 2006003784

A catalogue record for this book is available from the British Library.

The paper in this book meets the guidelines for permanence and durability
of the Committee on Production Guidelines for Book Longevity
of the Council on Library Resources.

For Imogen

Contents

Acknowledgments

Over the past ten years I have benefited from many conversations with professional colleagues, museum curators, librarians, craftsmen, and students on the topics of ancient and modern woodworking. My first study season in 1994 was immensely boosted with the kind permission of Stefano de Caro and his staff at the Museo Nazionale of Naples, which allowed me to examine iron and bronze tools from Pompeii stored at the museum. In that same year I was permitted to study traces of wood in buildings from Pompeii and Herculaneum by the Soprintendente of Pompeii, Baldassare Conticello. Similar access was granted to me by the Soprintendente of Ostia, Anna Gallina Zevi. The initial results of the fieldwork in Pompeii, Herculaneum, and Ostia were published in 1996 with the encouragement and support of Fred Kleiner, at that time editor of the *American Journal of Archaeology*.

A highlight of a tour of collections in Britain in 1995 was an invitation to examine the Roman tools held in the Verulamium Museum, under the stewardship of the Keeper of Archaeology, Vivienne Holgate. Shortly thereafter Francis Grew of the Museum of London graciously shared his office with me for a day, where I was able to study many Roman woodworking tools in near-pristine condition discovered in London. In 1999, additional work on carbonized doors and furniture was undertaken at Herculaneum with the permission of Conticello's successor, Soprintendente Pietro Giovanni Guzzo. My time at the site included a visit to the storerooms at Herculaneum to examine the singular pieces of carbonized furniture conserved there; these were published that same year in a comprehensive study by Stefan Mols, who would later allow me to include some of his drawings and photographs in the present volume.

I would also like to express gratitude to Christina Huemer and the staff of the library of the American Academy in Rome, and to the staff of the archaeological section of the Biblioteca del Castello del Buonconsiglio in Trento, Italy, as well as to Reinhart Sonnenburg and other reference librarians at Baker-Berry Library at my home institution of Dartmouth College, the last for accommodating near-daily interlibrary loan requests for the better part of two years.

Thanks are due as well to Alessandro Bettagno of the Instituto Nazionale di Archeologia e Storia dell'Arte; Robin Birley of the Vindolanda Trust; Edward Bradley;

Francesco Buranelli of the Vatican Museums; Andrea Bussmann of the Rheinisches Landesmuseum, Bonn; Enrico Cavada; Gill Clark of the British School at Rome (London office); John Clarke; Angelica Condrau of the Schweizerisches Landesmuseum; Francesca DeSimone; Sylvia Diebner of the Fototeca of the Deutsches Archäologisches Institut in Rome; Norman Doenges; Omar Foelsche and Susan Bibeau of Humanities Computing at Dartmouth; Brad Goedkoop of Kingfisher Woodworks; M. Cristina Guidotti of the Soprintendenza per i Beni Archeologici della Toscana; Kathy Hart of the Hood Museum; Hugh Kilmister of the Petrie Museum of Egyptian Archaeology; Joan Mertens of the Metropolitan Museum of Art; Bethan Mogford of the British Museum; John Oleson; Lorenza Mochi Onori of the Soprintendenza per il Patrimonio Storico Artistico e Demoetnoantropologico delle Marche; James Packer; Norm Roberson; Paola Rubino and Valeria Sampaolo of the Soprintendenza per i Beni Archeologici delle province di Napoli e Caserta; Susan Simon of Curricular Computing at Dartmouth; Anna Mura Sommella of the Musei Capitolini, Rome; Roberta Stewart; and Elizabeth Sullivan.

I am grateful to Harry Haskell, Keith Condon, Lawrence Kenney, and the rest of the editorial staff of Yale University Press for the support they have shown for this project. Rabun Taylor evaluated the manuscript for the press and offered many valuable suggestions. Finally, I would like to thank the office of the Dean of the Faculty of Dartmouth College, the Petitt Family Fellowship, and especially the Associate Dean of the Humanities, Lenore Grenoble, for offering financial support to underwrite the publication costs of this book.

Abbreviations of Ancient Sources

Apul. Apuleius, poet, philosopher, and rhetorician, born ca. A.D. 123
Fl(orida), *Met(amorphoses)*, *(De) Mun(do)*
English: Apuleius

Arn. Arnobius (of Sicca), Christian convert, third century
Adversus Nationes
English: Arnobius

Caes. C. Iulius Caesar, consul, general, dictator, 100–44 B.C.
(De) B(ello) Civ(ili), *(De) B(ello) Gall(ico)*, mid-first century B.C.
English: Julius Caesar or simply Caesar

Cato M. Porcius Cato, consul, censor, 234–149 B.C.
(De Re) Rust(ica) (= *De Agri Cultura*), ca. 160 B.C.
English: Cato

Cat. C. Valerius Catullus, poet, ca. 84–54 B.C.
English: Catullus

Cic. M. Tullius Cicero, statesman, philosopher, 106–43 B.C.
(Epistulae ad) Att(icum), *(Pro) Clu(entio)*, *(De) Nat(ura) D(eorum)*, *Orat(or Ad M. Brutum)*, *(In) Verr(em)*, *(De) Inv(entione Rhetorica)*, *(De) Leg(ibus)*, *(De) Off(iciis)*, *Parad (oxa Stoicorum)*, *(Orationes) Phil(ippicae)*, *(Epistulae ad) Q(uintum) Fr(atrem)*, *Tusc(ulanae Disputationes)*
English: Cicero

Columella L. Iunius Moderatus Columella, large estate holder, first century
(De Re) Rust(ica), *De Arboribus*, ca. A.D. 60–65
English: Columella

Curt. Q. Curtius Rufus, rhetorician, historian, first century A.D.
Historiae Alexandri Magni
English: Curtius

Dio Cassius Dio Cocceianus, historian, ca. A.D. 150–235
English: Dio

Enn. Q. Ennius, poet and playwright, 239–169 B.C.
Ann(ales), *scenica*
English: Ennius

Fest. Sextus Pompeius Festus, lexicographer, late second century
De verborum significatu cum Pauli Epitome, Lindsay edition
English: Festus

Gaius Gaius, Roman jurist, second century
Dig(est)
English: Gaius

Hor. Q. Horatius Flaccus, poet, 65–8 B.C.
Carm(ina), *Sat(irae = Sermones)*
English: Horace

Isid. Isidorus Hispalensis, bishop of Seville, sixth–seventh century
Etymologiae (or *Origines*), ca. A.D. 602–36
English: Isidore

Joseph. Josephus, historian, first century
B(ellum) J(udaicum)
English: Josephus

Juv.	Decimus Iunius Iuvenalis, poet, late first–early second century A.D. *Saturae* English: Juvenal
Liv.	Titus Livius, historian, 64 or 59 B.C.–A.D. 12 or 17 English: Livy
Luc.	M. Annaeus Lucanus, poet, historian, A.D. 39–65 *Bell(um) Civ(ile)* English: Lucan
Lucil.	C. Lucilius, poet, ca. 180–102/1 B.C. English: Lucilius
Lucr.	T. Lucretius Carus, philosopher, ca. 94–55 B.C. *de Rerum Natura* English: Lucretius
Man.	M. Manilius, poet, early first century *Astronomica*, early first century English: Manilius
Mart.	M. Valerius Martialis, poet, ca. A.D. 40–ca. 104 *Epigrammata*, mid-first century English: Martial
Ov.	P. Ovidius Naso, poet, 43 B.C.–ca. A.D. 17 *Ars Am(atoria)*, *(Epistulae ex) Pont(o)*, *Met(amorphoses)* English: Ovid
Palladius	Rutilius Taurus Aemilianus Palladius, poet, fourth century English: Palladius
Paul.	Paulus Diaconus, historian, eighth century *(Epitoma) Fest(i)* English: Paulus
Petron.	Petronius Arbiter, writer, first century *Sat(yricon)* English: Petronius
Plaut.	Titus Maccius Plautus, playwright, died ca. 184 B.C. *Aul(ularia)*, *Mil(es)*, *Mostell(aria)*, *Per(sa)*, *Poen(ulus)*, *St(ichus)* English: Plautus
Plin.	C. Plinius Secundus, historian, ca. A.D. 23–79 HN (*Naturalis Historia*), completed A.D. 77 English: Pliny (the Elder)
Plin.	C. Plinius Caecilius Secundus, writer of literary letters, ca. A.D. 61–ca. 112 *Epist(ulae curatius scriptae)*, published between A.D. 101 and 109 English: Pliny (the Younger)
Prop.	Sextus Propertius, poet, first century B.C. English: Propertius
Scaev.	Q. Cervidius Scaevola, Roman jurist, late second century *Digesta* English: Scaevola
Sen.	L. Annaeus Seneca, philosopher, poet, ca. 5 B.C.–A.D. 65 *Ep(istulae)* English: Seneca the Younger
Serv.	Maurus (Marius) Servius Honoratus, grammarian, fourth–fifth century *(In) Verg(ilium Commentarius)* English: Servius
Stat.	P. Papinius Statius, poet, ca. A.D. 45–96 *Silv(ae)* English: Statius
Strab.	Strabo, geographer, historian, ca. 64 B.C.–A.D. 21 *Geo(graphy)* English: Strabo
Suet.	C. Suetonius Tranquillus, biographer, ca. A.D. 70– ca. 140 *De Vita Caesarum* English: Suetonius
Tac.	Cornelius Tacitus, historian, born ca. A.D. 56 *Ann(ales)*, *Hist(oriae)* English: Tacitus
Tert.	Q. Septimius Florens Tertullianus, early Christian writer, ca. 160–ca. 240 *Apol(ogeticus)* English: Tertullian

Theophr. Theophrastus, Greek philosopher, ca. 370–ca. 285 B.C.
Hist(oria) Pl(antarum), *(De) Caus(is) Pl(antarum)*
English: Theophrastus

V. Flaccus C. Valerius Flaccus, poet, died ca. A.D.92
Argonautica
English: Valerius

Varro M. Terentius Varro, lexicographer, 116–27 B.C.
(De) Ling(ua Latina), *(Rerum) Rust(icarum Libri III)*, both published between ca. 43 and 37 B.C.
English: Varro

Verg. P. Vergilius Maro, poet, 70–19 B.C.
Aen(eid), *Georg(ics)*, late first century B.C.
English: Virgil

Vitr. Vitruvius Pollio, architect and engineer, first century B.C.
De Architectura
English: Vitruvius

Note: Citations in the text not listed here follow the format used in the *Oxford Classical Dictionary*.

Roman Woodworking

1 Introduction

The riches of the earth were hidden for a long time, and the trees and forests were considered to be the supreme gift bestowed by her to mankind. These first nourished him, their foliage cushioned his cave and their bark clothed him; there are still races that live this way.—Plin. *HN* 12.1[1]

Wood was arguably the most valuable natural resource utilized by the peoples of the ancient Mediterranean. Wood was a primary, and in some cases the only, component of tools, housing, household implements, modes of transportation, containers, and scaffolding. Many large public structures, from bridges to theaters, were built of wood. Wood was used as well for dyes, waterproofing materials, and pipes; it provided the sole source of energy for cooking, heating, smelting, and firing clay.

The cultivation, harvesting, transportation, and working of wood in a world without heavy machinery or power tools must have employed many people. Some of these were highly skilled. In the Roman world, the accomplished woodworker was a craftsman capable of carving fine moldings, inlaying and veneering exotic species imported from distant corners of the empire, and fastening disparate pieces of wood with an admirable variety of joints requiring no metal hardware.

Framing carpenters developed an intuitive knowledge of the structural capacities of different species of wood and, in departure from their Greek counterparts, recognized and exploited the enormous strength of trussed roofs and decked floors. Indeed, Roman builders understood that, with the exception of the dome, spaces just as vast and unencumbered by interior columns as those covered with barrel or groin vaults could be spanned with wood. Thick wooden beams were vital for the construction of scaffolding, construction forms, and lifting cranes, without which the most monumental of masonry buildings could never have been built.

Transportation in the Roman world depended upon the skills of the woodworker. Wood was the primary material used for all manner of carts, wagons, wheels, and, of course, merchant ships. Roman legions depended upon ready supplies of timber for military success. As a resource that was readily procurable in virtually all areas of the Mediterranean, wood was employed by the military for bridges, fortifications, and siege machines, including heavy artillery.

Timber not destined for the sawyer was employed for fuel. Without this single most important energy source, the Roman baths would have been inoperable, iron and bronze nonexistent, pottery, roofing tiles, and fired brick unobtainable.

The importance of wood in the Roman world is easy to overlook for an obvious reason. Modern discoveries of wood in Roman contexts are extraordinary; the survival of wooden artifacts requires unusual circumstances of environment. Carbonized beams, furniture, and everyday implements have been found embedded in the volcanic mud of Italy's Herculaneum. Plaster or concrete casts of decomposed wooden objects that have left imprints in fallen ash at Pompeii and Oplontis allow us to view ghostly images of wooden artifacts. The wet subsoils of northern Europe and Great Britain have preserved waterlogged timbers and tools; shipwrecks also yield valuable information. At the opposite end of the environmental spectrum are the dry climates of Egypt and the Middle East, where wooden objects from many historical periods have survived to the present day. While some of the most important wooden artifacts recovered in modern times have been lost—the immense wooden barges of Lake Nemi in Latium are a tragic example—other spectacular finds continue to be reported, and, just as important, the methods, if not always the money, now exist to preserve them for future study.

The scattered archaeological recovery of wooden artifacts and the evidence these finds provide are enhanced by related artifacts, such as tools, metal fittings, and depictions of tools, buildings, carts, boats, and wooden implements on wall frescoes, relief sculpture, and mosaics. The rectangular cavities left by wooden structural members in the walls of brick and stone structures can still be seen, as can the imprints of wooden planks used to frame vaults of concrete.

No less valuable for the history of Roman woodworking is the literary record. The importance and pervasiveness of wooden objects in Roman life are clear from the many references preserved in the extant corpus of Latin—and contemporary Greek—literature. Some of these references are technical and specific; best known is the commentary of Vitruvius, an architect and military engineer loyal to both Julius Caesar and, later, Octavian. Vitruvius devoted himself to the composition of his famous treatise *The Ten Books of Architecture (De Architectura)* at the dawn of the Roman empire, in the late first century B.C. Vitruvius's discussion of building methods, materials, and kinds of structures affords a wealth of terminology that would otherwise be lost, even if the ideals he espouses, in the form of proportional relationships or orientations of buildings and their constituent parts, seem to have been ignored by ancient builders more often than not.

For the types of trees and their uses Pliny is our most important Latin source. Although no botanist, Pliny studied and closely followed the works of pioneering Greek predecessors. No doubt the most important of these was the philosopher Theophrastus (ca. 370–285 B.C.), a brilliant and favored pupil of Aristotle. Theophrastus inherited his mentor's garden (not to mention the Lyceum itself) and wrote two books about trees and other plants (*Historia Plantarum* and *De Causis Plantarum*). Many, if not most, of Theophrastus's observations were based upon notes made by Aristotle, other written sources, and even student researchers. Born four centuries later, Pliny (A.D. 23–79) devoted six books of his *Natural History* to commentary on the typology of trees, woody plants, and their uses; much is derived directly from Theophrastus, but Pliny also provides a distinctly Roman perspective. Despite the achievements of such early naturalists, in the absence of canonical and universal methods of illustrating and describing plants, the transmission of botanical knowledge was severely hampered (Baker 1978, 20), and the modern scholar is commonly frustrated in trying to identify a specific tree when confronted with an ancient reference (Rackham 1996, 38).

Such shortcomings notwithstanding, a book such as this would be far the poorer without any access to the ancient voice. As Russell Meiggs pointed out in his landmark study *Trees and Timber in the Ancient Mediterranean World*, "There are indeed very few Greek or Latin prose authors who will not yield useful information (about trees and their uses) if diligently searched" (1982, 32). In the glossary attached to this book the focus is upon Latin terminology, and, as will be evident, the language of the Romans for the practices of carpentry was as rich as the applications of the craft were varied. Poetic sources are also important for terminology. Poetic metaphor can be used to reveal the function of an otherwise obscure technical term, despite Meiggs's caution that specific words may be substituted for one another to satisfy the needs of meter or literary allusion.

Interest in the technical aspects of Roman construction over the past century has been keen, to the degree that certain areas of ancient construction, such as the marble trade, have become subspecialties within the larger field of classical archaeology. The majority of attention has been focused upon masonry construction and the classification of Roman concretes and their various facings. Nevertheless, Roman woodworking has not been ignored. This particular field has enjoyed closer scrutiny over the past few decades as archaeologists have better learned to recognize and preserve fragile wooden artifacts, or at least document the remains carefully before they are lost. An increasing number of archaeological reports include an evaluation of the species of trees used for wooden artifacts; even microscopic traces of wood can now be analyzed and identified.

An early inspiration for this book was Hugo Blümner's four-part *Technologie und Terminologie der Gewerbe und Künste bei Griechen und Römern*, published between 1875 and 1887. Blümner's encyclopedic approach included documentation of Greek and Latin technical terminology of all types of Greek and Roman technology; he was also an acute observer of the physical remains extant during the mid-nineteenth century. Other modern scholars were interested in famous structures for which ancient descriptions exist but no actual remains have been discovered. Two examples come immediately to mind: Julius Caesar's bridge constructed over the Rhine River and the basilica Vitruvius himself built on the edge of the civic forum of ancient Fanum, a Roman port town on the Adriatic. Caesar's bridge was a structure built entirely of wood, while at Fanum wooden beams played an important role in the roofing of the basilica, as they would in virtually all similar buildings built over the course of the empire. From the ancient descriptions that have survived, precise for their time while elusive for the modern student, laborious analysis (for example, Saatman, Jüngst, Thielscher 1938) and ingenious restorations have been proposed.

Our knowledge of the timber trade and the types of trees used by Roman woodworkers was considerably enhanced by Meiggs's aforementioned book on trees and timber published in 1982. Meiggs was less concerned about technique than with the history of the timber trade in both Greek and Roman settings. His study includes not only an impressive compendium of ancient sources that address the topic of ancient woodworking, but vital commentary on the species of trees used and on the ambiguity that characterizes the categorization of trees provided by ancient writers of Greek and Latin.

The topic of ancient woodworking techniques has been addressed primarily by European scholars. Perhaps the most influential of these studies has been that of Jean-Pierre Adam, whose book on Roman building techniques was recently (1994) translated from French into English and published in the United States. Lesser known but also full of useful analysis, written from an engineer's point of view, is C. Giuliani's *L'edilizia nell'antichitá* of 1990. Both Adam and Giuliani, like their predecessors, treat woodworking primarily in the context of architecture and only as part of a larger overview of ancient building techniques.

Specialized tools developed for the woodworkers' trades during the Greek and Roman periods in many cases survived unchanged in form and function until the early twentieth century. Recent studies of Roman tools, such as Wolfgang Gaitzsch's *Eiserne römische Werkzeuge* (1980), have added new finds and analysis to a long tradition of interest in this subject, including the groundbreaking studies of W. M. F. Petrie (1917) and William Goodman (1964). Actual Roman tools, some in a pristine state with their original handles, can be seen in the regional museums of Europe and Great Britain.

Ongoing excavations in Britain, and especially London, have exposed tons of waterlogged timbers from the Roman period that have survived from both boats and buildings. This wealth of material has inspired a flurry of new research by archaeologists like Damian Goodburn and Peter Marsden. The sheer volume of the material discovered here and elsewhere (such as the prehistoric lakes region of northern Italy) means that there is much still awaiting careful study.

This book treats Roman woodworking from a broad perspective, building upon the works of predecessors and offering new analysis and evidence. Greater emphasis will be placed on tangible evidence than on speculative reconstructions. The topic is a vast one, and it would be presumptuous if not impossible to cover every dimension of the woodworker's trade in a single volume. At an early stage in the project it was decided not to integrate the subject of shipbuilding as a discrete topic. While some of the techniques of joinery and tools used by shipwrights are included in the study presented here, the techniques of lofting, constructing, and outfitting wooden ships are complicated enough to deserve a comprehensive treatment of their own. Even with these self-imposed limitations, the present volume aspires to examine the language and the practice of Roman woodworking through the types of literary and archaeological evidence just described, in a way that a classicist, historian, or modern woodworker can understand. The main text of the book consists of an overview of the tools, raw materials, and applications of ancient woodworking as well as the careers of its practitioners. To parallel the main text a comprehensive glossary has been organized so that words connected with woodworking and the timber industry can be consulted in either Latin or English. Definitions fall under the Latin terms, where they are known; English equivalents are cross-indexed. Each definition concludes with a sampling of quotations from Latin authors to illustrate how the term was used in an ancient literary context. Latin authors are listed alphabetically according to the conventional abbreviations of the *Oxford Classical Dictionary*.

II The Roman Woodworker

He distributed them [the body of the people] accordingly, by arts and trades, into musicians, goldsmiths, carpenters, dyers, leatherworkers, curriers, braziers, and potters. The remaining trades he grouped together, and made one body out of all who belonged to them. —Plutarch, *Life of Numa* 17

Plutarch, writing in his native Greek about the quasi-legendary second king of Rome, attributed the organization of skilled workers into associations, or guilds, to the earliest generations of Romans. Plutarch himself lived and wrote over seven hundred years after Numa was thought to have ruled (at the end of the eighth century B.C.), but the historian's notion that skilled workers were recognized and organized from a very early period was shared by other Latin writers such as Pliny. From such literary passages as well as inscriptions we know that Roman woodworking belonged to a larger group of manual skills practiced by craftsmen; Latin speakers called these skills arts (*artes*).

The practitioner of an *ars* was a person who, through training and experience, demonstrated a systematic knowledge of a technique to the point where he was able to make a living from it. The English derivative of *ars*, "art," can be similarly used in this broader sense. We refer to someone adept at the "art of diplomacy" or the "art of playing golf" to indicate one who has demonstrated a certain level of mastery. So, too, the Romans would refer to one skilled in the military arts (*artes militares*) or in the art of public speaking (*ars dicendi*) or practiced in poetry (*ars musica*). In these and other examples, the noun *ars* was paired with another word to specify the expertise in question.

The word *ars* was not used so freely that its emphasis on a learned skill was debased or trivialized. We do not expect, in modern English usage, to describe the digging of a ditch, in most circumstances, as artful, and Romans would have similarly drawn a distinction between what we may think of as skilled and unskilled labor. More specifically, we can understand through an examination of the Latin found in literary sources and funerary inscriptions that there were a number of subspecialties included within

Fig. 2.1. Two Roman legionaries drag a log from the forest, using a rope sling. From Column of Trajan, Rome (A.D. 113). (Anger, D.A.I. 91.112)

the practice of the woodworker. The slave or infantryman who dragged a log from the forest (fig. 2.1) need not have undergone the same training for his livelihood as the one who shaped the wood to make a finely crafted chest (the profession of the *arcularius*).

Such distinctions between skilled and unskilled can certainly be made, for the most part, on the basis of common sense and our own modern conception of craftsmanship. We would be straying too far, however, to infer any marked difference in social station between the unskilled and the skilled. For whether the worker was practicing a learned trade (an *ars*) or acted simply as a laborer in support of that trade made little difference on the global scale of Roman social standing. Highborn Romans and even rank-and-file free-born Romans with income were no more likely to become skilled carpenters than they were to choose professions in the fields of medicine, acting, or painting. Most Roman woodworkers, like their counterparts in other crafts such as metalworking, leatherworking, weaving, glass blowing, or inlay work, were slaves (*servi*), freed slaves (*liberti*), or landless freeborn plebeians. On the other hand, among these common folk it is easy to imagine a stratification based upon learned skills (*artes*), guild (*collegium*) memberships, and official positions within the guilds themselves (I shall return to the role of the guilds shortly).

That the Roman woodworker, like other craftsmen of the Roman world, took pride in his work is clear, even without the direct evidence of written testimony. Despite the disregard a Roman aristocrat like Cicero might have displayed for one who worked in a trade (although he nevertheless distinguished between the manual laborer and the skilled artisan),[1] those humble men who earned their living through the skill of their hands could take some comfort in remembering the names of predecessors celebrated in myth: legendary heroes like Daedalus, the *primus artifex* who designed the mysterious labyrinth on Crete but also was credited with inventing many of the very tools used by the woodworker, or the Greek Epeios, without whose skill (for it was he who designed and built the wooden Trojan horse) Troy would not have been taken. Homer imagines Odysseus himself describing the processes and tools by which he made his marriage bed from olive wood (*Od.* 23.195). The implications are that he who mastered his *ars*, at least in the Greek world, enjoyed some measure of respect. It follows that many of the names of artisans preserved on Roman-period funerary inscriptions are those of Greek freedmen.

From the archaeological record, nothing speaks more eloquently than the surviving funerary monuments that depict deceased Roman woodworkers with their tools. In the chapters that follow the evidence of these carved stones will be invaluable in reconstructing the appearance of the ancient woodworker's range of tools. Beyond this, however, the very fact that these images were carved so prominently on tombstones and altars and that they were sometimes the only images carved on such monuments asserts unequivocally that the skilled ancient artisan wished to be remembered as such by those that survived him.

Woodworkers were, like other traditional craftsmen (*artifices*, *fabri*), individuals who earned their living through handiwork. The product of their trade was a physical object, always utilitarian and occasionally of great aesthetic value as well. The *faber* exercised his corresponding *ars fabrica*, the creative skill required of his trade. In Roman towns where woodworking was of paramount importance—Rome's port town of Ostia is a good example—the term *faber* was synonymous with one who worked with wood. Nevertheless, the word *faber*, like "craftsman," in fact alerts us to the presence of a skilled worker but tells us little about the material being worked (for example, wood), let alone the specialty (a cabinetmaker). Thus the *faber ferrarius* is a man whose *ars fabrica* lies in the field of blacksmithing. The skilled carpenter was known by the general term *faber tignarius*, literally, a "beam craftsman," after the general term for a beam of wood (*tignum*).

From literary references, funerary epitaphs, graffiti, and dedicatory inscriptions it is possible to recover the names of a significant number of the subspecialties connected with woodworking; doubtless the titles of other specialists will never be known. The professions inscribed upon tombs are the most important, providing terms that occur nowhere else in extant Latin. The *faber carpentarius*, for example, from which the English word "carpentry" has been passed down to us, was a woodworker who specialized in the construction of the two-wheeled cart, or *carpentum*. A funerary inscription from Puteoli (*CIL* 10.1922) names a certain Gaius Atilius Fortunatus, whose trade was interior woodwork (*faber intestinarius*), someone more of a cabinetmaker than a common hammer-and-nails framer. *Fabri citrarii* specialized in working with inlays of exotic woods, among which the most precious was citrus or Thyine wood, a north African species, from which they derived their title. The *faber plaustrarius* manufactured wagons, the *faber pectinarius* crafted combs from close-grained boxwood, and beds and couches were fashioned by the *faber lectarius*.

From this brief sampling of subspecialties within the larger field of woodworking (for complete listings refer to the glossary), it is clear that the craft of woodworking was highly evolved, practiced by individuals focused upon specific applications. Our evidence for the lives of individual craftsmen in woodworking—or any ancient trade—is notoriously poor, usually limited to the terse lines of modest funerary plaques. Thus it is impossible to say how frequently, if ever, an individual craftsman practiced more than one specialty; epitaphs list only one, if any. From the little we know about the lives of such men and of the organization of ancient workshops, however, it is most likely that individual craftsmen devoted the majority of their energies to the manufacture of a single object or a closely related group of objects. Thus the *arcularius*, technically a "chest-maker," might have made all other types of furniture that were based upon the shape of a box with hinged doors or lids, like a cabinet for clothes (*armarium*). Tools were expensive, handwork was slow, and physical space for a workshop limited—craftsmen tended to work in a room of their dwelling place, if the excavations of Pompeii are any guide.

There is no evidence other than the documentable existence of professional guilds (*collegia*, *corpora*, and their meeting places, *scholae*) that woodworkers worked together in large numbers. The guilds provided members with social activities like feasts, a religious affiliation to a patron deity, and a promise of a decent burial. The collective purchasing power of the guilds resulted in some finely carved dedications to protecting gods. For example, the college of sawyers (*sectores materiarum*) of Aquileia, a thriving Roman colony in northern Italy where forests were intensively harvested, offered sacrifice to the rustic deity of the forests, Silvanus (*CIL* 5. 815). In Rome members of the *collegium fabrum tignuariorum* ("those who work with beams of wood," or "carpenters") set up an altar to their patroness Minerva, goddess of crafts; the stone surface of the dedication is intricately carved with the tools of their livelihood (fig. 2.2). While such guilds were a feature of artisan life in every Roman town, there is no indication that they oversaw standards of quality or involved themselves with issues of fair compensation. Indeed, we cannot even be sure that a given *collegium* restricted its membership to practitioners of a particular trade, despite its proclaimed affiliation (Cozzo 1928, 79); modern parallels come easily to mind.

Only the largest projects—military and commercial shipbuilding and public construction—would have required correspondingly high numbers of trained, coordinated woodworkers, and even these were probably manned by free men subcontracted for the purpose. Most Roman woodworkers, like other ancient tradesmen, tended to work in small workshops staffed by one or more master craftsmen and their apprentices, the latter most commonly sons of the master, his relatives, free citizens without land, or slaves (Burford 1972, 88). A particularly successful workshop is depicted in a battered relief that was found in Rome shortly before the Second World War, once part of a monumental altar dedicated to an unidentified deity (fig. 2.3). On the left side of the panel a representation of the goddess Minerva stands next to a male figure who may be the owner or master of the workshop. The center and right portions of the relief de-

Fig. 2.2. Marble altar dedicated to Minerva, probably paid for by members of the collegium fabrum tignuariorum. *Carpenters' tools dominate the lower half of the relief (from left): a bucksaw with toggle-stick, a large crosscut saw with handles on each end, a double-bladed axe, a small pickaxe, and an adze. H: 92 cm; W: (at base): 55 cm; D: 54 cm. Found in the Church of S. Giorgio in Velabro. Capitoline Museums, Rome, inv. 1909. Dated to the Augustan period (late first century B.C.—early first century). (Author [cast from the Museo della Civiltà Romana, inv. 3417])*

pict six craftsmen surrounded by their tools and workpieces. A group of three on the far right adds the finishing touches to a three-legged table resting upon a workbench. I will return to this relief for its important representations of woodworkers' tools.

Interestingly, modern Italy leads the rest of Europe in terms of those who regard themselves as self-employed, and the continuing tradition of sons following their fathers in trades echoes the old Roman way. The plethora of small workshops might lead one to expect a high degree of variation and creative impulse reflected in the wooden objects produced, but, as we shall see, one is struck by the general uniformity of work over time. As in the practice of other crafts in the ancient world, woodworkers probably worked in the vicinity of one another, in a particular quarter of a given town, where they saw one another's work and perhaps subcontracted subassemblies of their projects. The man who owned a lathe and knew the skill of turning wood (a *tornator*), for example, may well have performed tasks for a larger number of furniture makers.

The sudden destruction of the Campanian cities of Pompeii and Herculaneum has provided students of the Roman world a snapshot glimpse of Roman life as it existed in a prosperous seaside Italian town of the first century A.D. Excavations in the

*Fig. 2.3. Marble relief of a woodworkers' (furniture) shop, originally from an altar perhaps dedicated to Minerva, from Rome. Tools (from left): a frame saw attached to an apparatus, perhaps for ripping stock; a carpenter's square (*norma*), large calipers (*circinus*), and a bucksaw (*serra*). A lathe may be depicted on the stand in the center left of the relief and also before the seated figure to the right. Two craftsmen work on a three-legged table, visible on the right. L: 1.38 m, max. H: 0.58 m. Capitoline Museums (Montemartini) Rome, inv. 2743. Late first century(?). (Author)*

late 1920s unearthed a modest house now known as the House of the Craftsman (Casa del Fabbro, Pompeii I 10, 7), so named because of the large number of hand-tools found inside the dwelling. The tools and layout of the house offer a glimpse of how the small, independent workshop may have functioned at Pompeii.

Like other single-family Roman houses of the period, the Casa del Fabbro combined an assortment of interior rooms that lead to an enclosed garden, roughly square, that was open to the sky. The garden was bordered on one side by an airy covered porch, or portico (9.20×3.30 m). It was here that the majority of woodworking tools were found. These included fourteen iron chisels and gouges, six files, an adze-hammer, three hatchets, a saw, and a compass. Other rooms of the house yielded a folding foot-rule, drill bits, locks, bone appliqués (presumably for furniture), and latches. The main room of the house (the atrium) contained three heavy wooden chests (*arcae*) and a well-built armoire (*armarium*) with grated doors. Indeed, while this was one of the most modest houses found on the block, it held the richest assortment of tools and furniture.

The number and nature of the discoveries led the excavators to believe they had unearthed a craftsman's workshop, perhaps that of an *arcularius*, a maker of chests and furniture (Elia 1934, 292). If this interpretation is correct, then we can imagine such craftsmen working inside their homes, utilizing certain areas, such as the portico and open-air garden, as the "industrial" part of the residence. The precise function of the house may never be known. A large number of agricultural and medical tools were discovered here as well, prompting speculation that the house was in fact owned by either a tool merchant, a scrap metal dealer, or a jack-of-all-trades (Ling 1997, 162). Since the fine interior wall paintings had been allowed to fall into disrepair during the final years before destruction, it is thought that the last owner of the house was a man of modest means (Ling 1997, 150).

Learning a manual trade such as woodworking required apprenticeship; there were no ancient equivalents of technical academies. By the early fourth century Roman tradesmen were required to belong to their appropriate guild; this was one way the government could locate practitioners and impose the dreaded *collatio lustralis*, a crushing tax imposed every five years on merchants, moneylenders, and craftsmen. By this period the individual had lost even his freedom to switch occupations; the sons of Roman craftsmen were virtually bound to follow in their fathers' footsteps.

Romans of any social status did not choose to become woodworkers unless grave misfortune led them to apprentice themselves or their sons. Under such circumstances it is easy to imagine that the level of skills among woodworkers varied enormously. Yet from the evidence that has survived, it is clear that the master woodcrafter flourished alongside those who worked with precious metals, ivories, and other exotic materials so highly valued by the wealthy patron. Most forms of woodworking did not belong to the "higher" arts of painting, sculpture, or architecture. Without funerary inscriptions and chance literary mention, names of craftsmen like Gaius Atilius Fortunatus, Baebius Tertius, and Marcus Allius Apollonius would have remained forever lost. It is unlikely we will ever know the identities of the highly skilled men who made the towering doors of paneled cypress for the Temple of Artemis at Ephesus so praised by Pliny (HN 16.215) or the coffered ceilings of the Temple of Castor in the Roman forum eyed by Cicero while he stood in the forum and delivered his oration against Verres (Cic. *Ver.* 2.1.133).

III The Tools of the Trade

Like their modern counterparts, the tools of a Roman woodworker were precious assets. Each handmade instrument represented a significant investment. Some, as we shall see, were implements crafted with decorative flourishes to have their own aesthetic appeal. We have already noted that those Roman craftsmen who could afford a fine tomb included carved representations of their tools next to their names, yet the tools themselves were not generally included as grave goods. Presumably these were handed down from father to son as an important part of a simple estate.

Woodworkers' tools can be divided into two general categories. First are the measuring and marking devices: rulers, compasses, bobs, squares, levels, and chalk lines. Some of these were also employed in other trades, most notably that of the mason (fig. 3.1). Measuring tools were often fashioned from bronze, as strength was not an important factor and bronze provided a good surface for etching measuring marks. Cutting tools that chopped, sawed, smoothed, bored, and shaped wood fill the second category. The group includes various adzes, axes, saws, planes, knives, drills, files, rasps, and chisels; these implements consist of one or more cutting surfaces and a handle. In the historical Roman period, the blades were invariably of iron, which holds a sharper and more durable edge than bronze, and the handle of wood or some other organic material such as bone. Finally there are various tools belonging to neither category but frequently employed by the woodworker, including mallets for use with chisels, hammers for driving nails, clamps, and wedges.

Few tools from any single site have survived for modern study because they were rarely discarded. Blades could be resharpened or ground multiple times, handles could be replaced, and when there was little useful material left, the metal could be melted down and reused. Even at sites where a great number of finds might be expected—the buried cities of Pompeii and Herculaneum are the best-known examples—tools in good condition are few. In the Campanian cities just mentioned, for example, iron implements are badly corroded; thus the assemblage of cutting tools from here is paltry (fig. 3.2). Bronze measuring tools, in contrast, have been found in near pristine condi-

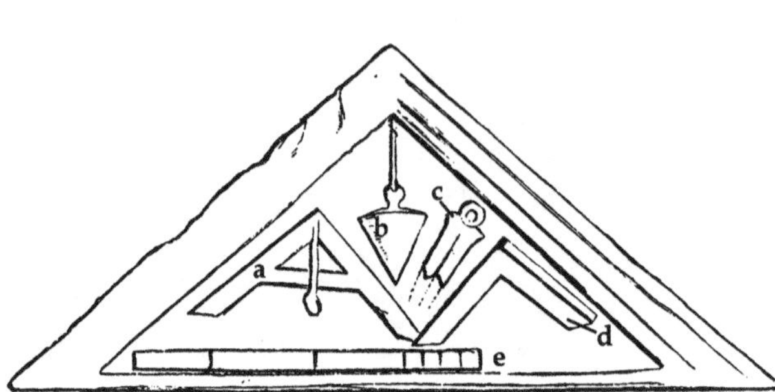

Fig. 3.1. A grave relief of Marcus Aebutius, freedman of the gens Aebutia, showing measuring tools used by both the mason and the carpenter: (a) a level (libella) with attached plumb line; (b) a plumb bob (perpendiculum); (c) compasses (circinus); (d) a square (norma); (e) a ruler (regula). Total H: 1.20 m. Travertine, originally from the Villa Mattei, Rome, now in the Capitoline Museums, inv. 2122. Second–third century. (After Blümner 1884 (3): 91)

tion. As hand-tools were both portable and valuable objects, we can imagine that many refugees from Pompeii and Herculaneum pocketed as many of them as they could carry.

There are additional worthy exceptions. We have already seen that the so-called Casa del Fabbro discovered at Pompeii was well stocked with woodworking tools when it was excavated, even though there was evidence that looters had broken into the house soon after the volcanic eruption in order to recover its most valuable furnishings. A discovery of a cache of tools in a house in Roman Augst (Germany) in 1964 unearthed a collection of twenty-two woodworking tools, including hammers, adzes, a rare rasp, plane blades, and chisels (Mutz 1980). Other fine examples of individual tools have been found in the Roman legionary outposts of northern Italy (for example, Aquileia), Germany (for example, Saalburg), and England (for example, Caerleon).

Among the best preserved of Roman woodworking tools are those which have been found in Roman Britain; important collections include those in the Museum of London, the British Museum, at Verulamium, and at Caerleon. Many of the tools have been discovered in waterlogged conditions in which oxidation has not ravished the metal surfaces. The tradition of working with wood and an abundance of fine timber in English lands may help to explain the impressive assemblage of tools found there.

Since finds of Roman tools tend to be haphazard, studies of a particular class of Roman tool, woodworking or otherwise, tend to draw upon examples found throughout the Roman world. (The appendix lists the tools that were studied for the following discussion as well as their find-spots and current locations.) From the relatively small samples of Roman tools that have been recovered, there does not seem to be a wide variation among a given tool type between different provinces (Gaitzsch 1980, 259). Within the period of the Roman empire, there is also little change over time in terms of physical form. Some tool types are quite rare. Fewer than twenty planes have been discovered in the entire Roman world. Twisted drill bits have been discovered at only a handful of sites. Since woodworking tools were long-lasting, portable, and disseminated widely by the carpenters who traveled with the Roman legions, we can assume that a given type of tool was, by the period of the high Roman Empire, known and used wherever the craft of woodworking was practiced.

Fig. 3.2. A plane recovered from the excavations of Pompeii. The iron elements of the tool have severely corroded. 21.3 cm long by 6.2 cm wide. Naples Museum, inv. 71964. First century. (Author, reproduced by permission of the Soprintendenza per i Beni Archeologici delle province di Napoli e Caserta)

We do not have to rely only on the finds of actual Roman tools to know what they looked like and how they functioned. As important as the discovery of actual tools is, much can be learned from the literary record and from artistic representations. Roman sources, among them Cato, Columella, Pliny, Varro, Vitruvius, and the Christian-period (seventh century) Isidore allow for a basic reconstruction of the names given to various tools and in some cases provide information concerning specific use. Unfortunately no treatise from the hand of an experienced woodworker survives, and our Latin sources do not divulge the specialized vocabulary that distinguishes within a certain class of tools (for example, different kinds of planes). In his *Origines*, Isidore comes the closest to giving specific definitions of tools according to their Latin names.

Ancient depictions of woodworking tools, both by themselves and in use, offer an invaluable perspective. As symbols of their profession, their pride, and their very identity, Roman woodworkers (as well as craftsmen from other fields) often had representations of their hand-tools carved in low relief on their tombstones. Such representations not only verify what has been found through archaeology, but also portray images of implements (for example, large rulers, levels, and sawing machines) that have never been found or otherwise identified through excavation. Components missing from actual finds, such as wooden handles or bows (in the case of the drill) can be restored on the basis of the representations. Finally, depictions of tools in use provide invaluable information about how implements were held and in what applications they were used.

It is through such chance archaeological finds, literary references, and artistic depictions that a fairly complete picture of the Roman woodworker's toolkit can be reconstructed. There remain nonetheless many unanswered questions. No complete depiction exists, for example, of a Roman lathe. One way of filling in such gaps is by examining tools and processes from post-Roman periods in Europe. Wolfgang Gaitzsch,

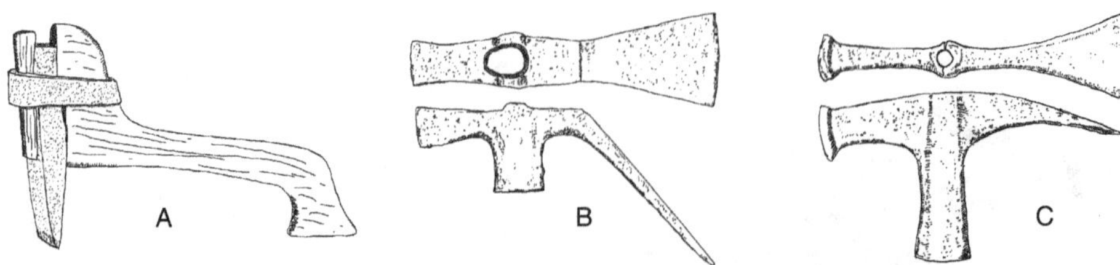

Fig. 3.3. The adze. (A) A flat blade is held in place by an iron collar and wedge. Two examples of the adze-hammer in the British Museum: (B) London, late first century, L: 22 cm; (C) Pakenham, Suffolk, 18.3 cm long, n.d. (Author—[A]: after Goodman 1964, 26; [B], [C]: after Manning 1985, pl. 9, B16, B17)

who has published the most comprehensive study of all Roman iron tools in recent years, suggests that the forms of such tools were fully developed in Roman times and did not change until the machine age (1980, 256). One only has to look at the form of a nineteenth-century wooden frame saw and its Roman counterpart depicted on a grave relief, or at an iron claw hammer found at Pompeii, to appreciate the remarkable continuity of form.

Roman Tools and Their Characteristics

THE ADZE (*ascia*)

Next to the axe, the adze is the oldest of woodworking tools. The sharp blade is attached so that the cutting edge is oriented to the handle like the blade of a hoe or mattock. Depending on the type of work to be performed, adzes were operated by one or both hands, heavier blades being attached to longer handles. Metal adze blades can be categorized into two major groups. The simplest (and older) blades look like thin metal wedges or even broad chisels. These unsocketed blades were attached to a bentwood handle and secured with a wooden wedge and sometimes a metal collar; the wooden elements of the tool have usually long disappeared (fig. 3.3). This simple form was known throughout the ancient Mediterranean and is documented from Sicilian contexts dating back to the Copper Age (Petrie 1917, 16); the basic assembly was derived from even earlier ancestors employing stone blades that were in use throughout Europe and Egypt from at least Neolithic times. A Roman-period variant in the form of a small double-handled adze has been identified by Gaitzsch as an *ascia-hobel* or "adze-plane." As its name suggests, it was a hybrid tool used for close smoothing work (figs. 3.4, 3.5, 3.6).

The second category of adze blades is characterized by blades with an attached socket. The simplest of these features a single blade, but Roman tools also commonly combined the adze blade with a second tool forged on the opposite side. These include the adze-hammer, with a cutting edge on one side and a striking head on the other, and the adze-axe (see the discussion of axes below). Variations in the design of the adze are

Fig. 3.4. The adze-plane depicted on the *funerary relief* of P. Ferrarius Hermes, from the vicinity of Pisa. In the depiction, the blade has been turned ninety degrees so that it is visible as if from the top. Note also the one-foot ruler marked with four digiti, plumb line, and square. A (boxwood?) comb is clearly rendered at the top center. H: 1.50 m; W: 65 cm. Archaeological Museum, Florence, inv. 1914. (D.A.I.: 4675)

Fig. 3.5. A relief in terra-cotta that depicts a craftsman, perhaps sharpening iron blades, from Ostia, Italy. Knives, a cleaver, and pruning tools are depicted in the upper right. Along the bottom of the relief are depicted (from left) four paring chisels, an adze-plane, and a tanged chisel. Two variants of the axe are shown as well: a double-bladed axe and a combination axe-pick (or axe with cooper's adze). 40 cm square. From tomb 29 of the cemetery of Isola Sacra (Ostia). Mid-second century. (Sansaini, D.A.I. neg. 65.277)

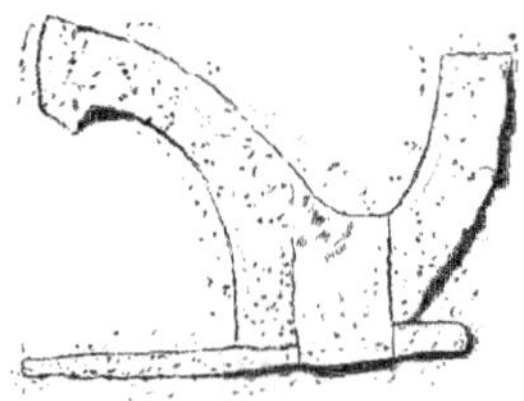

Fig. 3.6. Detail of an adze-plane from the funerary stele of P. Beitenos Hermes, a builder of couches. The tool is shown correctly in profile, with the broad but thin blade attached to the bottom of the handle. Louvre: inv. MA 934. (Author)

usually described in terms of blade width, curvature of the blade, and the angular relationship of the blade to the handle. Excavation reports have suggested specific uses for some recovered specimens (for example, a lighter, scooplike blade may have been intended for coopering [Evans 1894, 25]). Since the adze can be used in a wide range of applications, assignment of special uses to a particular tool is characteristically difficult.

Adze blades are generally thinner and therefore lighter than axe blades since less force is required in striking and the blade is never buried deeply in the wood. For rough shaping of large workpieces the craftsman stands over the timber and wields the adze with short strokes; note the stance of a Roman soldier standing over a timber as depicted on the Column of Trajan in Rome, which was dedicated in A.D. 113 to celebrate the emperor's victory over the Dacians (fig. 3.7). Large amounts of wood can be quickly removed. The adze would have been the tool of choice for hollowing entire oaken logs into the types of coffins used to bury the dead of Latium in the eighth through seventh centuries B.C. (Quilici 1979, 78) and to shape the dugout canoes rendered from single tree trunks that Pliny reports as being favored by German pirates (HN 16.203). On the farm the adze would hollow feed troughs for animals as well as all manner of simple wooden containers; such implements were common in rural Italy from Roman times through the early twentieth century.

In the hands of a skilled carpenter, the adze can be used to trim a thick tree trunk into a usable beam; timbers excavated at the Iron Age site of Glastonbury in the United Kingdom still bore the marks of this tool (Bulleid and Gray 1917, 373). Smaller adzes, wielded with one hand, were used for the shaping of wooden elements for furniture; a funerary relief from the Vatican collections depicts a craftsman shaping a table leg (fig. 3.8). The adze was indispensable for shaping the curved structural timbers of ancient ships; the grave relief of Publius Longidienus from Ravenna (early first century A.D.) highlights the artisan at work with his adze on a plank or interior rib destined for the hull of a boat under construction (fig. 3.9). In this same rendition we should note the lockable toolbox that Longidienus is using as a footstool. The adze is still used today by boatbuilders who build or repair wooden hulls.

THE AUGER (*terebra*)

The auger is a T-shaped boring and reaming tool; a vertical iron rod terminates in the bit, commonly shaped like a narrow spoon, with an upturned tip (hence a spoon bit). It

Fig. 3.7. A Roman soldier (lower right) *stands over a timber, which he steadies with his left foot as he wields a double-bladed tool; the curved blade probably is meant to indicate an adze. To the far right logs have been piled in alternating courses to form an abutment. The soldier at the portal* (top left) *appears to cradle a square* (norma), or libella, *under his right arm.* Column *of* Trajan, Rome (A.D. 113). (Anger, D.A.I. 89.730)

is the shape of the tip (and the solid, squarish tang) that helps to differentiate between an auger bit and a gouge (fig. 3.10). Twisted auger bits were rare if not absent from the Roman workshop. Some have argued that the twisted bit became common only in the eighteenth century (Salaman 1989, 30). Rare examples of twisted bits that have been found from Roman contexts in northern Italy and Gaul, however, would have worked well with augers; this is the only boring tool, as we shall see, that can be turned continuously in one direction (fig. 3.11). Large-diameter boring, whether for deeper holes or a series of overlapping shallow holes bored to prepare a workpiece for veneer or marquetry, required a drill that resembles a modern cheap wood-boring (or flat wood) bit.

Fig. 3.8. Funerary relief of a furniture maker from Rome. The craftsman shapes a table leg with a short-handled adze. The figure to the left smoothes a plank with a plane. Total H: 45 cm. Vatican, Museo Gregoriano Profano, inv. 3262. Mid–third century. (Vatican Museums, inv. 3262)

*Fig. 3.9. Relief of P. Longidienus, Ravenna. The adze is shown being used by a shipwright (*faber navalis*). The inscription on the small plaque in front of the craftsman proclaims, "Longidienus bustles on with his work." Total H: 2.66 m, W: 0.90 m. D: 0.44 m. Museo Nazionale, Ravenna inv. 7. First century. (Courtesy of the National Museum, Ravenna)*

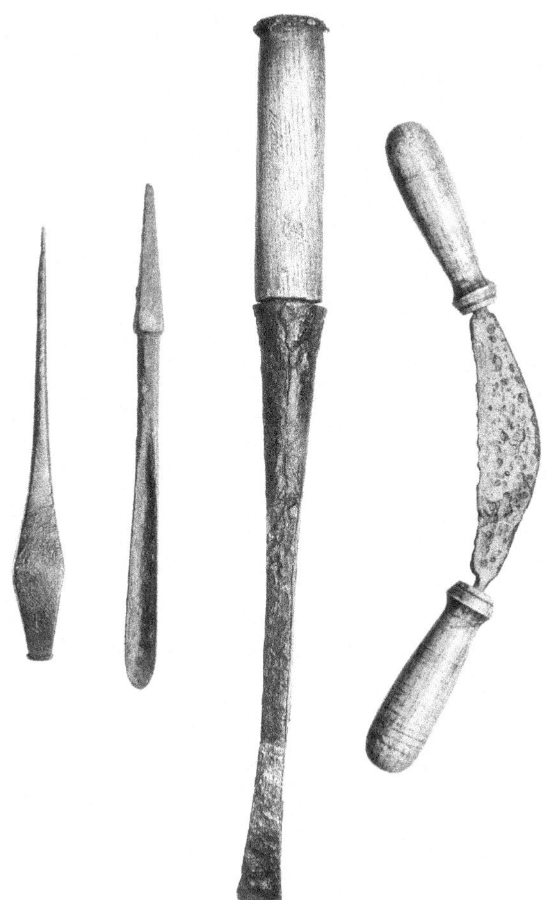

Fig. 3.10. Four Roman woodworking tools from the Museum of London. (from the right) *Drawknife (inv. 19167) with restored handles, blade length: 11.8 cm; socketed chisel (inv. 23317) with restored handle, total length: 26.3 cm, cutting edge 2.2 cm across; auger blade (or gouge; inv. 29.94/19), total length: 20 cm; one-piece fine mortising chisel (inv. A1893), total length: 17.5 cm—the narrow blade (.07 cm across) is seen from the side, and three small notches (owner's mark?) are visible on the right side of the top edge. (Courtesy of the Museum of London)*

A good example has been documented from Zürich (Neuberger 1919, 76). The Latin term, *terebra*, is also used to refer to a standard drill bit, discussed more fully below.[1]

According to Theophrastus (5.6.3), boring with an auger works best on seasoned wood, for the bit tends to become clogged when the wood is still green. In fact, modern craftsmen who use traditional tools can bore ably with a spoon bit when they are working with green wood (Langsner 1987, 175). The boring action of the auger is powered by the principle of leverage; great force can be applied when a long handle is employed. Thus the auger could be called upon to bore large-diameter holes; the difficulty of the boring, however, exponentially increases with diameter, as Vitruvius noted (10.16.5).

Pliny (HN 16.229) states that auger handles were best fashioned from olive, oak, box, holm oak, elm, and ash, that is, species most resistant to splitting. An iron spoon bit with a square tang has been restored with such a wooden handle in the Museum of London. The corners of the tang would resist slippage while the handle was rotated. No ancient depictions of a Roman using an auger have been identified (Goodman 1964, 165).

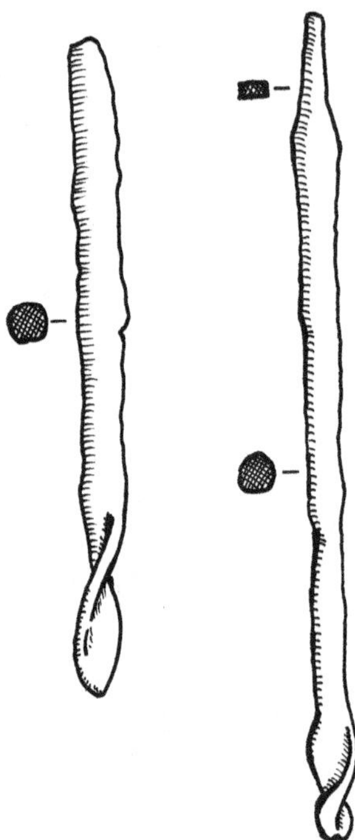

Fig. 3.11. Examples of spiral or twisted iron bits from Aquileia, Italy. (left) 12 cm long; (right) 15.2 cm long. Museo Nazionale (Aquileia). (Author, after Gaitzsch 1980, nos. 187, 188)

THE AXE (*ascia, dolabra, securis*)

The axe is the oldest of all woodworking tools; it has been used continuously since the Stone Age for tree felling and the initial preparation of timber. For the felling of trees the axe was better than the saw, for with an axe a man could work alone. Furthermore, saw-teeth could clog in green wood or the blade could become pinched and rendered useless by the weight of the tree. Moreover, axes had no match for rapidly delimbing fallen trees. Double-bladed axes (sing. *bipennis*) may have employed one sharper blade for cutting and a second, duller blade for splitting; it was also easier to sharpen a double-bladed tool in the woods, for one blade could be honed while the opposite side, embedded in a stump, held the axehead firmly. Romans used both single- and double-bladed axes, the most common being the single (fig. 3.12).

The first metal axes used in Italy, made of copper and then bronze, were little more than flat blades attached to a naturally bent piece of wood; the angle of blade to handle was determined by each piece of stock. These imitated the stone-bladed axes of their Neolithic predecessors. The metal blade could be attached through insertion into a slot cut into the wood or the wood itself was tightly fitted to a metal socket. The socket form required more skill on the part of the metalworker, but the resulting axe could be swung with greater force without splitting the handle.

By the tenth century B.C., the conventional woodcutter's axe closely resembled modern parallels; the wooden handle was fit at a right angle to the blade of the cutting head. Even earlier copper axes have been found in Italy (for example, at Monte Rovello)

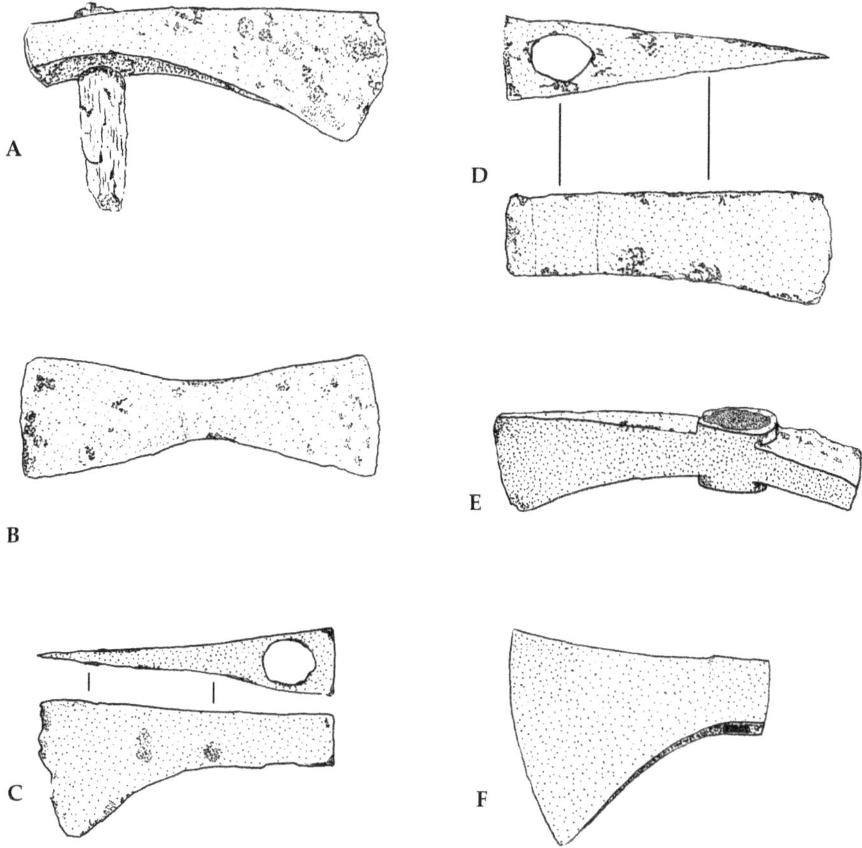

Fig. 3.12. Roman axes: (A) hatchet from Pompeii, with a fragment of the original handle, 19.20 cm long, Naples, inv. 286770; (B) bipennis from Pompeii, 28 cm long, Naples, inv. 286771; (C) axe, Combend (U.K.), 23 cm long, British Museum, inv. 1810.2-10.5/6; (D) axe, 19.1 cm long, Verulamium; (E) axe, 24.6 cm long, Silchester; (F) axe, 20 cm long, Pompeii. (Author, after Ciarallo 1999, 122, fig. 77, 123, fig. 78 [A, B]; Manning 1985, fig 3.2 [C]; Frere 1972, 165, fig. 60.7 [D]; Evans 1894, 147, fig. 11 [E]; Petrie 1917, pl. 10.34 [F])

with this fully developed form, but toolmakers were quick to appreciate the cutting properties of iron shortly after its discovery in the tenth century B.C. The iron head of the axe could be fashioned either by casting the tool in a mold or by hammering a thick sheet of iron to the desired shape. For the latter, a rectangular slab of iron was folded in half around a cylinder and hammered into shape; the socket, or eye, created by the gap was preserved by the cylinder. The process was easier and could be performed by a general-purpose smithy, although it produced an inferior tool.

A more durable socket was formed in the casting process. Axes were best cast from the butt, or poll, end. The thick iron poll developed incrementally over time as the casting process was mastered. The extra mass protected the open eye from deformation when the poll end was used as a sledgehammer. A wooden handle, or haft, was fitted into the socket, or eye. The diameter of the eye gives some indication of the species of wood that was used for the hafts. Egyptians, for example, may have used woods even harder than the European ash or oak and thus could reduce the size of the eyes of their iron axes (Petrie 1917, 11). In North America the wood of the hickory is now favored for tool handles and is exported around the world for this purpose.

The most common form of the Roman axe exhibits a flat poll, and the entire blade is often cast with a gentle curve (Goodman 1964, 24; Manning 1985, 15). If the blade is not curved, one or both sides of the front face sweep to provide a broader cutting edge. Axes are usually triangular in horizontal section; the eye is oval or, more rarely, round. Manning notes that while size and weight may vary, larger axes used for the heaviest work tend to weigh around 1.6 kg, twice that of lighter axes (or hatchets) (1985, 15).

Axes have been found at Roman sites throughout Europe, especially in Roman Britain, where a characteristic feature is a lug (or tubular sleeve) on either side of the socket to reinforce the attachment to the handle. The presence of the lug permitted the axe to be twisted free without distortion when embedded in wood (Petrie 1917, 12). Lugs are clearly visible on the two axes depicted on the grave relief of a knife sharpener from the Isola Sacra cemetery (see fig. 3.5).

Roman soldiers depicted on the Column of Trajan clear forests with a variety of axes. The most common military axe of the period (early second century) appears to combine an axe blade with an opposed pick (fig. 3.13). In Latin texts (for example, Tac. *Ann.* 3.46 and Juv. 8.248) referring to military activities, a distinction is drawn between the *securis* and the *dolabra*. The latter generally refers to the tool used for digging fortifications or attacking enemy walls and therefore most approximates the modern firefighter's pickaxe (Blümner 1875, 206). Actual specimens that are dead-ringers for the examples seen on the column were recovered from pits at the Roman fort of Newstead in Scotland (fig. 3.14). The lugged *dolabrae* from Newstead ranged in weight from 1.9 to 2.7 kg; one still held traces of its ash handle (Curle 1911, 278). Another form of *dolabra*, the *dolabra pontificalis*, refers to the special axes used for animal sacrifice and *dolabrarius* to the specialist who wielded the sacrificial axe (Fest. 319 M. 423 L.; Blümner 1879, 205). These special-use axes, which vary in form (some of them closely resemble the woodcutter's axe), are not considered here. An object that appears to be a double-axe with two broadly flaring blades is depicted along with other woodworking tools on a Roman-period funerary relief discovered at Priolo and now in the Syracuse museum (fig. 3.15). The object was first identified in the nineteenth century as a flour mill. That it is meant to show a hand-tool seems most likely, considering the other woodworking tools included on the relief and the prominent image of a ten-spoked wheel, which suggests the deceased, Eutyches, was a wheelwright or cart-maker. The image also resembles a type of "waisted" mallet we will see elsewhere. If indeed a double-bladed axe is represented, its appearance is strikingly similar to a well-preserved axehead recovered from the excavations of Pompeii (see fig. 3.12).

The small axe (hatchet, broad hatchet, bench axe) is probably indicated by the Latin *securicula*. This type of tool was meant to be used with one hand and was employed

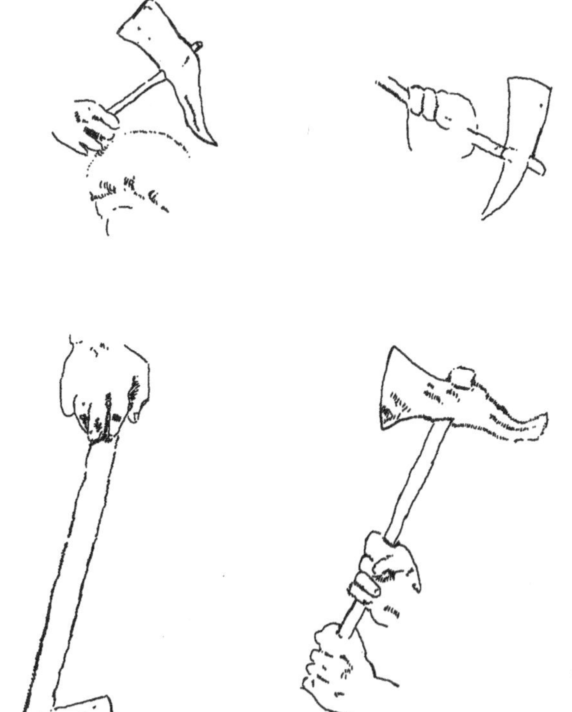

Fig. 3.13. Variants of axes shown in use felling trees, from Column of Trajan, Rome (A.D. 113). *The blade tends to be combined with an opposed pick, but the axe on the bottom left is clearly shown with a flat poll. (Author)*

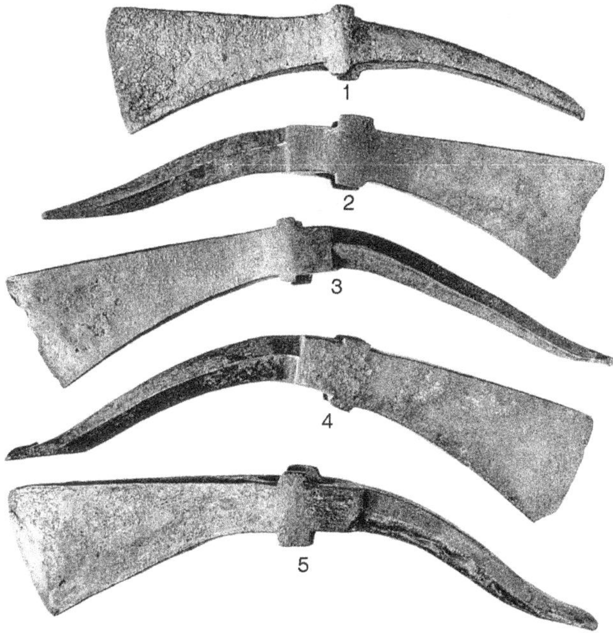

Fig. 3.14. Roman military-type axes (dolabrae) *from Newstead, Museum of Scotland, Edinburgh. Second century. (Curle 1911, pl. 57)*

Fig. 3.15. The funerary relief of Eutyches, from Priolo in Sicily. A double-bladed axe (or possibly a mallet) underneath which is a plane with discernible handles and low-angle iron. A wheel with ten spokes is also visible. The side of the relief is depicted in figure 3.21. H: 1.07 m. Museo Nazionale, Syracuse. Late third–early fourth century. (Fuhrmann, D.A.I. 37.651)

for hewing (shaping) smaller pieces of wood and removing branches from felled trees. Like its modern counterparts, it is distinguished from its big brothers mainly by its scale, not its form.

THE CHISEL AND THE GOUGE (*scalprum*)

The chisel is an essential woodworking tool where deep, angled cuts are required. It is therefore indispensable for shaping wooden joints: cutting mortises, forming tenons

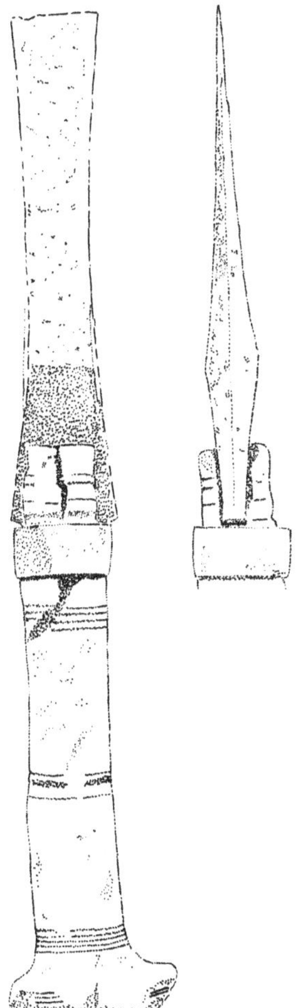

Fig. 3.16. The Aquileia chisel, with its original wooden handle. The metal ferrule serves to hold the tanged blade securely; (right) profile of the blade; (left) with handle: 28.6 cm; Museo Nazionale (Aquileia), inv. 80697. (Author after Gaitzsch 1980, pl. 38, fig. 181)

and dovetails, and removing stock for rabbets. The chisel blade can plane off imperfections from the surface of a board and scrape off excess drops of glue. A variation of the chisel, with a shorter, thicker, flat blade and a long wooden handle, was used to shape wood turned on a lathe. Theophrastus (5.6.4) observed that it is easier to work green wood with the chisel.

Roman chisels closely resemble their modern equivalents. Many ancient examples have been found, some in near pristine condition. A paring chisel recovered from the Roman town of Aquileia in northern Italy looks as if it belongs in a modern woodworker's toolbox; even the wooden handle, secured with a ferrule, survives (fig. 3.16). Several specimens displayed in the Museum of London remain in superior condition, with edges that would require only minor touching-up to return them to working condition.

Chisels were forged in a variety of shapes and sizes, some to endure heavy use, which included hard strikes with a hammer or mallet (see fig. 3.10). Others were designed to be used primarily with gentle pressure from the hand or shoulder. At the turn of the last century W. M. Flinders Petrie classified the many chisels found in Egyptian contexts by the manner in which the handle was attached to the cutting end of the tool (Petrie 1917, 19). Indeed, the method of handle attachment and profile of the blade can be good indicators of how a chisel was intended to be used (Manning 1985, 21). For the heaviest work, blade and handle were best forged from a single bar of iron. Other chisels were fashioned with a hollow socket opposite the cutting end into which a tapered wooden handle could be tightly fit. These, too, can accommodate heavy pounding without the handle splitting. Chisels were also forged with a tang that was inserted into a mortised wooden handle. This form of attachment was best when only light tapping or hand pressure was intended—nevertheless, some chisels intended for heavier use were of the tanged variety; it is this last type that Icarus uses to cut mortises in the famous fresco from the so-called Ixion Room in the House of the Vettii at Pompeii (fig. 3.17). Standard chisel blades were flat, with a taper from the handle toward the blade.

The paring chisel was used to make shallow cuts by means of pressure applied by the hand or shoulder. The blade width varied between narrow and broad. The blade itself was often formed by the gradual tapering of the iron shaft. At Verulamium handles and blades of paring chisels could be forged from a single piece of iron. Finds from other sites indicate that a tanged handle was common; the example of the Aquileia chisel cited earlier shows that the handle could terminate in a mushroom-shaped butt, thereby improving the grip for pushing the tool with the palm of the hand. The thickened profile of the upper blade could be firmly grasped in one hand while the mushroom-shaped handle fit comfortably into the palm of the other.

The "firmer chisel" (Manning 1985, 22) refers to a general-use category that was thicker and therefore stronger than the paring chisel and therefore could be struck with a hammer or a mallet. The handle and cutting edge were sometimes formed from a single piece of iron; a socketed handle was also favored (modern firmer chisels normally have tanged blades). The edge was formed by a continuous taper along the shaft of the blade; the blade itself was occasionally slightly splayed.

The mortising chisel was characterized by a thick, heavy blade and designed, as its name suggests, for cutting deep mortises in wooden stock. Normally the chisel was used to square off a cavity that had been initially bored with a bow drill (see below). The cutting of mortises in a tough wood like oak could stress the blade as much as if the workpiece material were stone. It is therefore possible to confuse these heavy woodworking chisels with the mason's cold chisel. The wooden handles of the mortising chisel were best seated in a socket to sustain hammering (although tanged examples have been found). The cutting edge was formed by an abrupt bevel at the end of the iron; it is occasionally curved slightly backward. Manning has observed that the mor-

Fig. 3.17. Icarus, the son of Daedalus, sits at a workbench and cuts mortises with a tanged chisel and a mallet. At his feet lie a bow and drill, and on the right side is an adze-plane. From the Ixion Room, House of the Vettii, Pompeii. Mid–first century (Courtesy of John Clarke and Michael Larvey)

tising chisel is the most common form of this category of Roman woodworking tools (1985, 23).

Gouges such as those found at Silchester were characterized by curved edges for making scooping cuts (fig. 3.18). The gouge was essential for making small objects with hollowed interiors or architectural moldings with concave profiles (such as acanthus leaves). The degree of curvature on a gouge is known as its sweep. Small Roman gouges that have lost their handles or blade tips can be confused with augers or large drill bits. The longest (31.5 cm) gouge from the iron "hoard" discovered at Silchester

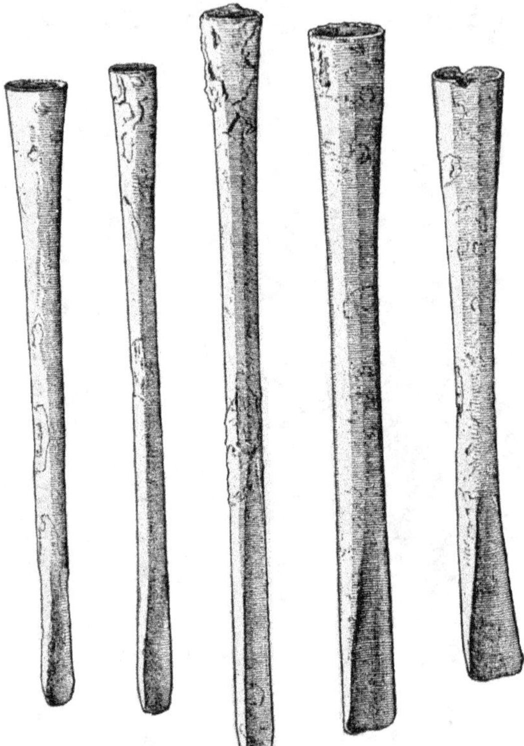

Fig. 3.18. A group of iron socketed gouges found at Silchester (U.K.). Lengths (from left): 26.1 cm, 27 cm, 31.5 cm, 30 cm, 25.8 cm. (Evans 1894, 150, fig. 16)

by Sir John Evans in 1890, sharpened on the inside edge, may have been used with the lathe (Boon 1974, 284).

THE DRILL (*terebra*)

Bow Drill

The bit of the bow drill was spun rapidly by the reciprocating motion of a thong wrapped around the shaft of the bit (or its wooden stock); the thong was kept in tension by the flexed stick, or bow. The right-handed craftsman steadied the bit and applied pressure at the top (using a piece of wood or hollowed stone known as a nave) in his left hand while operating the bow with his right hand. The alternating clockwise-counterclockwise rotation of the shaft explains the rarity and slow development of the spiral or twisted drill bit, which is effective only when rotated continuously in one direction. The low power of the bow drill could produce only small-diameter bores.

The principle behind the bow drill is of great antiquity. Egyptians of the third millennium B.C. understood the basic mechanics; it has been suggested that the Romans were the first to attach the thong to a bowed (not straight) stick (Goodman 1964, 161). Another, somewhat romantic view is that an actual bowstring wrapped around an arrow may have served as the first crude prototype of the tool (Petrie 1917, 39).

Fig. 3.19. Drill and stock of a bow drill from Hawara, Egypt. Total length: 55 cm. University College, London, inv. 27979. Roman period. (Petrie 1917, pl. 51)

The Egyptian (Petrie) collection of University College, London, holds a perfectly preserved Roman-period drill bit and wooden stock excavated at Hawara, Egypt (Petrie 1917, 39) (fig. 3.19). The wooden stock was turned on a lathe, its worn surface scored with parallel grooves to increase friction with the thong. The cylindrical nave was carved from two pieces of wood and then pegged along the seams. This allowed it to rotate around, yet be permanently affixed to, the wooden ball that was turned into the top of the stock. The shaft of the iron bit was square in section and terminated in a conventional diamond-shaped point. As a woodworking tool, the bow drill was especially useful for the initial hollowing out of mortises. The mid-first-century depiction of young Icarus cutting mortises with a tanged chisel in the House of the Vettii at Pompeii includes a rare representation of a bow and drill lying under his carpenter's bench; our impression is that Icarus has just finished the drilling step of the process (see fig. 3.17).

The appearance of the bow drill is known through a small number of additional ancient depictions. Its inclusion on funerary reliefs attests to the special skill associated with those who were expert at its use. One such relief, originally from near Rome and now in the British Museum, portrays two former slaves, P. Licinius Philonic(os) and P. Licinius Demetrius, along with smithing tools and, on the right border, as if hanging, a bow and drill (fig. 3.20). The funerary altar carved to honor the Sicilian wheelwright Eutyches includes the depiction of a bow and drill together with compasses and ruler. The handle of the drill is rendered as finely turned, with a prominent nave (fig. 3.21). The bow as a weapon instead of a tool can be discerned by its reflex profile and

Fig. 3.20. Funerary relief of P. Licinius Philonic(os) and P. Licinius Demetrius, from Frascati, Italy. The right border of the relief depicts a bow, a drill (with bit toward the top), a knife, an adze-hammer, and a tanged paring chisel or adze blade(?) (without the handle). The pedimental area depicts a hammer and tongs. The left border is adorned with a representation of the fasces. *H: 68 cm, W: 80 cm. British Museum, inv. 1954.12-14.1. Late first century. (© Copyright The British Museum)*

the presence of arrow shafts, as evident in a funerary portrait of Monimus, a Roman soldier stationed at Mainz in the first century (Selzer 1988: 153 cat. 80) (fig. 3.22).

The bow and drill were also shown in scenes other than those carved on tombstones. A gilt glass vessel found in the catacombs of Rome and dating from the fourth century offers a visual encyclopedia of craftsmen using woodworking tools, including the mallet and chisel, bench plane, bow and drill, adze, and frame saw (fig. 3.23). In the tiny scene of the bow and drill, the artist has depicted the craftsman incorrectly holding the tool by the middle of the stock (Goodman 1964, 161). A bit described as the *terebra gallica*, or "Celtic drill," suggests that Romans were ready to adopt unfamiliar tools they encountered as the military expanded into central and northern Europe (Mols 1999, 83). It is tempting to speculate that this variant refers to the rare bits forged with a spiral cutting edge, which have been found only in northern Italy (at Aquileia) and in Gaul. As noted earlier, however, these may have served best in the auger.

Strap Drill

The strap drill was a more powerful version of the bow drill. While the principle behind the operation of the drill was the same, the craftsman guided the iron bit while

Fig. 3.21. Bow, drill, compasses, and ruler from the side of a funerary altar erected for Eutyches, from the Priolo cemetery of Sicily, now in the Museo Nazionale, Syracuse. The facade of the altar is depicted in fig. 3.15. Third–fourth century. (Orsi 1891, 359)

an assistant pulled the strap wrapped around the wooden shaft in which the bit was mounted. In Roman workshops this tool was a mainstay for carving stone. But, as a reference from Homer shows (see below), the strap drill would have been useful also for boring deep holes in heavy timbers.

The best surviving physical representation of the strap drill is carved on the funerary relief of Eutropos, an early Christian sculptor of sarcophagi. The relief was found at the cemetery of Santa Helena on the outskirts of Rome and is now in Urbino (fig. 3.24). The seated sculptor holds the nave of the drill in his right hand and guides the bit with the aid of a rod in his left. An assistant (his son) operates the strap, which appears to be equipped with a short handle. The most graphic literary description of a strap drill in use is from one of the most dramatic scenes of Homer's *Odyssey*. Homer describes the blinding of the Cyclops Polyphemus by Odysseus and his shipmates, who plunged and twisted a sharpened stake into the monster's single eye, "just as a man bores a ship's timber with a drill while those below him twirl it with a strap they hold at either end, so the bit spins continuously" (*Od.* 9, 384–86).

Fig. 3.22. Funerary relief of the soldier Monimus, the son of Jerombalus. Monimus holds a bow in his left hand and arrows in his right. He died at age fifty. H: 1.35 m, W: 0.63 m, D: 20 cm. Mainz, Landesmuseum, inv. S166. First century. (Courtesy of the Landesmuseum, Mainz)

THE KNIFE (*culter, cultellus*)

Although knives play a small role in modern woodworking, the knife was traditionally an important tool for detail work and for carving smaller items such as latches, knobs, hinges, and tableware (Langsner 1987, 66). Cutting wood with knives is easier when the workpiece is still green. Woodworking techniques that make prominent use of cutting knives have survived to modern times in areas of central and northern Europe. The knife still plays an important role in barrel making in Switzerland, and in Sweden skilled carvers make a number of household utensils. Knives are also used to score cutting lines on stock or to lay out lines for veneer work. A marble relief depicting

Fig. 3.23. *The Vatican gilt glass vessel depicting the tools of the shipwright.* (clockwise from top right) *Mallet and chisel, bench plane, knife or adze(?), bow and drill (detail), adze or small hatchet (detail), frame (buck) saw. Museo Biblioteca Apostolica Vaticana, inv. 60788. Early fourth century. (Photo: Courtesy of the Vatican Museums; details: Author)*

Fig. 3.24. The strap drill; from the sarcophagus of Eutropos. A seated craftsman carves a marble sarcophagus with a double-handled tool while his assistant (his son) operates a strap. A container on the floor may hold additional bits. The sarcophagus was found on the Via Labicana (near S. Helena) in Rome and is now in the Museo Archeologico Lapidario, Urbino. Mid–fourth century. (Photo: Urbino, Museo Lapidario; detail: Author)

L. Cornelius Atimetus, a knife seller, with his wares offers a glimpse of the variety of the blades available (fig. 3.25).

THE DRAWKNIFE AND SPOKESHAVE

Although rarely used by woodworkers today, the drawknife, or drawing knife, was a versatile tool for the rough shaping of planks and boards in antiquity and was used with little change in design until the mid–twentieth century. The spokeshave works on the same principle as the drawknife; the main difference is one of size: the cutting edge of a spokeshave is about the same size as that of a modern single-edged razor blade. There is no surviving Latin term for the drawknife to distinguish it from a simple knife. The Roman drawknife was fashioned from a thick iron blade that was sharply beveled along one of its long, gently curved sides. The ends of the blade terminated in tangs that were inserted into turned wooden handles. The handles of modern drawknives are fixed at a ninety-degree angle to the blade. Roman-period specimens that have been found, such as those at Pompeii and from London, employ handles that were affixed at a lesser angle or even on the same axis as the cutting blade (see fig. 3.10).

As its name suggests, the drawknife cuts with a pulling force (although some craftsmen also push the tool for delicate work). Depending on the hardness of the wood, the drawknife can remove paper-thin shavings or thick strips of stock. The depth of the cut is controlled by the rotation of the handles while the workpiece is cut. Cuts

Fig. 3.25. Marble relief of a knife seller from the tomb of L. Cornelius Atimetus. The cabinet holds agricultural tools (top), knives (middle), and what may be racks holding gouges and chisels (bottom). H: 1.33 m; W: 1 m. Vatican Museums, inv. 9277. First century. (Anger, D.A.I. 1020)

can be made with the beveled (or basiled) side facing up or down. The drawknife is especially useful for the rough shaping of curved pieces of wood. A special variant of the drawknife was used to shape and gently hollow the planks used for making barrels. Drawknives could also quickly remove the corners from a squared timber being prepared for turning on the lathe. The spokeshave was used, as its name suggests, to shape smaller curved objects like the spokes of a wheel and also to smooth the rough surface left by the blade of the drawknife.

To date, no ancient depictions of the drawknife in use have been found. It does not appear on funerary relief sculpture. The existence of the tool is known primarily through archaeological finds of actual tools.

THE LATHE (*tornus*)

By attaching a piece of stock to a lathe and spinning it rapidly while holding a cutting edge to the wood, Roman woodworkers could create ornate bowls, cups, legs for tables and chairs, and even columns. Pliny credits the celebrated Greek architect, sculptor,

Fig. 3.26. A stone bed leg from a funerary bier in the Tomb of the Funeral Beds at Populonia (Italy). The stone has been carved to resemble a wooden leg turned on a lathe. Seventh–mid-sixth century B.C. *(Photo: Author)*

and metalworker Theodorus of Samos, who lived in the sixth century B.C., with the invention of this tool (Plin *HN* 7.198). Stone bed legs carved to look like wooden stock turned on the lathe can still be seen in a tomb from Populonia, an Etruscan site with stone and earth *tumuli* dating from the seventh century B.C. (fig. 3.26). Actual turned wooden vessels imported from Greece have been recovered from other Etruscan tombs (Rieth 1955, 6).

The appearance of the Roman lathe is known only imperfectly; it is likely that more than one type existed. Modern lathes powered by electric motors rotate the workpiece at a high speed in one direction; the axis of rotation is generally horizontal. The simplest of ancient lathes, however, used the same principle employed by the bow and drill to impart the rotational force. An Egyptian relief from Petosiris, third century B.C., illustrated by Adolf Rieth includes a pair of craftsmen—one operating a bow and a sec-

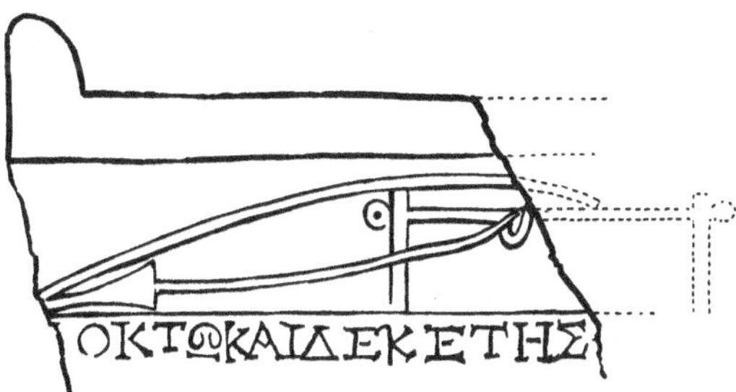

Fig. 3.27. A probable depiction of a simple bow-operated lathe from a Roman sarcophagus. Second century. (Author after Rieth 1940, 99, and Kontoleon 1890, 333)

ond holding a chisel—working on stock rotating on a vertical axis.[2] A Roman-period sarcophagus from Greece shows a similar apparatus (the bow lathe) minus its operators; the representation does not survive completely, but the bow and a simple frame for holding the workpiece in a more conventional horizontal configuration are recognizable (fig. 3.27). This appears to have been a tool well known in the premechanized eastern Mediterranean.

For a more sophisticated machine that could spin a workpiece with greater speed or be operated by a single craftsman we must turn to postclassical analogies. For centuries, for example, woodworkers in many parts of the world have turned wood on a simple machine known as the pole lathe. The concept is very simple. The workpiece is suspended horizontally between two uprights by pins so that it can rotate freely. A strap, wrapped once around the workpiece, is attached to a simple pedal below and a springy lath or tree branch overhead. The turner steps on the pedal, which rotates the workpiece, and applies pressure with his turning chisel. When the pedal is released, the workpiece revolves in the opposite direction and the process is repeated. Cutting takes place only on the downward stroke. Such simple lathes can be constructed just about anywhere and can be operated by one (coordinated!) person. The earliest depiction of such a tool in use dates to the thirteenth century, so there is no solid evidence that this version of the tool was known to the Romans (Aldred 1956, 232). Only slightly more complicated are lathes that employ a flywheel to supply a continuous turning action. The flywheel can be operated by an attendant or turned by a foot pedal.

A curious apparatus depicted on a relief of a carpenter's shop found in Rome in 1938 may afford a unique glimpse of an ancient lathe (see fig. 2.3). A circular object can be seen end-on resting upon an oversized stand built in the form of a sawhorse. On the floor, between the legs of the stand, rests another circular (or, if seen from the end, cylindrical) object. A craftsman stands to the left, apparently working on the piece. It is possible that the scene here is a rare, if not unprecedented, depiction of a lathe in use (Colini 1947, 23).

Even if the exact appearance of the Roman lathe is not known, its existence and use are clear from literary references. Relief sculpture and painting depicting turned chair and table legs as well as actual finds of objects turned on the lathe attest to its widespread use. The turned legs of a wooden bed excavated from Herculaneum are

Fig. 3.28. Turned legs on a bed depicted on a Roman funerary relief. The bed (lectus), holding a married couple, is backed on three sides. The turned legs of the bed stand in contrast to the carved legs of the three-legged table. From Isola Sacra (Ostia). Second century. (Author after Himmelmann 1973, pl. 24B)

good examples (Mols 1993, 489), as are the legs of banqueting couches shown on many a Roman funerary relief (fig. 3.28). Pliny (HN 36.193) tells us that the lathe was also used for glass making; even heavy stone drums were turned on massive lathes for the manufacture of columns (Plin. HN 36.90).

The small vessels fashioned by the specialist (perhaps a *tornator*) who mastered the lathe were highly prized, and the artisans themselves became minor celebrities among connoisseurs. Pliny (HN 16.205) mentions a certain Thericles who turned goblets from the wood of the turpentine tree (or terebinth); he is the only turner of wooden objects whose name is known to us from antiquity.

Carving a cup or bowl on the lathe presents a special challenge, for the pivot at each end of the workpiece interferes with the hollowing of the vessel. The Roman craftsman probably employed a practice later used by medieval turners. To make a bowl, for example, the woodworker would leave a central core intact for the pivot-point. This core would finally be cut away when the piece was removed from the lathe (Salzman 1923, 172; Mercer 1960, 220, fig. 190). Several Roman-period turned wooden bowls have been found in England (Pugsley 2003). One example, found during the rescue operations of 1969 at Fishbourne, is characterized by a rounded rim and foot-ring, a raised ring on the bottom to serve as a foot (Cunliffe, Henig, MacGregor 1996, 98, fig. 3). A similar bowl has been recovered from London (Wilmott 1982, fig. 32). A small cylindrical container (*pyxis*) and assorted circular lids, the latter of boxwood and all fashioned on a lathe, have been recovered from the Roman wreck from Comacchio of the first century B.C. (Desantis 1990, 264).[3]

If the turning chisels used with the lathe were given special names, they have been lost. Virgil (*Georg.* 2.449), for example, uses the general term for any blade, *ferrum*, to describe the knife used with a lathe. Turning chisels are usually flat and tend to be short (no longer than 15 cm) to reduce vibration. Curved cutting edges could be customized to create any desired profile. Bowl makers' chisels, shaped something like an oversized hook, are known from a pair of Celtic examples from Manchin, Germany (Jacobi 1974, 47); doubtless the form was well known to the Roman craftsman. Ro-

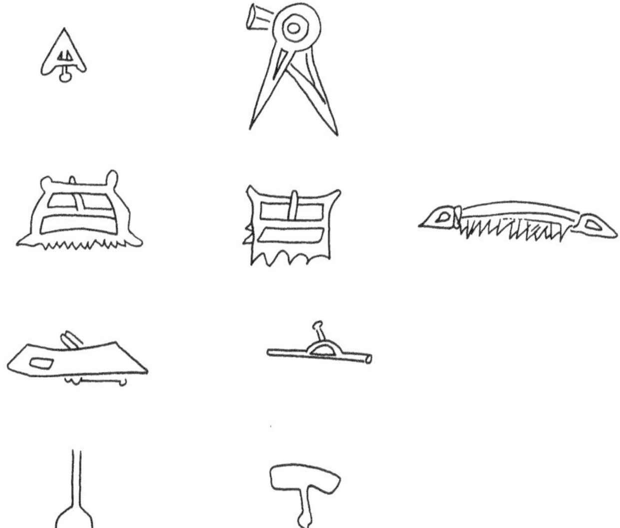

Fig. 3.29. Mint marks on late Roman republican silver coinage. (from top, left to right) *Level with pendulum, coin of M. Plaetorius Cestianus (69 B.C.); compass, coin of L. Papius (79 B.C.); two frame and one crosscut saw, from coins of L. Papius; plane, coin of L. Papius; plane, coin of L. Roscius Fabatus (64 B.C.); chisel (or plane blade) and auger(?), unknown moneyer.* (Author, *after Fava 1969*)

man turners would doubtless shape and size their own hardwood handles to fit both hands.

THE PLANE (*runcina*)

The plane (*runcina*) may have evolved from a small version of the adze, as already noted. With its blade, or iron, held at a fixed angle (known as the rake) and the cutting depth regulated by the use of a bottom plate, or sole, a plane pushed (or pulled) in the direction of the grain of the wood can quickly reduce the thickness of a plank or smooth and polish its rough surface by removing shavings no thicker than a piece of fine paper.

The *runcina* was apparently unknown in Pharaonic Egypt, where thin panels of wood were formed by the laborious process of rubbing adze-hewn planks with stones and sand. The proliferation of finely paneled doors in the Greek Classical and Hellenistic periods (fifth–second centuries B.C.) may have been possible by the invention of the plane. No Greek planes have been found, but the Greek *rhykane*, from which the Latin *runcina* is derived, indicates the existence of the tool (Varro *Ling.* 6.96).

The earliest physical evidence of the Roman plane exists in the form of diminutive mint marks on a remarkable series of silver coins minted during the first half of the first century B.C. Since the scale of these renditions is so small, few details are observable, but the schematic renditions include one or more handles, raked irons, and flat soleplates (fig. 3.29). The fact that variants exist in these depictions is an indication that the tool already existed in diverse shapes and sizes and that it was a common enough tool to be recognizable for what it was even when rendered schematically.

The range of planes in use by the imperial period was doubtless similar to that used by craftsmen well into the nineteenth century. All fall under three general cate-

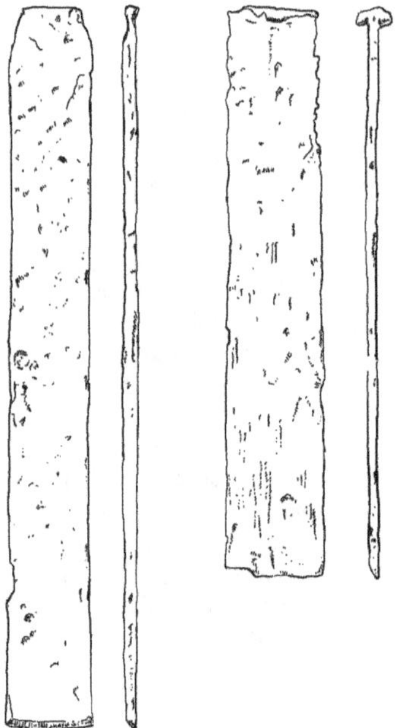

Fig. 3.30. Roman plane blades from Augst (Germany). Two irons are pictured from a set of four. (left) L: 23.5 cm; (right) L: 18.4 cm. *The bodies of the planes were not recovered. (Author, after Mutz 1980, figs. 3, 7–8)*

gories: all-purpose smoothing planes include the jack plane for quickly removing large amounts of wood; the trying or truing plane to further smooth the plank; the long joining plane, which smoothes and levels long boards in preparation for gluing; and a variety of small smoothing planes for closer and finer work. The second category consists of the molding planes, fitted with concave or convex blades and matching wooden soles in order to cut decorative moldings. Finally, there are the rabbetting planes, fitted with narrow blades, designed for cutting long grooves parallel to the grain of the workpiece. From actual finds of Roman planes, sculpted and painted representations of the tool, and finished wooden objects, it is clear that the Roman woodworker used planes from each of these categories, although extant Latin terminology does not preserve the special names used to distinguish them from one another.

The vast majority of Roman planes were probably completely wooden, with the exception of the blade itself; this explains the occasional discoveries of plane blades but no other traces of the tools. Any competent Roman carpenter would have had the skill to make his own hardwood stocks suitable for his specialty. An outstanding example of a set of Roman irons was recently found in Augst in Germany; here four irons of differing sizes neatly coincided with standard widths of the Roman measuring system (Mutz 1980, 126) (fig. 3.30). The shape of the iron's cutting edge is an important indicator of its function. Those irons that have been found still attached to their stocks vary only slightly in width, from 2.7 to 4 cm. A slight convex edge would cut deeply to remove large amounts of stock quickly. This shape characterizes the cut of the standard jack plane. Most blades of smoothing planes had straight edges. Toothed irons such as that

Fig. 3.31. Plane from the Roman Colonia Agrippina (now Cologne). L: 32.4 cm. Rheinisches Landesmuseum, Bonn, inv. 36.199. Second century. (Courtesy of Rheinisches Landesmuseum, Bonn)

found in a finely wrought plane discovered at Cologne in 1880 could both cut quickly and leave a roughened surface for subsequent gluing or veneer work (fig. 3.31). J. Curle illustrates an iron with a cutting edge ground for making moldings (1911, pl. 59).

One unusually well-preserved plane, found at the bottom of a well at the Roman fort at Saalburg, retained its stock of beechwood still attached to its iron sole. Such finds allow accurate restorations of planes that preserve only their metal fittings; the tool found at Verulamium is a case in point (fig 3.32). Beechwood, it should be noted, has been favored for plane stocks from Roman times to the modern era (Bealer 1972, 165). The blade of the Saalburg tool protrudes from the sole at an angle of fifty degrees (Liversidge 1976, 159), close to the forty-five-degree angle that characterizes the modern all-purpose plane. Angles of fifty to sixty degrees are preferred when use is intended for harder woods. Gaitzsch has found that the blades of extant Roman planes were all set at a relatively steep angle, ranging from fifty to sixty-six degrees (1980, 113).

As is evident from the Saalburg example, at least some Roman planes were fitted with durable iron soles. Nearly twenty other examples have been found throughout the Roman world (see appendix). With the single exception of a molding plane found from a Roman context in Egypt, these are all variants of smoothing planes, ranging from 43.9 cm to 21 cm in length. The oldest Roman planes (first century, before A.D. 79) were found at Pompeii in the nineteenth century (see fig. 3.2). The most sophisticated plane, in terms of design and decoration, is the example already mentioned from Cologne, dated to the second century. With a body framed in iron, the plane is characterized by fretted plates on the top surface which enclosed an otherwise standard wooden stock with front and rear grips. Interestingly, in the seventeenth and eighteenth centuries smoothing planes of northern Europe were similarly decorated with elaborate scrollwork, often carved directly into the hardwood stock (Bealer 1972, 171).

The discovery of a Roman plane in the summer of 2000 near Goodmanham in East Yorkshire, United Kingdom, also merits special mention. The Goodmanham plane

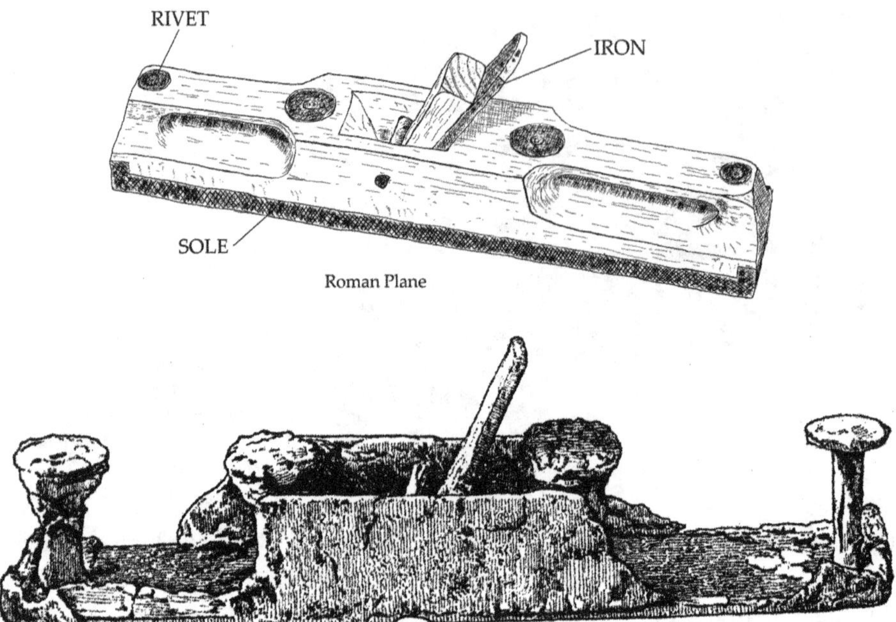

Fig. 3.32. (top) Plane from Verulamium, restored with its wooden stock. Only the iron sole and rivets were preserved. L: 43.9 cm. Verulamium Museum (St. Albans, U.K.) dated ca. A.D. *300. (*bottom*) Plane, iron sole, four rivets, side plates, Roman Calleva (Silchester). L: 34 cm. Reading Museum, inv. 07490. Fourth century. ([*top*] Author; [*bottom*] Evans 1894, 151, fig. 18)*

preserves not only its iron sole, rivets, and blade, but also, like the Saalburg example, its stock as well, which was rendered from a block of ivory. Why such a precious material was used is hard to say. The excavators speculate that the ivory may have been salvaged in late Roman times from another object or even that the plane may have been created as a presentation piece (Long, Steedman Vere-Stevens 2000, 19). The ivory stock preserves lateral slots that serve as hand-grips and a cylindrical iron crossbar against which a wooden wedge (not found) could be jammed into place to hold the blade firmly in position.

Ancient depictions of the plane supplement the archaeological finds. These fall in two basic categories: those depicting the plane as one of a group of otherwise isolated tools and those depicting the tool in use by a craftsman. Belonging to the latter type is a funerary relief in sandstone from Trier which depicts a Roman joiner standing at a bench using a jack plane, which he grasps with both hands (Goodman 1964, fig. 41); similar to this are the carpenters' guild fresco from Pompeii (fig. 3.33) and the Vatican gold-glass vessel (see fig. 3.23). Images depicting the plane floating on a blank background are generally found as part of a tomb's carved decoration. In all such examples the plane is shown with other tools but is usually the only one of the group that is exclusively used by woodworkers. The simple carved image of the plane in profile was considered recognizable enough to serve as a kind of emblem identifying the de-

Fig. 3.33. Fresco from Pompeii depicting a procession of carpenters. The covered float depicts (from left) Minerva (only her shield is visible), a man pushing a long bench plane, two men ripping a long plank with a frame saw, and a rendition of Daedalus, patron of carpenters, standing over the body of his son Icarus (or nephew, Perdix). The painting is believed to have belonged to a workshop (from Pompeii VI, 7.8-9), now in the Naples Museum, inv. 8991. H: 66 cm; W: 75 cm. First century. (Reproduced by permission, Soprintendenza per i Beni Archeologici delle province di Napoli e Caserta, Naples Museum neg. 1480)

ceased's profession. Important examples include a pediment-shaped relief from a tomb of a Roman carpenter now in the Museo Nazionale of L'Aquila, showing a jack plane and a *libella*; the rendition clearly shows both front and back grips and the projecting iron (Goodman 1964, fig. 42) (fig. 3.34). A virtually identical rendition is depicted on the funerary altar of Eutyches from Priolo (see fig. 3.15).

THE SAW (*serra*)

In the poetic imagination of Ovid, Perdix, the nephew of Daedalus, invented the saw by observing and copying the backbone of a fish (Ov. *Met.* 8.244–46). Toothed saws figured prominently at every level of woodworking, from the initial cutting and preparation of timber to fine veneering and jointing. The tool was employed in other industries as well. The quarrying of some soft stone, for example, was in part undertaken with special saws adapted for the purpose.

Fig. 3.34. Plane with front and rear grips depicted on a funerary relief. Above the plane a libella *(level) is clearly indicated. H: 38.5 cm. Museo Nazionale d'Abruzzo, L'Aquila, inv. 888. Dated to the late first or early second century. (G. Fittschen, D.A.I. 1007VW83)*

Whereas some heavy structural timbers, such as piles, could be used in their original round state or in a half-round (that is, split) state, the production of floorboards, furniture, doors, shutters, chests, or any objects formed from flat pieces of wood required good saws. Saw marks visible on Roman timbers excavated in Britain attest to the importance of the tool for cutting joints (Weeks 1982, 166).

From Pliny we learn that the teeth (*dentes*) of a saw were bent outward (known as the set) so as to expel the sawdust during the cut (Plin. *HN* 16.227; compare Theophr. *Hist. Pl.* 5.6.3). In fact, the most important function of the offset is to allow the kerf, the slit made by the saw, to be slightly wider than the sheet metal forming the blade of the saw, thus preventing the blade from becoming stuck in the wood (Goodman 1964, 116). This arrangement of the teeth also allowed the saw to cut while being pushed or pulled. The teeth of actual saws or saw fragments that have been discovered still reveal the offsetting that Pliny described, but examples in which this treatment is not employed also exist (Frere 1972, 166). The teeth of Roman saws, like those of their modern counterparts, could also be sloped in one direction, so that the actual cut took place on alternate strokes. Handsaws generally cut when pulled (the teeth slope toward the handle). Teeth were most easily formed by cutting notches in a flat strip of metal, invariably iron in the Roman period, with a file.

Ancient sawyers knew that wood that was still green and therefore contained high moisture content was difficult to saw because the wet sawdust clogged the saw and the kerf. Yet if the wood was too dry, according to Theophrastus (5.6.3), the saw would skip over the surface of the wood as if over baked earthenware. Sawdust might be saved for use as a packing material in ancient times and has in fact been discovered by archaeologists, who have been able to identify species of trees from the tiny particles recovered (Arena-Marcello 1963).

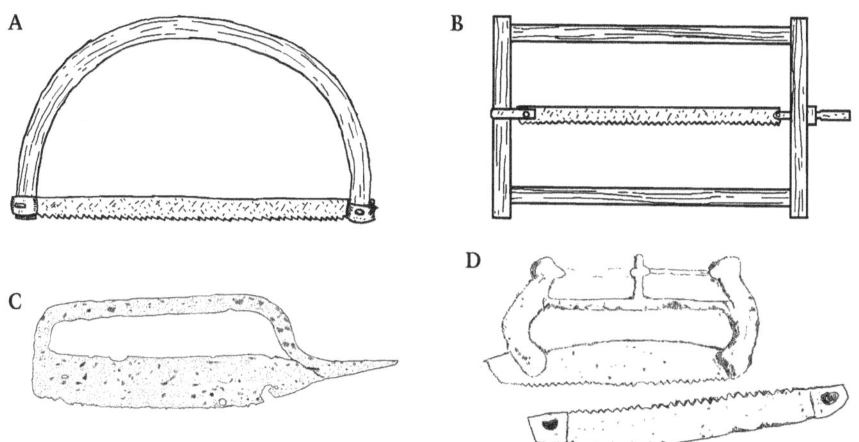

Fig. 3.35. The Roman saw. (A) Bow saw, bentwood frame, after an example from Fayum in the Petrie Museum, London; (B) frame saw; (C) a backed saw, with a tang for a (missing) handle, from the Landesmuseum, Zürich; (D) bucksaw and crosscut saw, from the altar to Minerva in the Capitoline Museums, Rome (see fig. 2.2). Drawings are not to scale. (Author)

Roman saw blades and their frames were made in various sizes depending upon the type (cross-grain or a ripping, that is, lengthwise, cut) and the fineness of the cut desired. Wolfgang Gaitzsch (1985) distinguishes five categories of Roman saws: bow saws (*Bügelsägen*), small frame saws (*Rahmensägen*), large frame saws for ripping logs (*Klobensägen*), crosscut saws for cutting heavy timbers and logs (*Bandsägen*), and small handsaws (*Stichsägen*; compass, or keyhole, saws) (fig. 3.35). The last of these (Latin: *serrulae*) were little more than serrated knives operated with one hand. The bow saw kept its blade in tension by the curved and springy wooden handle (something like the modern Swedish-style saw). Gaitzsch developed these categories by comparing ancient painted and sculpted depictions of the tool; extant Latin does not preserve the names of the specialized variants.

Frame saws of all sizes employed a thin, straplike blade kept under tension within a rectangular wooden frame. The Roman types described here remained in use, with only minor changes, until the early twentieth century. The smaller version of the frame saw (*Rahmensäge*, or bucksaw) could be operated by one man. The blade was kept under tension between two handles by a twisted strap of cord or sinew held in place by a toggle stick. The handles themselves are characteristically shown as bowed. The twisted cord was prevented from slipping off the frame by a decorative hook, in some cases a volute, that echoes the appearance of the ancient lyre. Some Roman depictions of the bucksaw, such as that on the marble relief of the furniture maker's shop in Rome (fig. 3.36), do not depict a toggle stick, while other more or less contemporary renditions of the tool include the device (see fig. 2.2). A blacksmith's funerary relief from Isola Sacra depicts a bucksaw with double toggles. William Goodman (1964, 122) suggests that the tensioning cords of some Roman saws of this type were simply twisted, wetted, and dried in order to maintain proper tension. In cases in which toggles are not

Fig. 3.36. Detail of the woodworking shop from Rome shown in fig. 2.3. A pair of calipers and a bucksaw hang on the wall behind the craftsmen. In the center a three-legged table with carved lions' heads on the legs is prominent. (Photo: Author)

shown, the artist may have ignored the feature, or possibly such details were painted and are no longer visible.

The largest frame saws, holding a blade in the center of a rectangular wooden frame, were employed vertically to rip logs or beams into boards. Unlike the configuration of the bucksaw, the teeth of the large frame saw are perpendicular to the plane defined by the rectangular wooden frame. The saw-pit, which was probably employed in Roman times, enabled one man to stand below the log (the pitman) and another to stand above (the top man), on the log itself. In this way a long log could be ripped into boards (Meiggs 1982, 349). Presumably the blade of these large frame saws was kept in tension by the same method employed by sawyers through the nineteenth century: a simple screw mechanism, fashioned from iron or wood, was attached to one end of the metal blade and passed through the handle of the frame. This tensioning device is clearly shown on the frame saw depicted on the relief of the woodworkers' shop from Rome already cited (see fig. 2.3). In this particular representation the saw hangs vertically from a long square beam, attached by a short strap. Clearly some kind of sawing machine is depicted, perhaps a way of allowing one man to operate the tool to rip long planks.

Roman-period depictions of craftsmen using their saws appear frequently in painted scenes of daily life. Images of two-man saws in action are known from both Pompeii and Herculaneum. From the latter a pair of cupids is depicted ripping a timber with a bucksaw, an inappropriate tool, as it turns out, except for a very shallow (or cross) cut (fig. 3.37). On the fresco depicting a procession of carpenters from Pompeii, two workers rip a board with a large frame saw; one stands on the board, which is supported by a vertical prop, while his mate works below (see fig. 3.33). A similar composition is known from a funerary relief found in Gaul, where a pit saw and its operators are shown about halfway through a cut (Meiggs 1982, 348). Similar depictions of

Fig. 3.37. Bucksaw in use, from a painting found in Herculaneum depicting two cupids operating the tool. The original painting is lost. This reproduction also depicts a carpenter's bench, toolbox, and wooden shelf. (Blümner 1875 [II] 346, fig. 59)

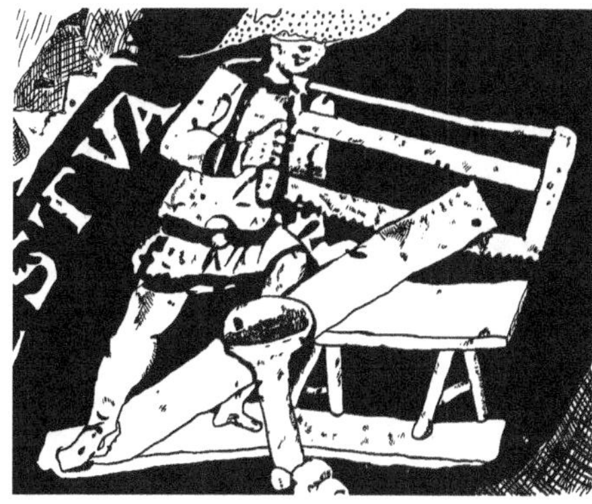

Fig. 3.38. Detail from the gilt glass vessel from the Vatican Museums (see fig. 3. 23). The bucksaw depicted, improbably, in a ripping operation. (Author)

craftsmen working alone with bucksaws (the smaller version of the frame saw, with the blade on one long side) exist, including the gold-glass vessel in the Vatican (fig. 3.38). In the Vatican rendition the craftsman is improbably using a bucksaw to rip a plank that he steadies with his right foot. Finally, a damaged marble relief found in the Isola Sacra cemetery near Ostia depicts a workshop scene dominated by two artisans, perhaps stonecutters, wielding an oversized bucksaw (or bow saw). Carved in deep relief, parts of the marble frame and the entire blade are now missing, yet the identification of the tool as a saw seems plausible (fig. 3.39).

There is no known depiction of a veneering saw in use. If postclassical examples are any guide, these saws looked like large frame saws (with the blade positioned in the center of the frame) but were wielded horizontally by two craftsmen to cut thin slices of exotic woods (Mercer 1960, 157). It is interesting that none of the depictions of saws in use represents the tools being applied to fell trees. On the Column of Trajan, where the clearing of forests is a repetitive theme, trees are always cut by axemen.

The sharpest saw blades were made of thin iron sheets. As might be expected, the state of preservation is usually poor. One is lucky to find a short fragment of a blade; the wooden handles and frames have not survived. Even so, from small fragments it is often possible to determine the basic type of saw represented by the find. The cutting

Fig. 3.39. Bucksaw in use, marble relief from the cemetery of Isola Sacra (Ostia). The large size of the saw in relationship to the artisans is unusual for this type of saw; perhaps the artisans are using the tool to cut stone. Ostia, inv. 138. L: 64 cm; H: 42 cm. Late first century. (Photo: Schwanke, D.A.I. neg. 80.3235)

and back sides of iron blades held under tension in a frame are usually parallel, while a handsaw that is to be pushed and pulled from one side generally exhibits a triangular configuration (Frere 1972, 166) (fig. 3.40).

Due to the scarcity of actual finds, most of our information concerning the appearance of the ancient Roman saw must come from artistic renditions. As is evident from the examples already cited, numerous depictions of saws carved on funerary reliefs and altars have survived. Bucksaws (along with planes, levels, and compasses) were also used as mint marks on Roman silver republican coinage (see fig. 3.29). A singular representation of a heavy-duty crosscut saw (other than a tiny representation used on a silver coin) with a handle attached to either end of a long blade is included in the collection of tools represented on the Capitoline altar to Minerva (see fig 2.2). Without such ancient representations little would be known about the appearance of the Roman saw and especially its handles and frame, even though important elements are often missing, such as toggle sticks. In addition, as we have seen, errors in depicting the correct tool for the job at hand can be cited.

THE WEDGE (*cuneus*)

Virgil's poetry refers to the ancestors of the Romans splitting wood with wedges before saws were widely employed (*Georg.* 1.144). The recovery of iron wedges from sites of the

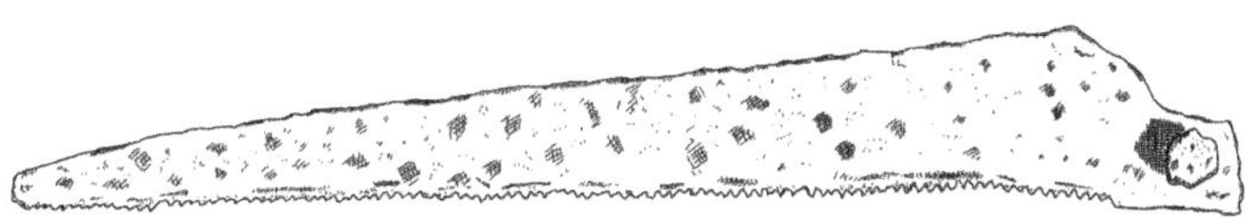

Fig. 3.40. Iron handsaw blade, from Verulamium. The wooden handle was attached by a rivet extant on the right side of the blade. The tang originally inserted in the wooden handle is broken. L: 41.7 cm. Ca. A.D. 160. (Author, after Manning in Frere 1972, 166 and fig. 61)

imperial period indicates that the wedge was always useful not only for the rough splitting of timber but as an aid in other cutting operations. Wedges inserted into the kerf of a piece of timber prevented saw blades from becoming stuck in the workpiece. In the workshop the wedge was indispensable when used with simple clamping mechanisms. Heavy wedges were certainly used for holding timbered centering in place during the construction of stone or concrete vaults. Wedges could be made of hardwood as well as of metal; if the latter, they were most commonly cast in iron.

STRIKING TOOLS: HAMMERS AND MALLETS (*malleus*)

Woodworking mallets were (and still are) customarily made entirely of hardwood; beech has been a favorite material for centuries. The mallet was used most often with wooden-handled chisels (*scalprum*, cf. Liv. 27.49.1), gouges, and small axes. It was also used to drive wooden dowels. The simplest mallets were fashioned from a single piece of wood and wielded like a small club or bat. The Egyptians realized that the life of the tool could be prolonged by fitting the head to a handle, thereby using the end-grain for the striking surface. Roman wooden mallets were characterized by a cylindrical head that was socketed and in profile something like an hourglass, a configuration sometimes described as "waisted" (Goodman 1964, 201). The appearance of Roman-period mallets is known through depictions in painting and sculpture. An outstanding example of a mallet in use is found in the painting from Pompeii of Icarus cutting mortises (see fig. 3.17). Mallets have also been identified on Roman funerary reliefs and are a common tool of the Roman soldier as portrayed on Trajan's column (fig. 3.41). The modern carver's mallet, characterized by a rounded, conical head, is not represented in Roman depictions.

Hammers with metal heads were used to drive nails and spikes of bronze or iron into wooden planks. By Roman times the hammer was developed to a point where it resembles the modern tool, that is, it had an iron head and in some cases a claw for the extraction of nails. Well-preserved examples of iron hammers from Pompeii are now stored in the Naples Museum. The heaviest iron hammers, or sledgehammers, were used when great force was necessary; this included the stunning of large animals, such as bulls, immediately before sacrifice. On Trajan's column a Roman legionary uses a short-handled sledgehammer to drive a pile into the ground for a bridge (fig. 3.42).

Fig. 3.41. A Roman soldier (center, top) is depicted striking a heavy mortising chisel (in his left hand) with a mallet. Column of Trajan, Rome, A.D. 113. (Anger, D.A.I. neg. 91.112)

MEASURING TOOLS: CALIPERS OR "DIVIDERS" AND COMPASSES (*circinus*)

Roman calipers look much like their modern counterparts. The basic model is characterized by two bowed legs joined at one end so that they can pivot. The caliper is used to transfer measurements between workpieces or between a plan (or a ruler) and a workpiece. A pair of compasses works on the same principle, but has straight legs to inscribe arcs and circles. The term *circinus* was apparently used for both types of tool. Its invention, like that of the saw, was attributed to Perdix, the nephew of Daedalus, by Ovid (*Met.* 8.247). Representations on Roman tombstones and on altars dedicated by craftsmen show both straight-legged (more common) and curved-legged examples (Burford 1972, pl. 8). Good depictions of the compasses, for example, can be seen on funerary reliefs now in the museums of Aquileia and Rome (see fig. 3.1; fig. 3.43); an actual straight-legged example in wood was recovered from the Nemi shipwrecks, dating to the first century A.D. An oversized device with curved legs is depicted hanging on the wall of the dedicatory relief to Minerva from Rome, already cited as an important physical document of woodworking practices (see fig. 3.36). This tool would have been of particular use for the turning of identically profiled legs. A rare example of a caliper gauge with one fixed and one movable jaw mounted on a straight wooden bar was recov-

Fig. 3.42. A Roman soldier hammers at a pile with a heavy mallet. Column of Trajan, Rome, A.D. 113. (D.A.I. neg. 71.2702)

ered in 1982–83 from the Giglio shipwreck off the coast of Tuscany (fig. 3.44) (Bound 1991, 31). Pins on the jaws of this tool permit accurate scribing on a wooden surface. The instrument may be of Greek manufacture from an early period (ca. 600 B.C.), but Roman versions undoubtedly existed as well.

The Plumb Line and the Level (*perpendiculum, libella*)

The *perpendiculum* appears to refer to the plumb line (*linea*) and bob used for establishing vertical lines, the *libella* to a tool for checking a horizontal surface. The *perpendiculum* looked little different from modern counterparts: a metal weight with a pointed end (shaped like a spinning top) to which a string was attached. Beautiful examples of bobs fashioned from lead-filled bronze have been recovered from Pompeii; the tool is also clearly represented on a relief in Rome dedicated to the god Silvanus, deity of the forest (fig. 3.45).[4]

The *libella* was used for checking horizontal surfaces. This compact instrument was known from Egyptian times (Goodman 1964, 200). Three pieces of wood connected by sheet metal braces were assembled in the form of the letter A (Adam 1994, 42). From the apex hung a line and plummet. When the line was centered on the horizontal piece of the frame, the surface upon which the instrument stood was level. The A-level was a popular motif carved upon the tombs of Roman freedmen, easily recog-

Fig. 3.43. Relief, perhaps in honor of a faber carpentarius *(wagon maker), depicting compasses, a two-foot ruler, and a wheel with eight spokes. The ruler appears to be divided into twenty-four* unciae *(fifteen are visible);* palmae *are marked with an X. H: 47 cm; W: 68. Aquileia Museum, inv. 1231. Late first–early second century. (D.A.I. neg. 82.415)*

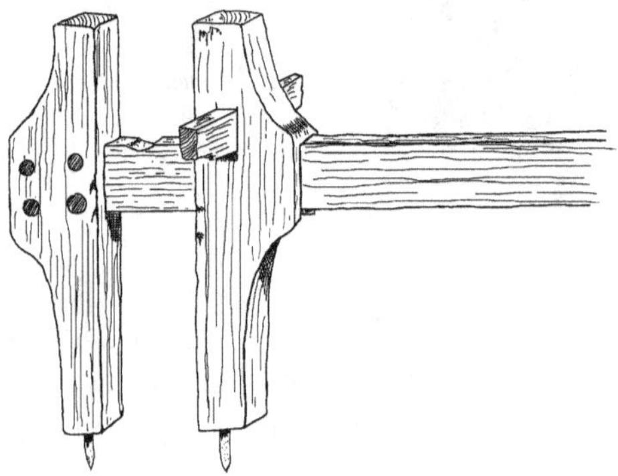

Fig. 3.44. Sketch of a wooden caliper gauge found on the Giglio wreck. Ca. 600 B.C. (Author, after Bound 1991, 31, fig. 65)

nizable to passersby; it signified the builder, stone mason, or carpenter and was frequently shown with other tools related to the given craft (see figs. 3.1, 3.34).

RULER (*regula*)

Like its modern descendants, the Roman ruler was a simple but essential tool consisting of a straightedge, usually of bronze, etched for measurements. The Latin *regula* is derived from the verb *rego*, which means "to fix the line." As far as we know, there were no numbers marked on Roman rulers, but there were lines or symbols to indicate segments. A folding *regula* one foot (29.6 cm) in length, for example, might be divided up into *palmae* (fourths, ca. 7.4 cm) and either *unciae* (twelfths, ca. 2.5 cm) or *digiti* (sixteenths, 1.85 cm). The increments were indicated by vertical lines, circles, dots, or crosses. Folding rulers of bronze have been found in various parts of the western empire; a comprehensive list was published by Goodman(1964, 190).

The most common form of bronze ruler discovered is the one-foot ruler with a hinge placed at the exact center so that when folded the ruler measures one-half foot (just under 15 cm) long. The folding rulers in the Naples Museum and in the Museum

Fig. 3.45. A statue base dedicated to Silvanus by P. Tartarius Chiaritus and his brother Atticus. The sides are decorated with tools used by builders. On the side shown: plumb line and bob, chisel, mallet, compasses, and square. Antiquarium del Celio, Rome. (Author)

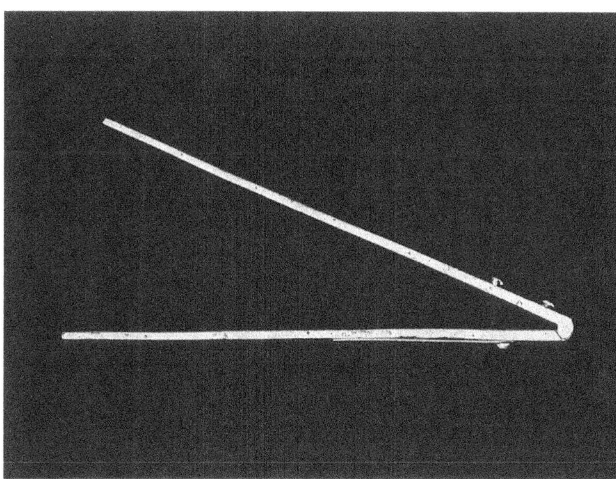

Fig. 3.46. Bronze foot-rule from Roman London. The ruler can be locked in the open position by the metal bar seen on the lower arm, which can be held in place by the two standing rivets seen above. Dots indicating unciae are also visible. Total length is one Roman foot (29.6 cm). (Courtesy of the Museum of London, inv. 29.94/7)

of London still work; the extended ruler is held open by a swinging latch (fig. 3.46). Rulers depicted on funerary reliefs tend to indicate tools of various lengths and markings. There is no indication that they were hinged (see figs. 3.4, 3.43). The foot rule of P. Ferrarius Hermes, for example, is marked with only four *digiti*. The rule on the Aquileia relief is a two-footer with *palmae* indicated with an "X."

SQUARE (CARPENTER'S OR FRAMING SQUARE) (*norma*)

The carpenter's square or framing square can be used to test right-angled joints, to scribe lines for right-angled cuts, to compute the pitch of a roof, to cut a rafter to the correct angle, and to mark out the rise of a stair. *Ad normae cacumen* can be translated as "right angle" (cf. Vitr. 4.3.5, in his discussion of the carving of triglyphs). In modern

Fig. 3.47. Square from Neftenbach (Switzerland); angles are 45, 90, and 135 degrees. (Courtesy of the Swiss National Museum, Zurich, inv. A-5392, neg. P-5950)

Fig. 3.48. Bronze square from Pompeii. First century. (Courtesy of the National Museum, Naples, inv. 76689, neg. 112251)

parlance, the shorter end of the square is called the tongue and the long edge the bar. Pliny credits the same Theodorus of Samos posited as the inventor of the lathe as the creator of the *norma* (Plin. HN 7.198).

Isidore (*Orig.* 19.18.1) describes how a simple square can be made from three wooden battens fixed to one another in the form of a triangle. An outstanding example of such a tool is an iron framing square discovered at Neftenbach, now in the Landesmuseum at Zürich. In this case the sides (ca. 6, 8, and 10 inches long) form a right-angled triangle; a fourth bar provides an additional 45-degree corner extension (Goodman 1964, 200) (fig. 3.47). The most common type of *norma*, represented by the two bronze specimens recovered from Pompeii and now stored in the Naples Museum, consist of two arms forming a right angle (fig. 3.48). The smaller of the Pompeian *normae* has arms of equal length (16 cm); both terminate in a crescent profile. The broader of the two arms (1.9 cm; the thinner is 1.2 cm across) is fitted with a flange or fence 2.3 cm wide, which enabled the *norma* to stand upright without additional support or served as a guide to push against the edge of a plank.

Most of the squares depicted on Roman relief sculpture are of the two-armed variety represented by the Pompeian examples (see figs. 3.1, 3.4, 3.45). Such representations are found on funerary reliefs and dedicatory bases. The tools carved on the monument of P. Ferrarius Hermes include a *norma.* A military architect depicted on the Column of Trajan cradles a *norma* under his right arm (see fig. 3.7). Although the tool is partially obscured, a slightly curved plate formed the hypotenuse of the two arms, suggesting that the *norma* could be combined with a *libella.* The English words "norm" and "normal" are derived from the Latin *norma.*

Miscellaneous Tools Related to the Field of Carpentry

THE JOINER'S DOG

The dog is a metal device, usually of iron, that acts as a clamp or cleat to steady a wooden workpiece on the surface of a workbench. Similar in appearance to a large staple, one leg of the dog is anchored in a hole in the workbench while the other grips the wood. Examples of iron dogs have been found at the excavations of Fishbourne (Cunliffe, Rudkin 1996, 230, figs. 1, 2).

CLAMPS

Clamps are essential to most methods of cabinet and furniture making. They are particularly important when gluing is employed to join planks edge to edge (see chapter 4). Since the components of the clamps were probably fashioned entirely out of wood, it comes as no surprise that ancient examples of clamps are not extant. And the utilization of the clamp does not appear to be a subject of interest to the ancient painter or sculptor. One exception is the previously cited painting from Herculaneum, now lost, that depicts two *erotes*, or cupids, wielding a bucksaw (see fig. 3.37). A device depicted at the end of the table resembles a modern C-clamp. But without the original painting to examine, the accuracy of this rendition must be questioned.

A simple, effective clamp can be made from a wooden board and a wedge or two. An opening corresponding roughly to the desired width of the finished work can be cut in the board, boards can be laid at right angles within the gap, and the assembly tightened with wooden wedges (fig. 3.49). Most clamps, however, depend upon the tightening action of a screw. We know that the principle of the screw was understood and applied by Romans in mechanical devices. In all such cases the screw was fashioned from a wooden blank; the invention of the device itself has been attributed to the Greek mathematician Archytas of Tarentum (died ca. 350 B.C.). Actual Roman examples include a wooden press, now carbonized, used for drying cloth which was

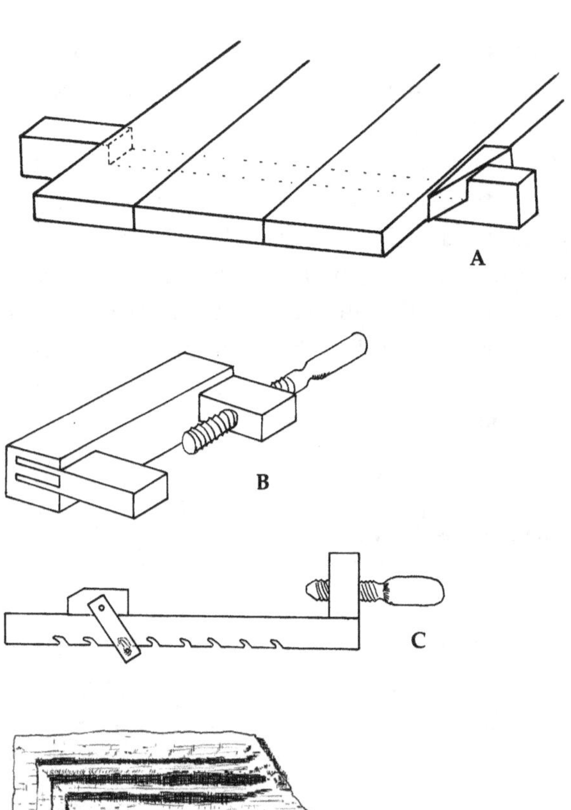

Fig. 3.49. Sketches of clamping mechanisms: (A) simple bar clamp made from a notched beam of wood and a wedge; (B, C) all-wooden clamps used in preindustrial northern Italy (after examples in the Museo degli Usi e Costumi della Gente Trentina, San Michele all'Adige, Italy). Note how the bars in example B are attached with a multiple slot-and-tenon joint. (Author)

Fig. 3.50. Depiction of a capstan of an olive (?) press from Aquileia. Note the threads of the screw. National Museum, Aquileia. First half of the third century. (Author)

found at Herculaneum and the image of a screw-press with a capstan tightening mechanism carved on a relief discovered at Aquileia dating to the third century (fig. 3.50). As has been mentioned, the wooden screw would have been used as a tightening device in large frame saws holding a central blade. Thus it is not difficult to imagine that Roman clamps were similar in appearance to the mechanisms made from hardwood by Italian woodworkers through the early twentieth century.

IV Joints

The fundamental skill of the woodworker throughout the ages has been measured by his ability to join securely—and with elegance—two pieces of wood (hence the English *joiner* and *joinery*, derived from the Latin *iunctura*). Sound wooden joints do not require metal fasteners, but the use of nails and metal plates, such as gussets, makes the task easier. The woodworker needs greater skill and expends more time in making strong joints of wood alone. In the ancient world, wood-to-wood joints were often made without the use of glue, nails, or clamps, although all these aids were known; many fine woodworkers today prefer to employ few if any nonwooden devices for joints.

Two or more wooden pieces were nevertheless commonly coupled with metal fasteners including pins (*fibulae*), nails, and spikes (*clavi*). These aids were normally associated with connecting heavy timbers: the framing of floors and roofs, the building of bridges and ships, and the fashioning of utilitarian objects like carts, barrels, wheels, and wine presses. Julius Caesar writes of iron nails used in ship construction having the thickness of a man's thumb (*BGall* 3.13.3). Even in these utilitarian applications, complex bevels, mortises and tenons, and scarfs were used in conjunction with metal; modern analogs are numerous. Slender metal rods were also used on finer examples of carpentry. The moldings of the Roman-period wooden sarcophagi excavated on the shores of the Black Sea in the late nineteenth century were affixed by pins that anchored the carved strips to the frame of the casket (fig. 4.1). The metal wood-screw, a mainstay of modern carpentry, was not employed by the Roman craftsman, despite the fact that the principle of the screw was understood and used in many industrial applications, as we have seen.

The use of flax or hemp twine to lash or sew wooden planks together must have been widespread, though archaeological evidence for such methods is understandably rare. Such sewing was presumably one of the first methods utilized for joining wood. The practice has been thoroughly documented with the construction of ancient boats, both large and small, and was shared by ancient Mediterranean cultures. Examples have

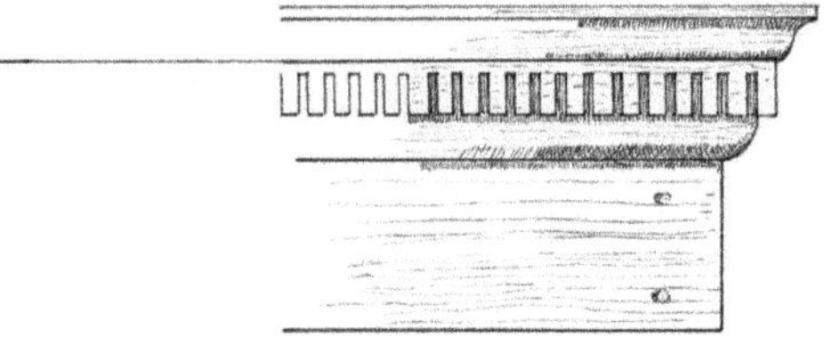

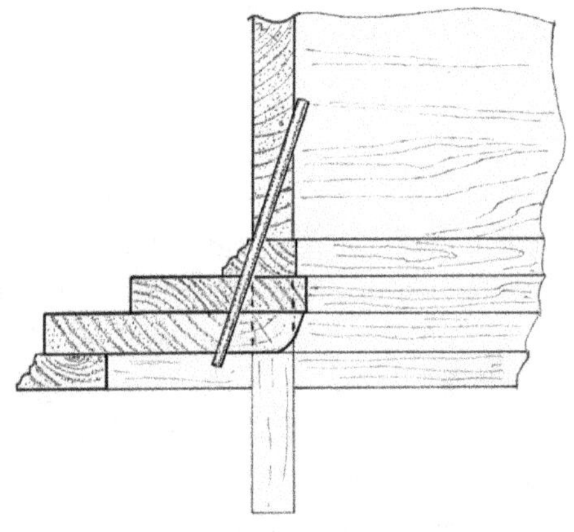

Fig. 4.1. The use of metal pins to attach the moldings of Roman-period sarcophagi, excavated in and around the area of Kertch, on the Black Sea. Both from the Hermitage, St. Petersburg. (top) *Crowning molding, inv. P.1910.21;* (bottom) *base molding, inv. Pan. 1692.* (Author, *after Wasowicz* 1974, 110, *fig.* 51; 111, *fig.* 53)

been found in England (McGrail and Denford 1982) and on the Italian coast (Bound 1991, 31). Pliny refers to the sewn ships (*sutiles naves*) of Homer's day (HN 24.65), but even in the historical Roman period (and beyond) boats were still built with sewn planks, especially along the Adriatic coast and the banks of the river Po. Commonly the holes used for binding did not penetrate the planks entirely but were bored at an angle, from the inner face to an inside edge, so that the bindings were not visible on the outside of the hull (Bonino 1985, 91). Bronze models of ships found in tombs from central Italy, like the so-called Noah's Ark discovered at Vetulonia (seventh century B.C.), include details such as lashing of the forward stem and figurehead of the ship to the hull (Göttlicher 1978, no. 428). The oldest, most spectacular example preserved is surely that of the cedarwood boat built for the Egyptian pharaoh Cheops in the mid-third millennium B.C. (Lipke 1985).

The type of wooden joint used in a particular application depends on both the stress that will be placed upon the wood and the direction of the grain in the piece to be joined. Equally important is choosing a joint that will not crack or split the wood during the normal cycles of expansion and contraction caused by changes in atmospheric humidity (Azevedo 1957, 13). In structural applications, for example, a mortised joint is better for a vertical connection (for example, separate beams for a vertical framing element, where the wood is in compression, with force parallel to the fibers), while a scarf joint is more suited for a horizontal join subjected to tension (for example, a keel of a ship). Other joints, such as the draw-tongue joint (a type of mortise and tenon), are designed to yield somewhat under torsion (twisting). The give necessary for the planks that make up a wagon wheel or the hull of a ship is not desirable for the roofing of a building (Weeks 1982, 164). Thus a joint that is good for one purpose may be unsuitable for another.

Types of Joints Used by Roman Woodworkers

The types of joints employed by Roman woodworkers are more easily comprehended through drawings rather than physical descriptions (fig. 4.2). Moreover, the *names* used to distinguish different kinds of joints can be confusing, since there is no standard of usage in English, and one joint may have several different names or be given different names when employed in diverse applications. As Jane Weeks writes in her study of carpentry joints observed primarily in timbered, Roman-period wells in Britain, "As in all areas where archaeology overlaps with other disciplines, the archaeologist is faced with a ready-made technical vocabulary and the problem of whether to apply it wholesale, and thus render the subject more abstruse than necessary, or whether to use a simpler series of terms, and run the risk of longwindedness" (1982, 159). In the summary offered in this chapter, the English terms used for woodworking joints are those in common usage among craftsmen who work with wood. There are some minor variants between American and British usage, and variant terms for the same joint are included when practical in the following discussion.

Latin terminology is problematic in a different way. Surely there was a rich vocabulary for the variety of joints used by the Roman woodworker, but since no treatise of ancient woodworking has survived we have only texts written by (or intended for) nonspecialists to consider. These do provide us with some of the essential vocabulary but tend to disappoint where specifics are concerned. By far the most important evidence about what Roman joints looked like is owing not to written accounts but to actual wooden objects recovered from a number of contexts in the Roman world.

The simplest form of joint is known in English as a *butt joint*. The edges of the workpieces are simply pushed together and nailed or glued. The butt joint is a mainstay in rough carpentry, a term used to indicate the basic practice of framing a structure. Butt joints can be quite strong when braces or gussets are used. They are especially suitable for temporary structures which require long vertical timbers. Imprints of heavy timbers placed vertically to support the wooden forms of the concrete foundations at the Temple of Venus and Roma in Rome (second century) show evidence of the butt joint. This site will be considered in more detail in the following chapter.

A more elegant and stronger way to attach two lengths of wood to create a longer element is with a *mortise and tenon* or a *scarf*. In the mortise and tenon, a tongue of wood (the tenon) is inserted into a recess (the mortise) to make the joint. Usually, once the join has been made, both mortise and tenon are hidden. The tenon (*subscus*) can be

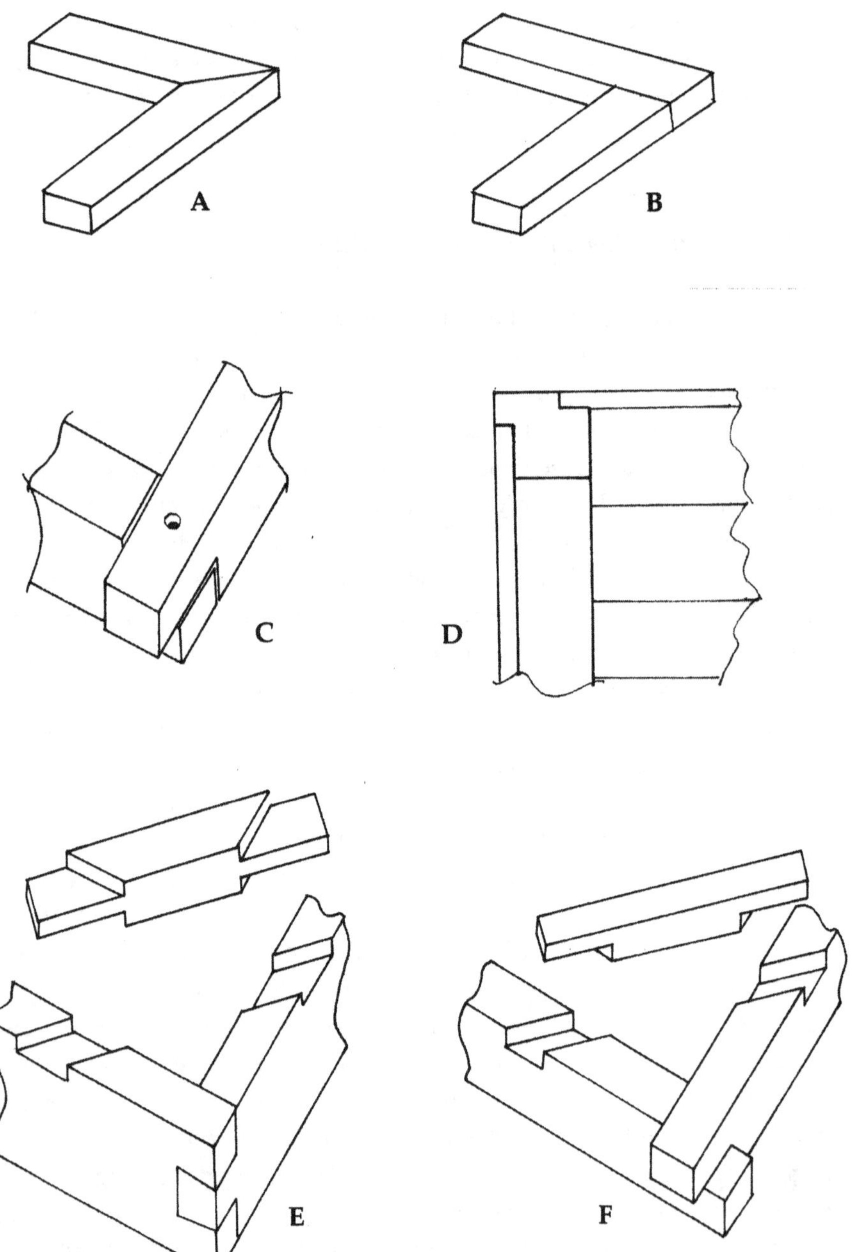

Fig. 4.2. Roman woodworking joints. (A–F) Corner joints: (A) simple miter; (B) butt joint; (C) half-lap (York); (D) rebate in a corner post for attachment to side planks (London); (E, F) corner braces with half-lap and saddle joints (York). (G) mortise and tenon; (H) tongue-and-groove joint hides end-grain of parallel glued boards (Herculaneum); (I) half-lap (saddle). (J–M) Dovetail joints: (J) asymmetrical (half-) dovetail (London); (K) dovetail; (L) lap dovetail or dovetail halving joint (London); (M) edge-halved scarf with one dovetail butt (London). (Author, adapted from: Adam 1994, 101 [G, I, K]; Goodburn 1991, 199 [D, L, M]; Mol 1999, fig. 22 [F]; Raines in Carver 1978, 23 [E, F]; Weeks 1982, 158 [J])

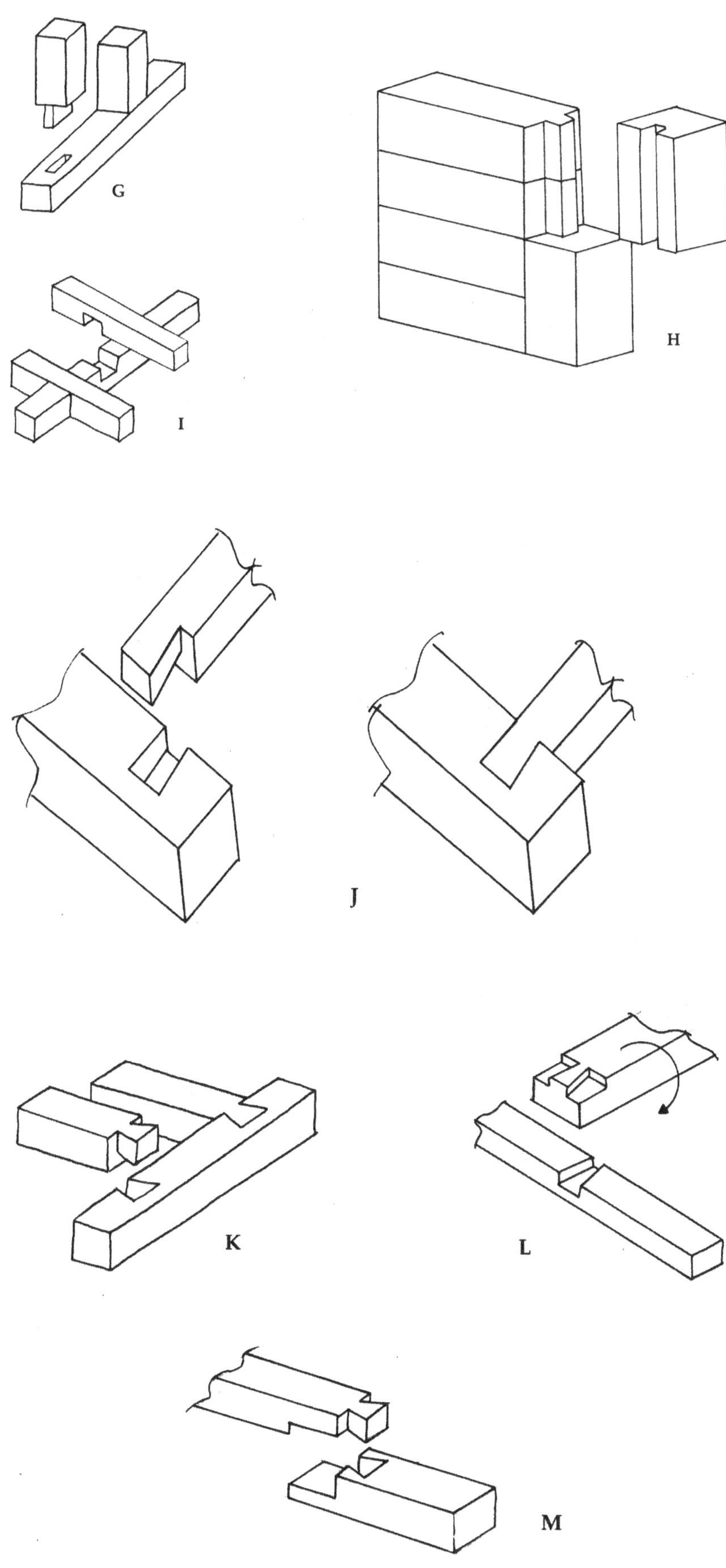
G
H
I
J
K
L
M

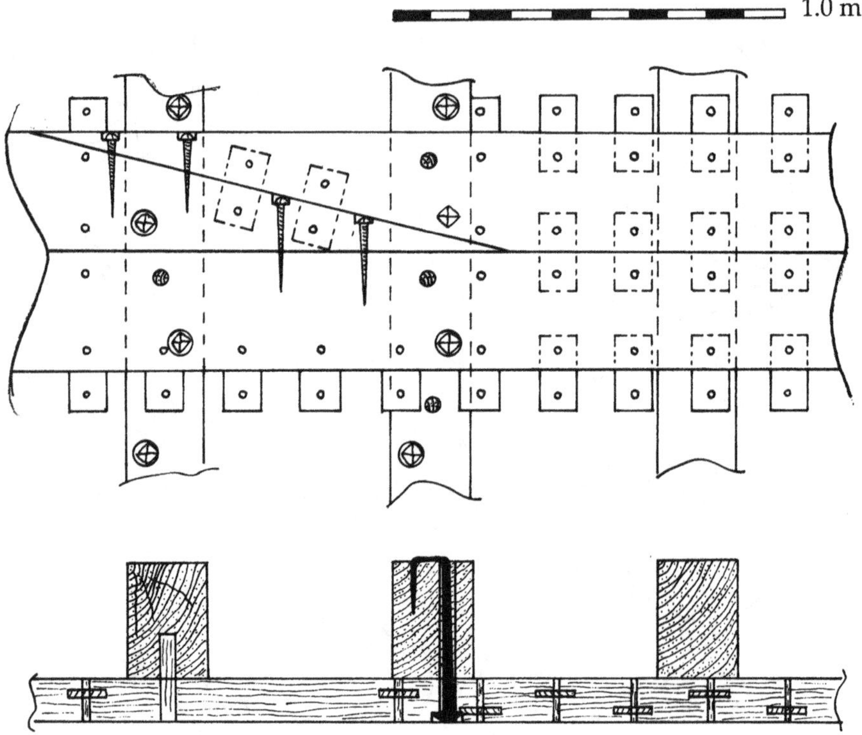

Fig. 4.3. Mortises and tenons used to connect the strakes of the Lake Nemi barges. Note how scarfs are reinforced with iron nails. (bottom) *An edge-on view of the planks. Note the staggering of the mortises. Early first century. (Author after Ucelli 1950, 153, fig. 153)*

cylindrical or rectangular in shape. The mortise joint was of great antiquity, known, for example, to the builders of Stonehenge in England (last phase, second millennium B.C.). It was universally employed in Roman carpentry; many a Roman apprentice must have spent long hours with hammer and chisel cutting mortises to his master's specifications (see fig. 3.17). Thousands of mortises were cut into the planking of a single Roman boat to join the strakes of the hull edge to edge; usually each tenon was reinforced with an additional dowel, as was the case of the sunken imperial pleasure barges discovered at Lake Nemi seventy years ago (fig. 4.3). Doors and window shutters and all types of furniture relied upon the mortise and tenon. In timber buildings, uprights were securely fitted to horizontal wall plates and sleeper beams.

The most distinctive form of the mortise and tenon is the *dovetail joint*, so described because of the distinctive fan-shape, or tail, of the tenon. The Latin *securicula* is best translated as an "axe" or "hatchet" blade joint. Dovetail joints were used in both heavy framing jobs and fine furniture work. Dovetails might tie the ends of heavy timbers, such as floor or ceiling joists, at right angles to a longer timber. Such practice has been observed in the well-preserved floor decking of a riverside building on the Southwark waterfront, London, dated by dendrochronology to the mid-second cen-

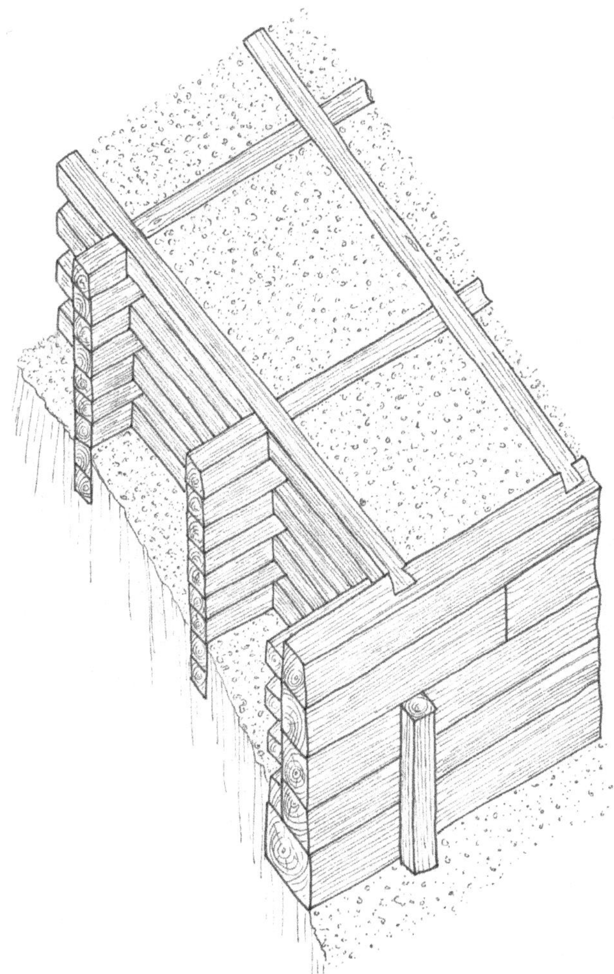

Fig. 4.4. A retaining wall along the London waterfront, created by half-lapped heavy timbers and dovetails. Second century. (Author, after Marsden 1994, 30)

tury (Brigham, Goodburn, Tyers 1995), and for contemporary pier construction along London's Roman port (Marsden 1994, 28 [fig. 4.4]).

The fan-shaped tenon of the dovetail joint must be cut with the grain, that is, at the end of the beam or plank to be joined, while the mortise can be cut either with or across the grain. The tenon itself can also be cut as an independent element, fan-shaped on both ends so that it has the silhouette of a butterfly and used as a fastener for two elements with opposing mortises. Oaken dovetails, for example, were used to hold the stone blocks of the Forum of Augustus in Rome (2 B.C.) in place during construction (Ganzert 1996, 124). The simplest way to cut the dovetail was to remove only one side from the stock, leaving an asymmetrical tenon sometimes referred to as a half dovetail. Well-preserved examples have been found cut on the timbers of a lined well in England (Weeks 1982, 158).

Another variant of the dovetail was employed to join heavy squared timbers side

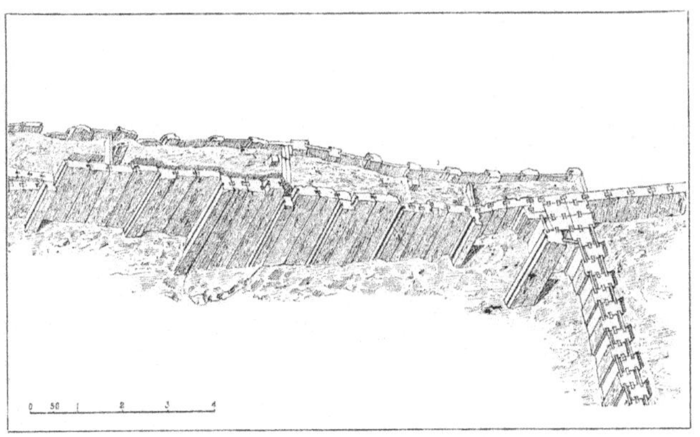

Fig. 4.5. Retaining wall built of squared beams and flat planks. Discovered at Lake Nemi. The excavators reported that the beams were connected with dovetailed slats. (Giammiti in Ucelli 1950, 121, fig. 126, courtesy of the Istituto Nazionale di Archeologia e Storia dell'Arte, Rome)

by side in order to create a robust retaining wall. At the site of Genzano, for example, on the shores of Lake Nemi near Rome, long dovetailed tenons were inserted into grooves cut into the shorter faces (20 by 30 cm in dimension) of rectangular beams to create a continuous palisade (Ucelli 1950, 126) (fig. 4.5). This practice has also been documented in Rome, where similar dovetail joints joined the oaken piles that formed a stout mole along the Tiber River in the Roman imperial period. This discovery will be discussed in more detail in the following chapter.

Cabinetmakers used the dovetail as a strong corner joint for boxes, chests, and drawers. A simple variant, using right-angled rather than fan-shaped mortises, is known as a finger or box joint (fig. 4.6). Roman use of the dovetail for connecting the sides of boxes has also been documented through archaeological finds. In 1976 a small wooden box, or *loculus* (45 cm long by 37 cm wide and 34 cm high), that employed both dovetails and iron corner straps was found in a late first-century villa excavated at Bradwell (Buckinghamshire, United Kingdom). Here the pins, or segments of wood visible between each dovetail, measured ca. 1.20 cm at the base and ca. 2.60 cm at the widest part. The congruency between the greatest width of the pin and the thickness of the wooden sides of the box corresponds to modern practice (Keepax and Robson 1978, 38). Another outstanding example, assembled without the use of any metal plates, was recovered from the Comacchio wreck. This *loculus* is significantly smaller in overall dimensions (6 cm wide by 4.5 cm high by 10.4 cm long), and exhibits finely cut dovetails and mitered corners (Desantis 1990, 266) (fig. 4.7).

The *scarf* is yet another option for joining two pieces of wood end-to-end to make a longer continuous wooden unit. The most common method of scarfing was to cut the end of each plank to be joined on a bias so that the surface area of the connection would be maximized (the resulting join is sometimes referred to as the splayed scarf). Nails could be driven through the edges of the plank to strengthen the scarf, as was done with the strakes of Caligula's pleasure barges excavated at Lake Nemi (fig. 4.8). At Nemi, to prevent the planks from splitting, each nailing point was first bored and filled with a softwood dowel to receive the spike (Barrett 1989, 202). This time-consuming

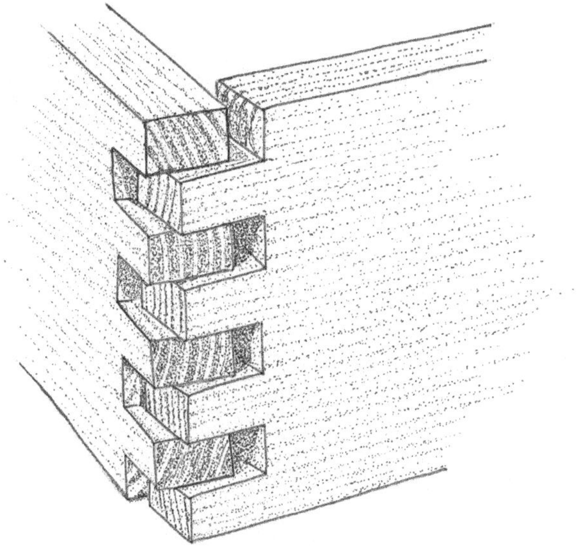

Fig. 4.6. The box or finger joint, shown in the process of assembly. (Author)

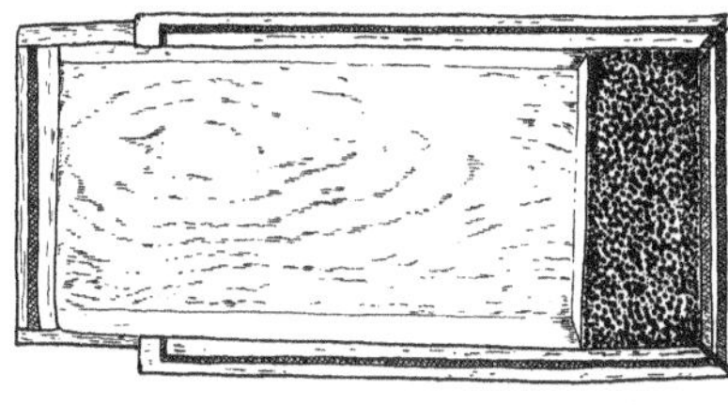

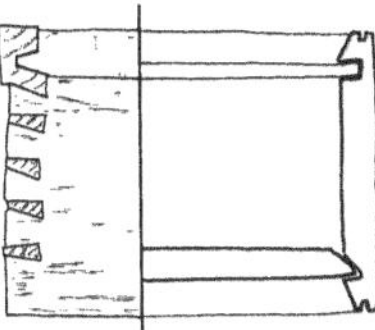

Fig. 4.7. A wooden box (loculus) from the Comacchio wreck. The beveled edge of the top slides into grooves cut into the sides of the box. The corners are joined with dovetails and top miter joints. H: 4.5 cm; L: 10.4 cm. Museo della Nave Romana di Comacchio, inv. 55075. First century B.C. (Author, after DeSantis 1990, 265, no. 237)

practice was not confined to the emperor's pleasure craft; the identical use of such trenails (also treenail, or trunnel) has been documented in the planks of the working boats of the harbor at Ostia, Rome's port (Scrinari 1979, 48). Indeed, the most complicated scarfing documented has been found in the timbers used for the keel of an Ostian ship from the third century, which employs offsets and keys to form a strong, rigid bond that will resist torque caused by the currents and waves upon a hull (fig. 4.9).

As might be deduced from the examples just cited, the scarf is best for horizontal connections; it is still the single most important joint for wooden boat construction for keels, clamps, and planking. Serpentine scarfs cut in the form of a curve and countercurve were used to create the long cedar strakes of the famous Egyptian ship found buried at the foot of Khufu's pyramid in the 1950s. The scarf is vulnerable when

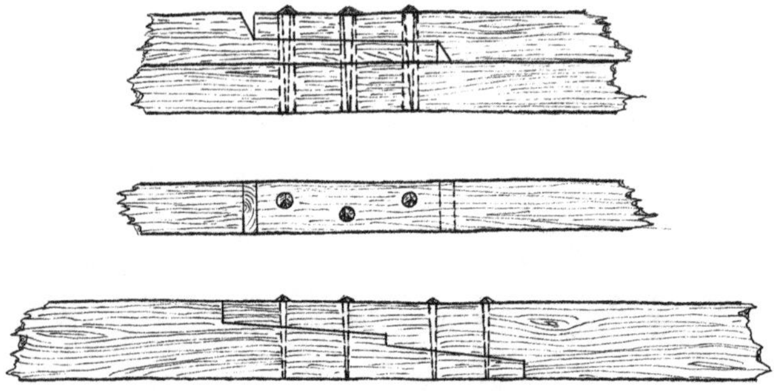

Fig. 4.8. Scarf joints used on the Lake Nemi barges. (top and middle) *The scarf is nailed to a lower beam to strengthen the join. Note the predrilling for the staggered spikes; the bored holes are filled with soft wood;* (bottom) *the scarf, with offsets, is nailed. First century. (Author, after Ucelli 1950, 156)*

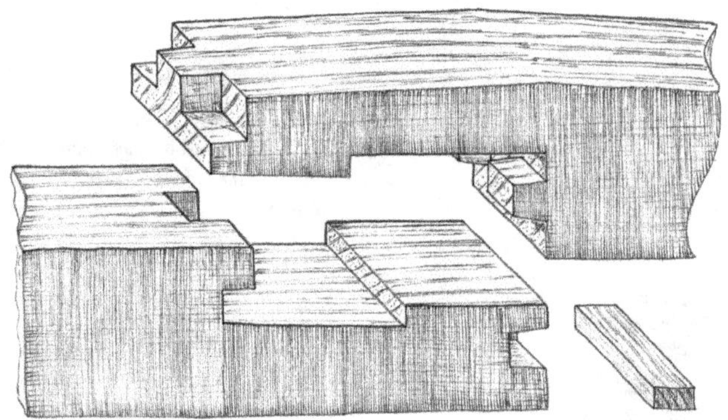

Fig. 4.9. Scarfing of a keel from a cargo ship excavated at Ostia in 1962. The scarf is held tightly in place by the small rectangular key. H: 17 cm; W: 12 cm. Museo delle Navi, Ostia (Rome), inv. 37797. Third century. (Author, after Ricciardi in Scrinari 1979, 42, fig. 18)

under tension (that is, a pulling force), as would be the case if used as a tie-beam. There is not much point in using a scarf for vertical elements in compression; the extra work to make the joint is not usually worth the trouble. Although Romans apparently avoided using scarfed timbers as tie-beams (even though modern reconstructions of ancient buildings sometimes include this feature), scarfing would have played a role in construction in such applications as the purlins used on a large roof or even the main ridgepole, the latter a massive timber in archaic-period buildings that supported the primary rafters. Obviously scarf joints are not seen in wooden objects that do not require long timbers. It is not, therefore, a joint associated with furniture making. The more elaborate the scarf, the greater skill and time required to make it, and in many cases, particularly if the join was hidden from view, the use of iron gussets would have been easier and just as effective.

Another common way of joining two pieces of wood of the same thickness, for either linear or angled connections, is with the *half-lap joint*. In this case half of the thickness of the stock is cut away at each end of the two planks to be joined. The half-lap differs from the standard diagonally cut scarf in two major ways. First, it is a suitable joint for timbers that are being joined vertically. Second, it is a useful way of joining two planks at right angles to one another to form a rectangular frame (also known as a corner halving joint). Since the half-lap joint does not hide end-grain, it is used where strength is more important than aesthetics.

Fig. 4.10. A wooden window grill exhibiting the use of the saddle joint. Similar wooden grates were used in Roman doors and windows or as barriers between columns. From Castel Sant'Elia, Lazio, Italy. (Photo: Author)

Closely related to the half-lap join is the *saddle* or *halved* (or cross halving) joint. This is nothing more than a half lap that is cut at a point other than the very end of the piece of stock, so that two planks can cross and continue past the joint. This joint would have been a mainstay for the woodworker making wooden grills for windows and doors or the X shaped braces used for the wooden balustrades of bridges, balconies, decks of ships, and military towers. Such wooden grills can still be seen in the windows of older buildings in the province of Latium in Italy. An example from Castel Sant'Elia is illustrated here; each joint has been reinforced with a nail (fig. 4.10). With heavy timbers left round or roughly squared, the saddle joint could be used to build stout structures similar in appearance to the modern log-built house. These would be used throughout the Roman period for retaining walls, artillery emplacements, cofferdams, or well linings. In 1877 Pigorini discovered at the site of Castione, near Parma, features he described as *gabbioni*, large boxes or pens, skillfully made from squared timbers joined at the corners with saddle joints. Filled with clay and debris, the structures formed a perimeter wall around a late Bronze Age village (late second millennium

Fig. 4.11. Saddle-jointed heavy timbers arranged as a retaining wall at the Bronze Age site of Castione, northern Italy. Second millennium B.C. *(Pigorini 1883, pl. IV)*

B.C.), built by the so-called Terremare culture of northern Italy (Säflund 1939, 96) (fig. 4.11). We will see that this practice was to have a long history. Saddle joints both elegant and crude were employed to form timbered well-casings, as is apparent from the joinery observed at Skeldergate (York, late second–third century) and Camulodunum (Colchester, first century) (figs. 4.2, 4.12).

For right-angled joins where strength is not paramount and end-grain is to be hidden a *mitered cut* is suitable. Two pieces of stock are cut at a forty-five-degree angle and connected to form a right angle. Despite the relative weakness of the miter, it is the only practical solution for moldings, so that the decorative profiles can extend unbroken around each corner (it is for this reason that modern picture frames and molded door panels use mitered joints). The small box found intact at Comacchio, already cited, exhibits easily discernible miters on its upper corners.

Glued Joints

Glues were just as important a means of joining wood in Greek and Roman times as they are today. The parallels of specific usage between ancient and modern applications are striking. The most characteristic is the use of glue for joining two or more planks side by side to create a large, flat panel. Thus we can imagine the gluing of boards to make tabletops, door leaves, and the sides of a chest. Mols (1999) has considered several probable examples of gluing in the surviving pieces of wooden furniture he has studied from Herculaneum. Although traces of the glue had not survived, Mols surmised that in the absence of any other detectable methods of attachment, glue was the only solution (1999, 96).

Fig. 4.12. The saddle joint used with thin oaken planks to form a well casing. From Camulodunum. First century. (Reprinted by permission, Society of Antiquaries, London [Hawkes 1947, 127])

Such modern studies are substantiated by the unequivocal comments of Pliny and others that glues were essential to the craft of woodworking. Pliny credits Daedalus as being the inventor of glues (*HN* 7.198), distinguishes between fish and bull-hide glues, and even recommends the choicest parts of the bull for making glues (the ears and genitals, *HN* 28.236). The boiling of a cow's horns and hooves creates a sticky substance that remains liquid as long as it is kept hot. Pliny also notes that some types of wood are more suitable for gluing than others and that gluing is essential to veneer work. His reference to the gluing of the constituent parts of the doors of the Temple of Artemis at Ephesus (Plin *HN* 16.215) reveals that ancient woodworkers understood the necessity of tightly clamping glued pieces together while the glue set. From a chance reference in Lucretius (6.1069) we hear that a well-glued joint would remain fast even if the board itself cracked somewhere else. Thus while Romans did not have access to the strong synthetic formulae available today, their two essential concoctions clearly worked extremely well.

V Foundations

A revealing passage concerning the role of wood in everyday construction, in this case a farmhouse (*villa rustica*), is included as a kind of model contract in Cato's *de Re Rustica*, written ca. 160 B.C.: "If you are contracting for the building of a villa from the ground up, it is the contractor's obligation to make the following: all the walls, as has been specified, of stone and mortar, piers made from cut stone, all the wooden beams which are needed for the job, thresholds, (door)posts, lintels, supports, and (floor?)boards" (*Rust.* 14.1).[1]

In this passage Cato, a conservative Roman veteran and politician who was himself of common stock, manages to convey both the actual construction sequence of this hypothetical farmhouse and the two most important skills needed to realize the project. The stonemason was entrusted with the foundations, the walls, and supporting piers. Most of the masonry had to be in place before the work of the carpenter could begin: upper floors, ceilings, lofts, door and window frames, and roofing. Cato labels the person responsible for undertaking this work simply as *faber*. In this context the *faber* is probably the equivalent of a general contractor who will employ specialists to carry out the various phases of the job, even though the term often means, in other contexts, a "craftsman."

Cato's attention moves on to the furnishings and fittings of the house: the required number of tables, stools, benches, windows, and so on. He does not specify either the kinds of wood to be used or their dimensional requirements. Instead, he lays the responsibility for the choice of raw materials selected for the project on the shoulders of the client, who was expected to furnish the timber in its raw form as well as a good saw and plumb line (*Rust.* 14.3). These last would be used to rip the raw timber into planks, preliminary tasks performed under the supervision of the contractor or foreman (now described as a *conductor*). Ideally the timber was to be felled on land attached to the new farmstead; we see that Cato advises that the *conductor* should not only saw the timber into planks but fell the trees as well. At the very beginning of his treatise on agriculture Cato includes a stand of wood in his list of nine features of ideal farmland.

It is easy to guess what kinds of wood Cato had in mind for the villa in question.

Having concluded his remarks on estimating the price of construction and the dimensional requirements of the villa's foundations (14.4–5), he recommends the best felling times for oak, pine, and elm. All are species one expects to find on the arable land in the vicinity of Rome and Tusculum, the rural village where Cato was born. A full discussion of species and their native locales can be found in chapter 10.

I now turn to the various phases of construction and the use of wood in buildings large and small. Structures are built from the ground up; the phrase has not changed from Cato's day (*ab solo*). The first phase of any building project is the construction of the foundations. Here, too, as we shall see, wood plays an important role.

Foundations (*substructiones* or *fundamenta*) anchor a building to the ground. The method employed can vary considerably from place to place, depending upon such factors as availability of building materials, technology, building type, soil composition, and the local water table. For most buildings, where durability is a priority, the best foundations are those made of strong, rot-resistant materials such as stone or concrete. In Roman times, however, as is true today, wooden foundations were used in some applications where stone was unavailable, impractical, or unnecessary.

Piling

In Roman Italy foundations of wood were most commonly based on the principle of the pile (*palus* or *sublica*), a vertical beam (usually left round) pounded by hand or rammed into the ground mechanically to which the walls, floor, or decking of a structure were attached. The sharpened ends of the piles could be sheathed with iron (Jackson and Ambrose 1976, 44; Marchetti 1891, 51). Beams that were placed in predug holes or trenches will be referred to as posts (see below).

Wooden piles were especially suitable for wet, unstable substrates. They were indispensable for the construction of docks, bridges (military and civilian), and waterfront warehouses. In places where the water table was high or flooding a nuisance, entire villages stood virtually suspended upon piling, their decked "ground" floors several feet off the ground. Strabo (5.1.7), whose *Geography* was published in the late first century B.C., observed that low-lying Ravenna, rising from the Adriatic coast of northeastern Italy, was a river-girt city built entirely of wood. The town was flooded at times by the same high tides that threaten nearby Venice today. In such an environment piling was a necessity; Vitruvius (2.9.11) adds that every building at Ravenna, public and private, used this method of construction. Trajan's column depicts an enemy Dacian village of planked buildings supported on piling being fired by Roman auxiliaries in the opening years of the second century A.D. (fig. 5.1).

Waterlogged wood does not rot; timbers below water level can last for millennia, as recent archaeological finds have amply demonstrated. A few examples suffice to illustrate the longevity of wood when maintained in waterlogged conditions. The oaken

Fig. 5.1. Column of Trajan, Rome. Dacian buildings on wooden piles are shown on fire. In the foreground a planked fence is depicted. Note the figure in the lower left corner who appears to be holding a chisel or a drill. A.D. *113. (Anger, D.A.I. neg. 91.103)*

planks and stakes of a track dating to the end of the second millennium B.C. have been excavated in southern England (Orme, 1982). The Bronze Age Lake Dwellers of northern Italy (ca. 2000–1500 B.C.) inhabited dwellings built on piles (*palafitte*) driven into the ground along the shoreline. Major sites that preserve substantial remains of these old timbers include Fiavé, Fontanello, and Molina di Ledro. Closer to Rome, and testifying to an early indigenous tradition in central Italy, are the piles on the shores of Lake Mezzano in northern Latium, which date to the late Apennine period (ca. 1200 B.C.). North of the Alps similar piles of oak have been excavated at such sites as the Late Bronze Age village of Auvernier-Nord (ca. 850–830 B.C.) in Switzerland (Arnold, 1982).

Romans themselves would later use the same species of wood that had been found over centuries of trial and error to be best suited to wet conditions. Both Vitruvius and Pliny, for example, lauded the qualities of the alder (*alnus*), larch (*larix*), and oak (*quercus*), all species that have been found in prehistoric settings. According to Pliny (HN 16.219), the wood of the alder lasted forever when driven into boggy ground and was capable of sustaining great loads. Other woods could be treated to resist decay. Vitruvius (5.12.6), for example, says that charred piles rammed into pits packed with charcoal could support bridges, theaters, and city walls. Techniques used to protect wood from decay are considered in more detail in chapter 10.

Post Pits and Post Trenches

The term *post* is generally used to refer to a vertical wooden structural beam that is smaller in dimension than a pile and may also play a role in defining and supporting the superstructure of a building. Furthermore, for the purposes of this discussion, piles have been defined as those structural members that have been driven into the ground by pounding. Posts, like piles, were often set into the ground, but only after a hole or trench was dug first; the post was then set therein. Even after the wood has long since deteriorated, careful excavators can expose cavities, or postholes, that reveal the original footprint of the lost structure; in Latin the general term for such a hole is *foramen* (Varro *Rust.* 1.14.2). Modern technical reports may make a distinction between postholes and post pits. The latter term is used to indicate a hole substantially larger than the timber post placed within it. Gravel or other packing material could be tamped around the post to anchor it securely (Maxwell 1976, 34).

Post or pile construction was especially attractive in situations in which speed of construction was of the essence and mature trees were abundant. Wooden posts and piles could be set into place quickly from trees logged right at the site. For this reason posts and piles were frequently used for the gates, towers, walls, and interior buildings of Roman military encampments (*castra*). Forts such as the military post of Vindolanda, situated along a lonely stretch of Hadrian's famous wall in northern England, are instructive examples. Here Roman soldiers were protected by a high wall of turf that surrounded the encampment, topped by a wooden parapet built on a framework of piling driven deeply through the turf wall into the subsoil (Birley 1977, 161). The excavation of a Roman signal tower at Beattock Summit, Lanarkshire (Scotland), revealed post pits that measured between 60 and 45 cm in diameter and between 76 and 96 cm deep dug for posts 20 cm square (Maxwell 1976, 34).

To ensure safe storage of grain and other perishables for a legion stationed far from home, Roman military engineers constructed granaries supported above the ground on wooden *columnae*, or posts, so that drying air would "blow through them from all sides, and even from below" (Plin. *HN* 18.73). Dozens of such granaries have been identified in England and Germany. Geoffrey Rickman (1971, 214) has pointed out that such wooden granaries were doubtless built wherever the Roman army was present and there existed an exploitable local supply of timber (fig. 5.2). In England alone more than fifty such structures have been excavated (Manning 1975, 105). The excavations of the Romano-British warehouses have produced enough data to offer some interesting evidence about the standardization of timber-framed structures, at least as far as first-century Roman Britain is concerned. The rows of posts that supported the wooden floors of the granaries were most commonly spaced about 1.5 m (5 Roman feet) apart; the posts themselves were also often spaced at five-foot intervals in each row. The posts were usually circular in section and of more or less constant size in a particular granary. Small warehouses used posts about 13 cm in diameter, medium-sized grana-

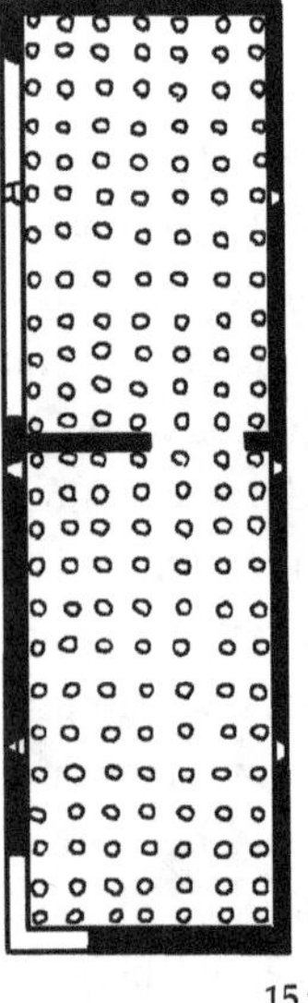

Fig. 5.2. Plan of a Roman military granary from Niederbieber, Germany. The floor was supported by wooden supports placed in postholes arranged in twenty-five rows. Overall dimensions: 53 by 16 m. Second century. (Author, after Rickman 1971, 244, fig. 53)

ries used a more robust support measuring either ca. 18 or 23 cm in cross section. The posts were set up either in long trenches or, more rarely, within individually dug pits.

The placement of individual posts along the bottom of a long dug trench (post trench) was quicker than digging individual postholes and facilitated the laying out of long, straight lines of supports. The disadvantage of this method was some sacrifice of stability; archaeologists have found that posts were placed snugly against one wall of the trench to aid in bracing the uprights firmly in the ground (Hanson 1982, 170). This is the principle, as we will see, employed to anchor the posts used for Iron Age (ca. 800 B.C.) buildings in early Rome. One way to help strengthen the posts was by adding a horizontal sill plate, or sleeper beam.

Sleeper Beams

The sleeper beam is probably a good example of the Roman builder adopting a local building practice to suit his own needs. In this case, the use of sleeper beams was widely practiced in northern Europe and England. The sleeper itself connotes a heavy timber placed in a shallow trench; this beam supports the framework of the building's walls and, if a granary, the elevated floor of the structure as well (fig. 5.3). Clearly the technique evolved from the simpler post trench, just described. The sleeper beam itself may be considered the foundation proper of the building; the posts connected to it are more accurately described now as wall studs (*arrectaria*).

Although the employment of a sleeper beam requires more skilled labor, it does have some clear advantages over post or pile construction. In especially unstable

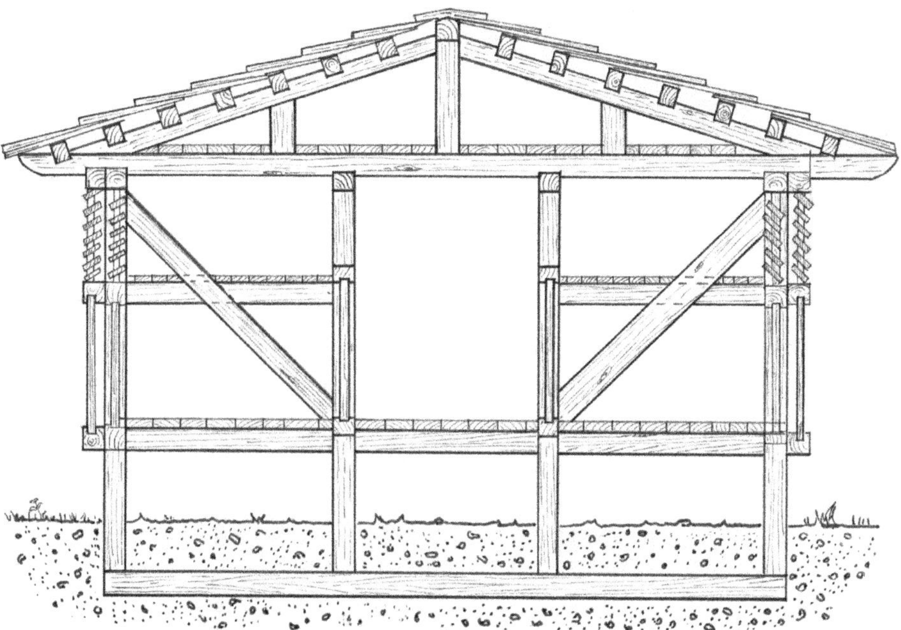

Fig. 5.3. Section of a timbered Roman military horrea *(warehouse) from Fendoch, Scotland. The superstructure is hypothetical, based upon the pattern of supporting posts. The floor was supported by posts on sleeper beams. First century. (Author, after Rickman 1971, 237, fig. 47)*

ground, the load of bearing walls is distributed evenly along the beam. If the sleeper is set upon a masonry dwarf wall, rot can be avoided (in this case the sleeper beam is exposed to the air and is more accurately described as a sill plate). Finally, there is the possibility of prefabricating sections of wall before they are set into place. The existence of sleeper-beam construction is often missed by excavators, who confuse this type of construction with that of the post trench. In Roman Britain, use of the sleeper beam was primarily a first-century practice.

At Vindolanda, for example, R. Birley found foundation timbers still *in situ*; the major uprights (*arrectaria*) of a building in the legionary camp were observed mortised into slots cut into the sleeper beam; the interstices were filled in with panels of wattle and daub (Birley 1977, 110). Birley has reported another variation from the same site, where the *arrectaria* were mortised into individual short lengths of beams: each segment thus acted as an independent footing; here the features are dated to the early second century. Rickman cites the use of sleeper beams to support flooring at the military warehouses found at Pen Llystyn and Fendoch in Scotland (1971, 215).

Similarly, at Verulamium, S. Frere (1972, 6) reported that oaken sleeper beams 30 cm wide by 6 cm thick were installed to support walls about 20 cm thick. Oaken uprights measuring 15 by 7.5 cm were placed 35 cm apart. In another application the uprights measured 10 cm square and were spaced 55 cm apart. These and other examples indicate that the spacing of the uprights was variable. At Valkenburg (I) in Britain, for example, uprights were placed, on average, 80 cm apart (Hanson 1978, 304). Further discussion of such uprights and how they were filled in to form walls will be considered in chapter 6.

The First Bridge Over the Tiber: The Pons Sublicius, or "Pile Bridge"

The most venerable of pile-built structures at Rome was the Pons Sublicius (the "pile bridge") over the Tiber. According to Livy (1.33.6), the Roman historian who lived during the dawn of the empire, the Pons Sublicius was one of the earliest large public monuments of the city. Romans believed the bridge had been first constructed in the seventh century B.C., a date that, if correct, makes the bridge even older than the Temple of Jupiter Optimus Maximus (dedicated in 509 B.C.). Livy also tells us that the patron of the bridging project was one of Rome's early kings, Ancus Marcius, who desired to unite the east and west banks of the Tiber (another story attributed the bridge to Numa, the grandfather of Ancus Marcius). Two additional ancient passages, written by Plutarch (*Numa* 9.2–3) and Dionysius of Halicarnassus (3.45.2), claim that the bridge was built without metal fasteners and that the pontiffs of Rome took responsibility for its maintenance and replacement.

As implements of bronze and iron were fully accessible to the seventh-century builder, the oath not to employ metal fasteners was apparently in deference at first to strategic concerns and later (for the old wooden bridge was maintained or replaced as necessary for centuries) to hallowed tradition. Pliny informs us that the ban was put in place so that the bridge could be dismantled quickly in times of war (see Vitruvius's similar advice for contemporary wooden military walkways, 1.5.4); legend told of the difficulty defenders of the city had in tearing down the bridge when Lars Porsenna, the Etruscan king of Clusium, marched upon Rome at the dawn of the republican period (the battle is said to have taken place in 508 B.C.). Other sources, including Plutarch, assert that the avoidance of metal fasteners was based on the advice of an oracle. The pile bridge, perhaps of oak or, less plausibly, chestnut (Cozza 1928, 188), was located to the south of Tiber island. Here the breadth of the river is greater but the current less forceful. No traces of the Pons Sublicius have been found, and its precise location is yet unresolved (Galliazzo 1994.2, 25).

It is not merely wishful thinking to hope that someday traces of the ancient Pons Sublicius, in at least one of its early manifestations, might be found. Roman-period wooden piles belonging to moles and quays along the Tiber have in fact been discovered, and in a remarkable state of preservation. Perhaps the most notable of these artifacts was the timber cladding of a mole used to unload marble from barges towed up the Tiber to Rome. Domenico Marchetti, who published the discovery in 1891, described massive oaken piles between 6 and 8 m long, squared to 55×50 cm, dovetailed to one

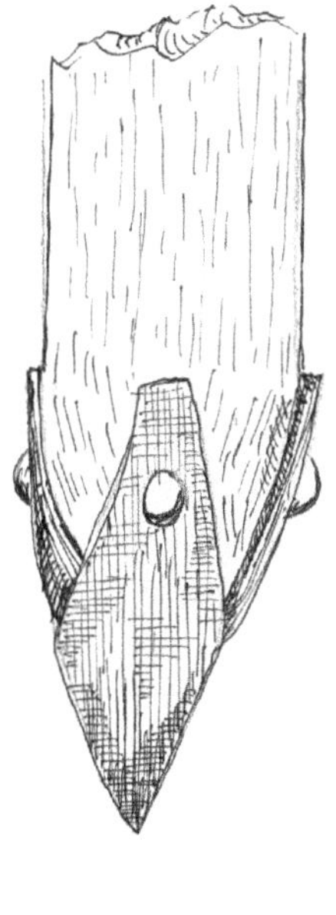

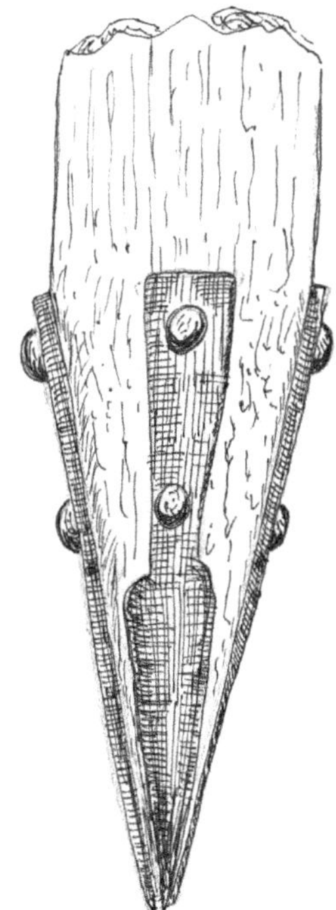

Fig. 5.4. Methods used to sheath oaken piles with iron before ramming them into place. Examples excavated along the Tiber River, Rome. (left) Iron plates 32 cm long, weighing 6–7 kg per plate; (right) iron plates 58 cm long at 13.4 kg. each. (Author, after Marchetti 1891, 51)

another along their lengths to form a strong, durable wooden shell that protected the perimeter of the mole. The piles were sharpened and sheathed with plates of iron to facilitate ramming them into place—while protecting them from splitting (fig. 5.4). Magnificent oaks were felled to create piles of such length and thickness. Similar piles, also of oak, with sharpened, ironclad points have been documented from the Roman-period bridge (first century) at Trier, spanning the Mosel, in northern Germany (Cüppers 1969, 47). In this application the piles were hidden from view, hammered into the riverbed in a compact grid to form a stable platform upon which the superstructure of squared stone blocks was placed (fig. 5.5).

The discovery in early 1998 in Rome of a first-century fresco depicting a city, possibly a fantastic interpretation of the capital itself, revealed an image of a covered bridge spanning a broad river. As a timber bridge, the Pons Sublicius may have been covered by a timber shed, not unlike the way in which bridges in New England were traditionally protected to prevent the thick planks used for the roadbed from rotting.

The Pons Sublicius is apparently the bridge depicted on a medallion issued by Antoninus Pius in A.D. 140–44, in celebration of a restoration of the old monument. The piles themselves are represented as clusters of three vertical lines; the roadbed is horizontal on either end and arched in the middle. Whether the image represents the appearance of the original structure or even offers an accurate sketch of its much later reconstruction is hard to say (Galliazzo 1994, 26). There exists no ancient description

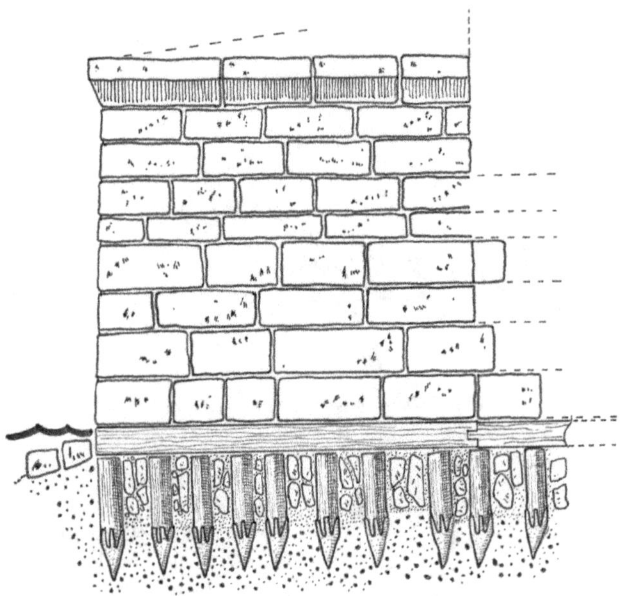

Fig. 5.5. Piles used for bridge construction in Germany. The ends of the sharpened oaken piles were sheathed with iron; most were ca. 2 m in length. Note the scarf joint observed in the heavy horizontal timbers. First century. (Author after Cüppers 1969, 51, fig. 36)

of how Rome's original Pons Sublicius was in fact built, but a striking parallel example can be called upon to give us some idea: Julius Caesar's famous wooden bridge built over the Rhine.

Julius Caesar's Bridge over the Rhine

One of the most illustrative of literary passages concerning the use of wooden piling is found in Julius Caesar's account of the wooden bridge he ordered built in 55 B.C. so that he and his army could cross the Rhine, as Caesar says, with dignity. The crossing could have been achieved by boat or even pontoon bridge, but the construction of what must have been the first permanent bridge over the river would in fact mirror Ancus Marcius's achievement in Rome nearly six hundred years before, and itself anticipates the emperor Trajan's feat of spanning the Danube about a century and a half later.

Caesar's detailed description of the construction of the bridge attests both to the skill of the engineers who traveled with the army and to the commander's fascination with the construction process. We do not know if Caesar wrote the passage unaided or incorporated a description supplied by one of his engineers (cf. Granger 1985, 281). The account is of great importance for several reasons. Not only does it attest to the speed and skill of the Roman military engineer (Caesar boasts that the entire project, including the procurement of the timber, required a mere ten days), but the details included in the description are especially useful for determining the definitions of a number of technical terms used by the Roman carpenter. It is worthwhile quoting the account in full:

Caes. *BGall.* 4. 17: Caesar decided to build the bridge on the following plan. He coupled together at an interval of two feet a pair of piles [*tigna*] one and one half feet thick, sharpened a little way from the base and measured to accommodate the depth of the river. These he lowered into the river by means of cranes [*machinationibus*] and set them fast, and pounded them in by pile drivers [*fistucis*], not vertically, in the customary way of piles [*sublicae*], but leaning forward at a uniform slope, so that they inclined in the direction of the stream. Opposite these were driven two more piles coupled [*iuncta*] in the same fashion, at a distance of forty feet from base to base, slanted against the force and rush of the stream. These pairs of piles were subsequently connected by beams two feet wide [*bipedalibus trabibus immissis*], whose ends fitted perfectly into the spaces between the two piles of each pair. The pairs were kept apart from each other by braces [*fibulis*] on the outer side of each end. So, as they were at once held apart and affixed together, the stability of the structure was so great and its character such that, the greater the force and thrust of the water, the tighter were the connections [*illigata*]. These trestles [*derecta*] were interconnected by woodwork [*materia*] laid over at right angles, and floored with long poles [*longuriis*] and wattle [*cratibus*]. In addition, piles [*sublicae*] were driven obliquely on the side facing downstream, projecting below like a buttress and joined with the whole structure, so as to absorb the force of the current. In the same way other piles [were placed] a short distance above the bridge, so that if the trunks of trees [*trunci arborum*] or boats were launched by the natives to break down the structure, by these defenses the force of such things would be lessened lest they harm the bridge.[2] [Figs. 5.6, 5.7]

From Caesar's full description the modern reader can visualize easily the general appearance of the bridge: a series of trestles with splayed legs carried a deck of timbers that in turn supported a roadbed of wattle (woven branches) and probably a clay surface.

Although the main points of Julius Caesar's description are clear, the details of the bridge's construction have invited debate. Were the braces (or pins) (*fibulae*) Caesar describes used to bind each pair of the trestle's legs or to attach each paired set to that standing opposite? A *fibula* is a word commonly used to indicate a metal pin employed to fasten a garment; did Caesar have a metallic fastener in mind? The rapid pace of construction may have precluded or at least limited the use of metal fasteners. Vitruvius's use of *fibula* to describe the connection of the ends of a lifting crane (10.2.1) clearly indicates a brace used to bind the ends of two heavy beams together, although in 1.5.3 he compares the long tie-beams installed transversely in a fortification wall to *fibulae*. Thus while the sense of the term can be assayed without much difficulty, its exact nature remains elusive. Here, too, Caesar indicates a pile by both *tignum* (a general term for a wooden beam) and *sublica*. In sum, even though this well-known description of the Rhine bridge offers an unusually clear description of a Roman structure for which we have no archaeological clues, the terminology, while perhaps perfectly clear to the ancient architect or engineer, leaves us less than certain.

Fig. 5.6. Reconstruction of Julius Caesar's bridge over the Rhine. The model is in the Museo della Civiltà Romana, Rome. Originally built mid–first century B.C. *(Author)*

Caesar claimed that the wooden piles used to support his bridge measured 1.5 feet in diameter (a standard measurement described as *sesquipedalis*). This corresponds to a metric measurement of 45 cm, which, when compared to a sampling of dimensions reported for other piles (or postholes) that have been excavated, seems right in line with analogous applications (table 1).

Caesar's first bridge over the Rhine (there is record of a second construction in a later campaign) represents the most basic form of the military bridges that spanned the rivers controlled by Rome. It was built to serve an immediate purpose; longevity of service was secondary in concern to speed of construction. In a modern study of pile bridges built in Roman Germany, Eckart Mensching (1981) sketches four general categories that rely increasingly on the durability of quarried stone, yet even the most advanced solutions were achieved through the utility of wooden piling. The all-wood construction of Caesar's bridge ("der Pioniersteg") could have been swept away by a turbulent current swollen by winter or spring storms. A stronger bridge, also entirely of wood, was built upon piles spaced closely together in parallel rows ("die Pfahljochbrücke"); the dense clustering of the piles could bear more weight and resist floodwaters more effectively. Better still was the bridge made from stone piers that rested on a series of submerged wooden piles pounded into the riverbed and packed with rocks ("die Pfahlrostbrücke"). Such a bridge was durable as well as strong and could be built without directly accessing the bedrock of the riverbed. The actual spans supported by the masonry piers could be constructed of either stone arches or wooden beams.

Fig. 5.7. Section (hypothetical) of Caesar's bridge over the Rhine. The flow of the river is indicated with an arrow. The diagonal brace is not explicitly described by Caesar, unless it represents one of the fibulae mentioned in the Latin passage. (Author)

Table 1. Examples of excavated post or pile dimensions

Location	*Type of Building*	*Thickness of Pile*	*Source*
Palatine Hill, Rome	Iron Age hut	20–45 cm diam.	(Puglisi 1951, 71)
Beattock Summit, Lanarkshire (U.K.)	signal tower	20 cm. square	(Maxwell 1976, 34–35)
Northhamptonshire (U.K.)	bridge	24–28 cm square	(Jackson/Ambrose 1976, 43)
Mosel River, Trier, Germany	bridge	23–43 cm diam.	(Cüppers 1969, 47)
Tiber River, Rome	wharf	50–55 cm square	(Marchetti 1891, 51)
Lake Nemi, Italy	retaining wall	30 × 20 cm (rect.) 25 × 52 cm (rect.)	(Ucelli 1950, 133)

Finally, there was the bridge whose stone piers rested directly on the bedrock of the riverbed (die Steinpfeilerbrücke). Wooden piling nevertheless played an important role in construction, for wooden dams fashioned from closely spaced vertical piles enabled engineers to drain and excavate the riverbed in order to seat the stone foundations.

It is tempting to see these four variants in terms of a linear, evolutionary development from less to more sophisticated. It is probably more apt, however, to see the types as complementary solutions applied to varying political and environmental exigencies. On Trajan's column, for example, erected over 150 years after Julius Caesar crossed the Rhine, we continue to see representations of simple pile bridges that exist coevally with others built with the highest technical skill (fig. 5.8). Of particular interest here is the importance of the wooden pile over centuries and the extent to which abundant natural resources in the form of mature stands of timber were apparently readily at hand. I shall return to the issue of the superstructures of large timbered bridges in a later chapter.

The Use of Wood for Framing Foundations of Concrete

Foundations made entirely of wood, represented by the employment of piles and sleeper beams already discussed, provided an important role for wood in Roman building. Most large Roman buildings in Italy, however, were footed on foundations of stone or concrete; for the construction of the latter, wood was still of great importance. Concrete foundations, used in Italy from the late third century B.C., could be fashioned simply

Fig. 5.8. A simple military bridge built of piles and planking. Column of Trajan, Rome. A.D. *113. (Deichmann, D.A.I. neg. 41.1705)*

by dumping the freshly mixed concrete into open trenches in the ground. A more accurate and less wasteful method involved using a form for the concrete. The form itself, a rigid outer skin built to hold the wet concrete, could be built of stone or even fired brick, but for many projects builders constructed envelopes of wooden beams and planks; the practice is often referred to as shuttering (Taylor 2003, 77) (fig. 5.9). Such forms, depending on the material used, were left wholly (as in the case of stone and brick) or partly (in the case of wood) in place after the concrete had set. Today the imprints of beams and planks left on the surfaces of concrete foundations provide an important record not only for the practice of building wooden forms for foundations, but also for the dimensional characteristics of ancient sawn lumber. As ancient liter-

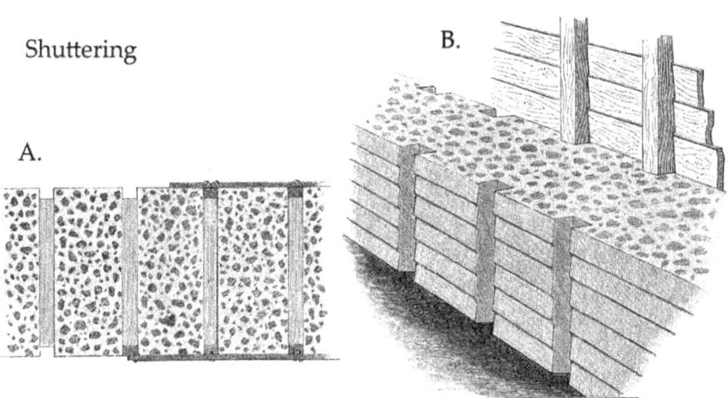

Fig. 5.9. Reconstruction of shuttering. (A) Overhead view; transverse timbers have been used to reinforce the vertical posts and outer boards; (B) foundation wall with vertical beams and exterior boards; in the foreground the cured concrete face as it would appear after the wooden elements have been removed or have deteriorated. (After Middleton 1891, 48, fig. 3)

ary sources do not describe this method of construction, the archaeological record is of vital importance.

An accessible and easily located example of the use of wooden forms for massive concrete foundations can be seen in the center of Rome. The north retaining wall of the huge artificial platform (145×100 m) built to support Hadrian's Temple of Venus and Roma (dedicated in A.D. 135), now flanking the modern Via dei Fori Imperiali, illustrates how wooden forms were suitable for the largest of public building projects (fig. 5.10). Here heavy posts, most measuring about one Roman foot thick (28–30 cm) and spaced about 3.5 to 4 Roman feet apart (1.03–1.18 m), framed a wall of planks of varying width (plank width is more difficult to determine; where good impressions survive the width is ca. 24 cm). The planks were nailed to the outside faces of the uprights, so that the latter were embedded in the concrete mixture.

The outer skin of the formwork, made up of the planks themselves, could be recovered for reuse once the concrete had set, perhaps even for the same project at a higher level (Roman concrete walls, particularly foundations, tended to be built up in layered sections). To increase the strength of the wooden form, the framework of timber uprights could be joined by transverse beams that would, as the foundation work advanced, become completely encased by the wall save for the ends; cavities left by these beams after the wood has decomposed can still be observed. This was a common practice in the Roman world wherever concrete was used in foundation work. There is no evidence for the use of transverse beams in the foundation work just described. The builders must have used heavy diagonal props to hold the shuttering in place as each layer of concrete was added.

Another remarkable example of a timbered grid required for concrete foundation work was discovered in 1982 in the submerged remains of the Roman harbor at Caesarea Maritima, located in modern Israel. The construction of the harbor at Caesarea, an ambitious provincial project under way during the principate of Rome's first emperor, Augustus (27 B.C.–A.D. 14), included the challenging task of creating artificial breakwaters to protect ships in the harbor and to provide additional dockage. Some

Fig. 5.10. The north retaining wall of the Temple of Venus and Roma, Rome, begun in A.D. 121, finished ca. 135. Imprints of the shuttering used for the foundations are clearly visible. A butt joint is apparent in the vertical channel at center. In this example transverse tie-beams were not used. (Author)

of the lessons learned at Caesarea Maritima may well have been applied to Rome's own artificial harbor at Ostia, another project requiring massive offshore seawalls, which would be built some thirty years later.

In 1982 archaeologists exploring the northwest tip of the northern breakwater at Caesarea discovered immense blocks (the largest measured 12×15 m) that had been created by filling wooden forms, or caissons, with concrete. The forms themselves, essentially cavernous wooden boxes, were reinforced with a carefully crafted matrix of interior wooden beams that would eventually become embedded in the finished blocks (fig. 5.11). Not only did the cavities left by these timbers survive, but segments of the sleeper beams used to frame the base of the form could be observed in their original state. The discovery is important not only for the history of maritime architecture, but for our understanding of the carpenter's trade in an innovative and specialized application:

> The formwork, best preserved at the northwest corner, was built on massive pine (*Pinus*) and fir (*Abies*) sleeper beams ca. 0.29 m square that ran along the base of each face of the block and interlocked at the corners with simple lap-joints. A series of pine and possibly fir (*Abies*) uprights (ca. 0.12–0.15×0.23 m) mortised into the horizontal beams at 1.60 m intervals, carried an external and internal wall of horizontal pine planks 0.08 m thick and 0.14 m wide. The lowest of the planks on the inner face of the formwork was inset slightly into the upper surface of the large sleeper beam and fastened with mortise and tenon joints. In the examples recovered, the tenons were of oak (*Quercus*) and poplar (*Poplus*). . . . The planks must have been nailed to the uprights, but it was not possible to examine any surviving joints for verification. The wood species are not indigenous to this area. (Oleson 1989, 129)

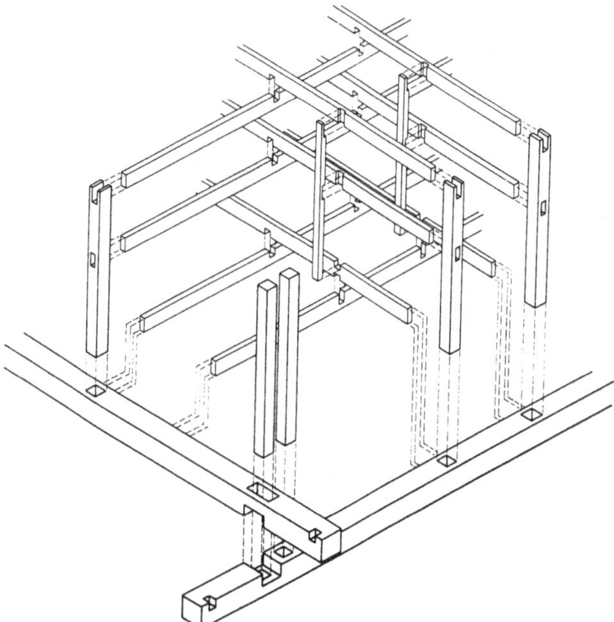

Fig. 5.11. Caesarea Maritima (Israel). An underwater formwork (caisson) framed with timber. Hypothetical reconstruction. First century. (Oleson 1989, 409, by permission)

Presumably this formwork, once it had been constructed on dry land, was floated into place, sunken and anchored at its intended location, and then filled with concrete. Apparently planking was not needed on the bottom of the form; the wet concrete rested directly on the sea floor. Timber was clearly imported to the worksite. The use of wooden forms for underwater concrete structures is not limited to Caesarea Maritima; the practice has been documented at other ancient Roman Mediterranean harbors (Yorke and Davidson 1985).

Wooden Retaining Walls

Retaining walls constructed entirely of wood are closely related to the assemblies created for concrete foundations just described. Such walls are commonly found along the steep banks of lakes or rivers and were thus particularly important for the construction of quays. Depending on the materials available and the strength required, the retaining walls were built with a combination of squared timbers and planks or simply heavy squared timbers alone. Both methods may be found side by side at the same site.

At New Fresh Wharf in Roman London, for example, a line of upright squared timbers supported horizontal rows of planks to square off the banks of the Thames and provide good dockage. The construction method was essentially the same as that used for shuttering and concrete construction, except that the fundamental components are reversed. Thus the planks are on the inside face of the retaining wall, held in

Fig. 5.12. Alder logs bored to serve as water pipes. From Vindolanda. Average length: 75 cm; W: 25 cm; bore of 6.5 cm. A slab of oak used as a junction box is visible in the background. Early second century. (Photo: Courtesy the Vindolanda Trust)

place by the external vertical piers. The date is roughly the same as that of the Temple of Venus and Rome (mid-second century). Nearby, at the Custom House site, the quay was formed only slightly later by building up boxes of half-lapped squared beams that were dovetailed into an exterior wall of heavy squared trunks (Marsden 1994, 30) (see fig. 4.4). This method of isolating cubical volumes of earth within interlocking rectilinear walls was also known and used by builders of temple *podia* (immense artificial platforms) and tombs constructed entirely with concrete or stone masonry.

On the shore of Lake Nemi a wooden retaining wall was found built of squared piles of oak and fir, interspersed with oaken planks 5 cm thick that were fitted into grooves, or rabbets, cut along the length of each pile (Ucelli 1950, 163) (see fig. 4.5). Such walls facilitated the creation of flat terraces along the steep shore of the lake.

Wooden Pipes

The existence of wooden water pipes is known through both written testimony (for example, Plin. *HN* 16.224) and archaeological finds. They are common in what A. T. Hodge calls "isolated systems" (1992, 111), that is, systems not connected to main aqueduct lines, such as villas and forts, but also used in urban settings, particularly those of northern Europe and Britain. At Aachen, Germany, for example, beams of oak 30 cm square were bored and attached to one another by cylindrical metal collars pounded into the flat face at each end of the pipe (Hugot 1982, 167); a center ridge in the metal cylinder ensured that each end of the connector penetrated the pipe to a uniform depth. In Roman Britannia similar wooden pipes or their metal fittings have been found at sites like Caerleon (Zienkiewicz 1986, 318), Fishbourne (Cunliffe 1971, fig. 55), London (Wheeler 1930), Silchester (Boon 1974), and Vindolanda (Birley 1994, pl. 11).

Excavations at Silchester, for example, revealed the vestiges of a wooden pipeline over 200 m in length. The wooden pipes had vanished, but the iron collars that once connected them still lay *in situ*. From the spacing of these collars, each 11.5 cm in diameter and 4 cm in length, it was estimated that the individual pipes were roughly 2 meters long and carefully drilled with augers of increasing sizes (Boon 1974, 88). Similar evidence has been found at Caerwent (Hodge 1992, 112).

In addition to oak, alder was considered a good material for wooden pipes (we have already seen that alder was valued for subterranean applications). Recent discoveries at Vindolanda included a still-functioning water line of alder pipes that supplied the Roman fort from a nearby spring. The line was constructed without the use of the more common metal collars. Instead, the end of each alder log was tapered, torpedo-like, and inserted into a hole bored into opposing faces of a block of oak (fig. 5.12). According to Pliny (*HN* 16.224), less durable woods such as pine could also be bored for piping. Even though softer pinewood would deteriorate more rapidly, it was easier to bore than oak or alder and, if buried in moist soil, could last for a very long time.

VI Framing and Walls

From the tree we raise buildings.—Pliny HN 12.5

Wood played a vital role in ancient construction long before the utilization of mud brick, stone, and fired brick. Even when fired brick, masonry, and concrete construction prevailed during the first through third centuries A.D., stout framing timbers were still essential structural and decorative elements in most Roman buildings. Raising walls of concrete or cut stone required wooden scaffolding. Vaults were built upon semicylindrical wooden forms. For agricultural and military buildings, wood was always an indispensable resource; many such structures were built entirely of timber. Large public temporary structures, including theaters and amphitheaters, were built entirely of wood, and wooden framing later became a significant component of the uppermost stories of the most magnificent stone arenas. Wooden floors were employed in most Roman buildings in all periods, even if the joists and subfloors were ultimately hidden under layers of concrete, mosaic, or tile. In addition to the floors, the framing of a building included door and window frames, partition walls, ceiling, and roof.

The First Timber-Framed Buildings

Wherever forests grew in the Mediterranean basin, architecture began with the exploitation of wood. Thick timbers were used to make the skeletal frame of the building, which Vitruvius calls *compactio*. Where straight logs were plentiful and easily procured, the entire structure could be robustly constructed from heavy unsawn timbers (roundwood) to withstand climatic conditions, a method still used in North America today. Whether such log cabins ever played an important role in Italy is hard so say, but it is unlikely. Vitruvius (2.1.4) is able to describe such construction but has no word for it;

he considers it a barbaric practice, crude and unskilled, and attributes its use to the race of the Colchi in the heavily forested regions of the Black Sea (an area that produced superb woodworkers in historical times, as we shall see). An exception that exists well into the imperial period is the use of logs, either round or squared, laid in horizontal rows, each layer placed perpendicularly to the one below it, to build stout platforms for military artillery during sieges. Such structures are depicted clearly on Trajan's column (A.D. 113), and we can imagine that they had useful civilian applications as well, such as for the construction of bridge abutments and cofferdams (see fig. 3.7).

A varied Latin vocabulary existed to describe vertical and horizontal wooden beams, that is, the two essential components of the timber-framed building. Equally significant, the archaeological discoveries of early Italic buildings, particularly from the Italian Bronze Age (from ca. 1800 B.C.) and Iron Age (from ca. 1000 B.C.), indicate that the structures we consider as precursors to Roman were framed by a grid of timber beams filled with thin panels of woven twigs and mud, animal hides, or bark. Such structures, while on one hand primitive, never fell completely out of use and, in some parts of the Roman world, such as the agrarian communities of northern Italy, remained the dominant forms of architecture.

The first such timber frames consisted of unhewn trunks and thick limbs. Uprights were seated in postholes, and horizontal timbers were lashed to them with twines of bark fibers and sinew. This simple formula established the fundamental trabeated look of Greek and Roman architecture for a thousand years, and thoughtful Romans who cared to consider the origins of their buildings were conscious of the ancient pedigree of the architectural spaces they inhabited. A particularly beautiful example of this awareness is exhibited by the Augustan poet Ovid, who describes the miraculous transformation of a primitive hut inhabited by two peasants into a temple (*Met.* 8.700). At the climax of the miracle, "columns took the place of the forked wooden props."

The forked wooden prop (*furca*), familiar to any rural child who has attempted to build a fort in the woods, was certainly one of humankind's earliest architectural innovations. The V-shaped end naturally holds a crosspiece in place. The stout, forked trunks of Rome's native evergreen holm oak (*ilex*) were more useful to the primitive builders on the Palatine Hill than the tall, straight shafts of mountain pine and silver fir so highly prized by later generations.

No modern student of the topographical and archaeological history of the ancient city of Rome is unaware of the remains of the Iron Age huts discovered in the early twentieth century on the summit of the Palatine Hill. The first traces of these were observed by Dante Vaglieri in 1907, but it was not until the campaigns undertaken by Pietro Romanelli and Salvatore Puglisi in the late 1940s that the old buildings were brought to light; publication followed in 1951.

Puglisi described the remains of foundations of two simple huts, and by drawing upon evidence from other sites in Italy and from Iron Age funerary urns fashioned

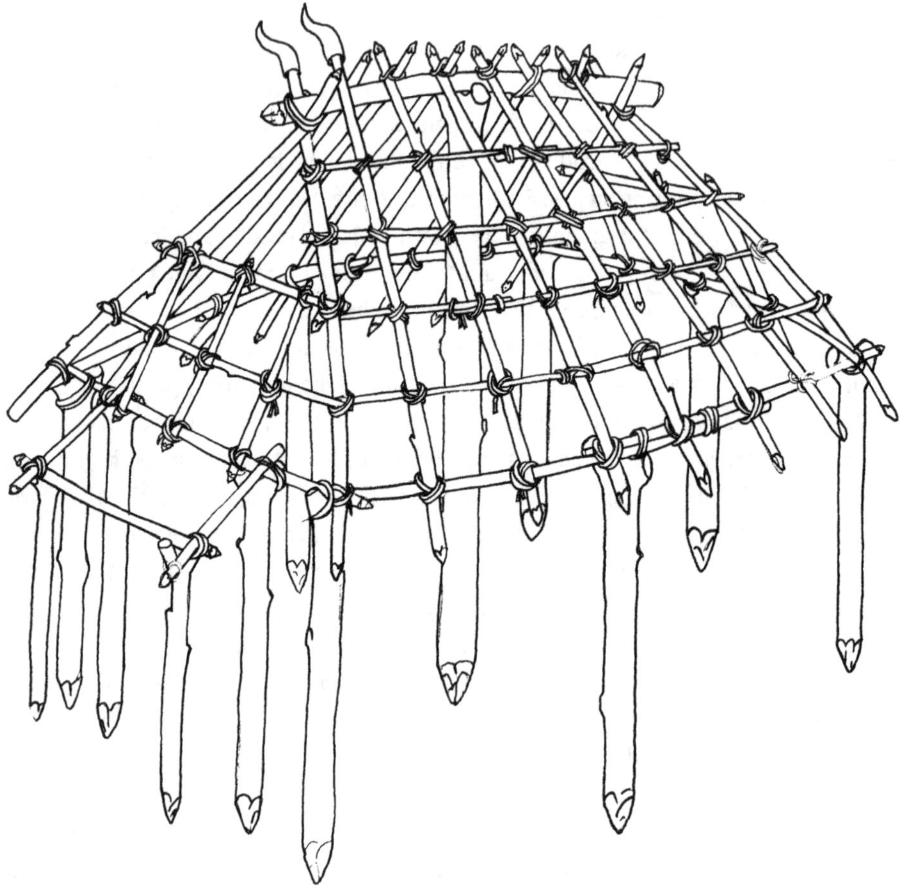

Fig. 6.1. Hypothetical reconstruction of the timber framing of Roman Iron Age huts, based upon the foundations discovered on the Palatine Hill, Rome. The lashing of the timbers would have been the easiest way to stabilize the structure. Eighth century B.C. *(Author, after Bartoloni 1987, pl. 99)*

to look like contemporary dwellings, he was able to provide a plausible description of the wooden frame employed by the builder (fig. 6.1). These simple structures would continue to be used by farmers for thousands of years and even now can occasionally be observed in remoter areas of the Italian countryside (fig. 6.2).

It is assumed that each of the two oval structures on the Palatine Hill housed a single family, and that the huts represent the presence of a village there dating back to the eighth century B.C. The floors of these huts were sunken slightly below ground level, so the perimeter of each dwelling can still be easily traced. The most distinctive features observed are the well-preserved postholes that anchored the upright timbers that supported the walls and roof. These remains, however paltry, allow us to understand how wood was prepared for construction—it was not, for example, squared before use—and to determine the sizes of the timbers employed.

The walls of the best preserved and most soundly built of the Palatine huts (4.90×

Fig. 6.2. A farmer's hut made of poles and thatch observed in 1998 near the Etruscan necropolis of San Giuliano (Lazio). The structure had been standing for some time and has since disappeared. (Author)

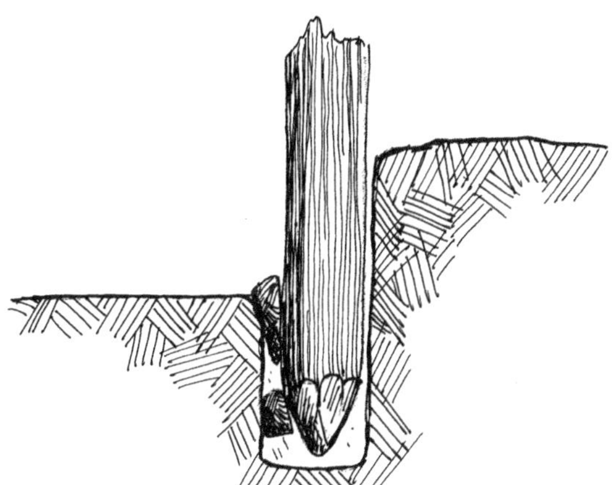

Fig. 6.3. Method of anchoring wooden posts in Iron Age dwellings in Rome. The combination of the posthole and higher exterior ground level reinforces the upright. (Author)

3.60 m) were framed by three upright timbers (probably *furcae*) on each long side, three across the back (counting the corner timbers twice), and two pairs of smaller timbers at the entrance, which apparently framed a simple door prefaced by a small porch. Great attention was given to securely anchoring the uprights. The postholes were deep (up to 55 cm). These were placed around the edges of the floor, which, as has been mentioned, was sunken up to 90 cm below ground level (fig. 6.3). Thus the posts were not only seated snugly in their holes but were further braced against what would become the lower wall of the house (the same principle behind the post-trench of the historical era described in chapter 5). This area is where bracing would be needed the most, for as far as we know these early huts used no interior horizontal timber ties to counteract the lateral pressures generated by a pitched roof.

The roof itself was a precursor to what is conventionally known as the prop-and-lintel, or in this case, more accurately, a prop-and-ridgepole. Virtually the same technique would, within 175 years, support the roofs of the first archaic Greek temples and,

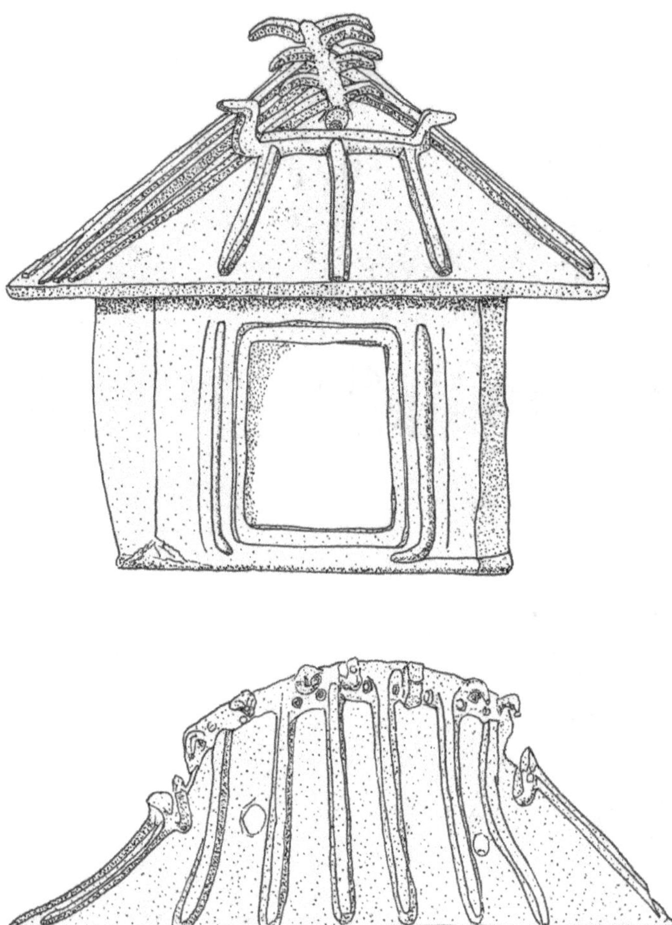

Fig. 6.4. An Iron Age "hut urn" of thick fired clay (impasto). The artisan has included details that reflect the framing of a contemporary house. From the Necropoli delle Arcatelle, Tarquinia, excavated in 1881. H: 39 cm; diameter at base: ca. 30 cm. Civitavecchia, Museo Nazionale, inv. RC638. (Author, after Bartoloni 1987, cat. 77; fig. 41)

along the Po and Tiber watersheds in Italy, Etruscan shrines. Here, on the Palatine of Rome, the ridgepoles of the Iron Age huts were supported by a thick, centrally placed, upright beam. The placement of this beam and the evidence from contemporary cremation urns fashioned as small replicas of the huts, such as the urn illustrated here from Tarquinia (fig. 6.4), indicate that the roof beam was not as long as the building itself; thus the light timber framework must have been sloped on all four sides (in other words, a simple hipped roof). The presence of rafters on the exteriors of the roofs of the cremation urns suggests that the sheathing was composed of thatch and held in place in part by superimposed branches or poles.

The timber framework of the Palatine huts is rather overbuilt; the timber uprights are stouter than they need be. The central posthole of the larger of the two Palatine huts measured about 45 cm (1.5 Roman feet) across. Even if the wooden post that was fitted into this hole was slightly smaller in diameter (and made fast with small

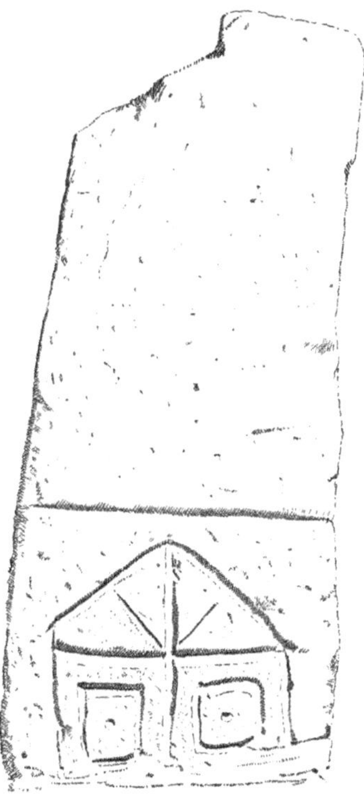

Fig. 6.5. Sandstone grave marker (stele) excavated at the Porta San Vitale necropolis (tomb no. 793) near Bologna. H: 55 cm; W: 25 cm. Museo Civico Archeologico, Bologna. Villanovan culture (ninth century B.C.). (Author, after Pincelli 1975, pl. 332)

wedge-shaped stones, as postulated by the excavators), it would still have been an impressive, trunklike support (Puglisi 1951, 71).

The phenomenon of overbuilding, or using more materials than is structurally necessary, is a pervasive characteristic of Roman construction. Later, Roman carpenters, as will be shown, also overbuilt upper floors by using thick, closely spaced joists. In our Iron Age Palatine example, the explanation for this practice is simple: the trunks of unhewn mature trees (roundwood) were used for the interior support and exterior walls. Practically speaking, time and effort were saved by not squaring the logs. Raw materials were plentiful. We may speculate, however, that such massive trunks contributed a sense of well-being and pride to these simple houses, and it is not unreasonable to assume that iron nails may have been driven into the central support post to hold prized possessions out of harm's way or even to showcase them to visitors.

Perhaps the single most remarkable representation of an Iron Age wood-framed structure has been found carved on a sandstone grave marker (*stele*) excavated at the Porta San Vitale necropolis near Bologna (fig. 6.5). The roughly rectangular slab of stone (55 cm high by 25 cm wide), although broken across the bottom, clearly depicts a building with a pitched roof and two incised squares indicating doors or windows. Of great interest is the rendition of a vertical center pole and two diagonal braces that reinforce the rafters. It is tempting to identify this rendition as the earliest existing interpretation of a triangular trussed timber roof, a form that would dominate the roofing of large Italian structures six hundred years after the San Vitale slab was carved (Bartolini 1989, 62). The fact that the center pole extends from the ground to the main roof beam, however, suggests that a prop-and-lintel, not a true timber truss, supported the structure. An obvious point, but no less remarkable, is the significance of the framed wooden structure as a symbol of the afterlife: the carving was the only representational work of sculpture recovered from the San Vitale cemetery (Pincelli, Morigi Govi 1975, 501).

The roundwood framing of the early Palatine huts and of the homes of those who dwelt in the vicinity of San Vitale was covered with screens of wickerwork smeared with mud and protected by deep, overhanging eaves. The use of such sheathing had a long history in Roman construction and was itself dependent upon the woodworker's trade. Such wattle and daub construction, as it is called, is indicated in Latin by the term *cratis.*

Wattle and Daub Construction (*cratis*)

Hardened fragments of mud bearing the impressions of woven reeds and twigs are the only traces remaining of the walls of the earliest Palatine huts. Such finds indicate that wattle, along with sun-dried brick, was one of the earliest materials used for wall construction. Vitruvius (2.1.2) speculates, somewhat romantically, that primitive builders learned the technique by observing the nests of swallows, and thus the first walls of Rome were in fact "woven."

Wattling was also occasionally used for the construction of floors and other applications that would seem ill-served by the flimsy nature of the material. At the Roman fort at Vindolanda, for example, Robin Birley found a drainage ditch with walls and covers of wattle; the soggy conditions of the soil and the waste water prevented the seemingly fragile matrix from decomposing for nearly two millennia (1977, 111). Romans arriving in Britain would have found that the native population along the Thames floored their simple huts with wattle. At Brentford, archaeologists have uncovered substantial traces of such floors (Wheeler 1930, 35). We have already seen that Julius Caesar's bridge over the Rhine employed wattle work for its roadbed; Caesar also describes covering military ramparts with roofing of wattle covered with clay (*BCiv.* 2.15.2) and laments how easily such surfaces can be punctured by enemy ballistics (*BCiv.* 2.2.1).

Wattle work required strong fingers and springy wood. Staves, round or squared (easily cleaved from heavier stock) and sometimes as thick as one or two inches, were woven basket-fashion to form a wattling panel, which was supported between two uprights (Vitruvius 2.8.20 calls these *arrectaria*). Wattles were gathered from the forest or by the practice of coppicing. In northern forests, birch branches were a favored source (Birley 1977, 111). Usually wattles were woven vertically and wrapped around horizon-

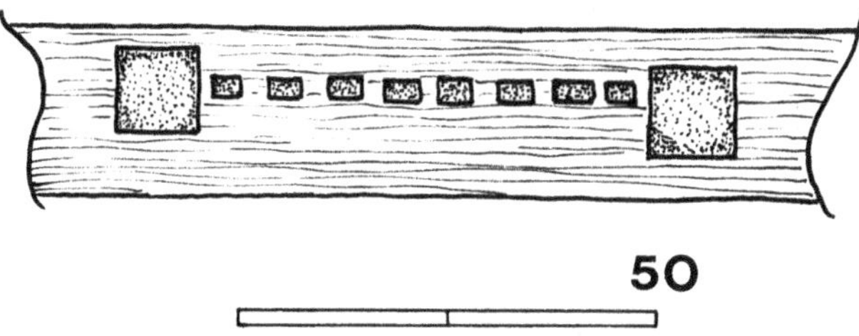

Fig. 6.6. Reconstruction of wattle work found at Verulamium, Britain. The sleeper beam and framing elements were of oak. Period II D (ca. A.D. *150). (Author, after Frere 1972/I, 8, fig. 4)*

tal rods (*transversaria*), which were themselves fitted into holes bored into the uprights (*arrectaria*) on either side. At Verulamium some wattling panels were in fact nailed to the timber frame. If a foundation trench (*fossa*) was used, the ends of the wattles were buried and stabilized in the trench; a better method anchored the bottoms of the vertical wattles into mortises cut into a sleeper beam. Well-preserved examples have been found in Britain at Verulamium and elsewhere (fig. 6.6). Gaps in the wattling could be filled with mud or clay—hence the term wattle and daub—and a finishing coat of plaster. In some areas of the Roman Empire, particularly in the northern provinces, the use of wattle had a particularly long history. In England, panels of wattle enclosed the walls of Roman military towers (Maxwell 1976, 36); from the same region well-preserved examples from as late as the tenth century have been found (Hall 1982, 237).

Opus Craticium (Half-Timber Construction)

Opus craticium evolved from—but never fully supplanted—wattle and daub construction. Whereas simple wattle work was based upon the principle of creating walls by the construction of a light framework of timber sheathed by a grid of wattling panels, the more durable variant known as *opus craticium* employed squared structural timbers and

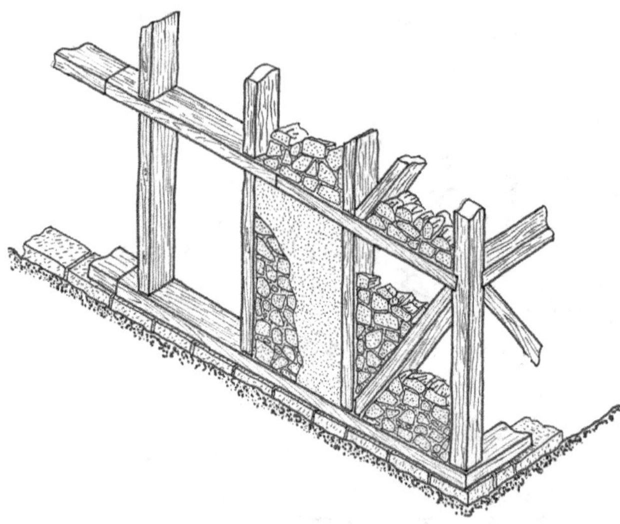

Fig. 6.7. Early use of proto opus craticium *from Lavinium, Italy. Seventh century* B.C. *(Author, after Guatoli 1981)*

thicker, tougher infill for the panels. The most popular filler was concrete and rubble. The resulting cheap form of half-timber construction was employed frequently. Roman builders were not concerned about the homeliness of such walls, for if time and expense permitted, a thick layer of painted stucco would cover both the timber frame and the concrete filler.

Vitruvius did not approve of the construction of *opus craticium*. He grudgingly admitted that the method was fast and economical, but he worried that the exposed wooden beams of walls so constructed were like torches (*faces*) ready to catch fire (2.8.20). The passage makes it clear that many builders and owners never got around to applying the finish coat of plaster or did not bother to repair walls that lost their original coating, not surprising considering that the construction method was used to save time and money. Vitruvius pointed out that walls built this way were ultimately unsuitable for plastering because the timber framing would swell from the moisture in the plaster and then, as it dried, its shrinkage would crack the plaster surface. He did admit, however, that the light weight of *opus craticium* provided a means of building partitions in upper stories that did not require underlying piers.

Excavations have revealed numerous examples of walls constructed of *opus craticium*. Early Italian examples of such walls, using squared horizontal and upright timbers, the latter mortised into sleeper beams, date from the seventh century B.C., too early for the use of a proper concrete infill. At Lavinium excavators have identified packed stones and clay covered with plaster as the fill material (Guatoli 1981) (fig. 6.7). Despite this early date, it is not known if the technique of *opus craticium* was of Italian invention. Variations have been found throughout the ancient Mediterranean. Valerio Papaccio has suggested that the practice was imported into Italy from Asia Minor (1993, 611).

Fig. 6.8. Opus craticium *in a house from Herculaneum, now restored. The so-called Casa a Graticcio is located in Insula III, nos. 13, 14, and 15 at Herculaneum. First century or earlier. (Photo: Author)*

Numerous surviving examples of *opus craticium* have been found at Pompeii, but the practice seems to have been particularly favored at Herculaneum. A well-known house excavated at Herculaneum between 1927 and 1933 was so abundantly characterized by the use of this method—visible even in an exterior room on the second floor projecting over the sidewalk—that the excavator, Amedeo Maiuri, named it the Craticium House (fig. 6.8). Here the main structure of the house is supported by piers of *opus incertum*[1] and brick, but numerous curtain walls of *craticium* (*parietes craticii*) subdivide the interior. The squared timber uprights (*arrectaria*), now carbonized wood, were sturdier (8–12 cm thick) than the horizontal *transversaria* (6–8 cm). Each panel measured between 50 and 80 cm per side. These were filled with concrete and rubble. Another well-preserved wall in the Augustan shrine on the *decumanus maximus* of Herculaneum employed squared *arrectaria* 8 cm thick and *tranversaria* 5–6 cm thick. Panels 78 cm wide varied in height between 83 and 92 cm. The horizontal timbers are carelessly butt-jointed, possibly without the use of fasteners, suggesting that the uprights were put into place first and then fill was added until it was time to "cap" a panel with a horizontal timber.

Upper floors at the House of the Menander at Pompeii, thoroughly studied by Roger Ling (1977), employed a timber framework that created panels 50–70 cm on a side. In all cases, concrete and rubble were the fill materials.[2] The similarity between the examples at Herculaneum and Pompeii is notable, although apparently no rigid rule was followed by the builders in terms of panel size.

Despite Vitruvius's censure of this popular method of wall construction, many walls of *opus craticium* survived the violent earthquakes that accompanied the eruption of Vesuvius in A.D. 79 to be excavated virtually intact in modern times. Papaccio has pointed out that the flexible wooden framework used in such a wall would act similarly to the iron rods built into a modern wall of reinforced concrete. *Opus craticium*, then, may have been chosen not only for its reputation as an economical alternative, but as a time-tested method of building in a region prone to frequent devastating seismic shocks.

Planked Walls

Not surprisingly, there is little extant evidence of walls made entirely from wood from the Roman period. Depictions of such walls, however, attest to their existence and, in some areas of Roman life, predominance. Nowhere is this clearer than on the reliefs on Trajan's column documenting the exploits of the Roman and Dacian armies. The numerous wooden structures depicted on the column, while rich in detail and a necessarily important component of any study of Roman woodworking, must be considered with certain caveats. First, the context is a military one; second, the places depicted are all far from Rome (let alone Italy); and third, many of the details appear to represent Dacian, not Roman, customs. Nevertheless, many of the best woodworkers in the Roman world were employed in the army, where their skills were called upon daily for any number of projects. Many of these same men, specifically those lucky enough to retire, would no doubt eventually ply their trades in civilian settings. It is not far-fetched to imagine that such craftsmen would bring back skills they had learned or observed on foreign soil.

One of the most common forms of planked wall pictured on Trajan's column is a type of defensive, open-air wooden fence. Some of the fences depicted are of roundwood staves placed vertically and connected with horizontal beams; others are clearly made of sawn planks. Planks are most commonly shown placed vertically, often with pointed tops, and bound by horizontal rails (see fig. 5.1). Less commonly the planking itself is horizontal, nailed to upright posts. It is hard to believe these planked walls would be constructed of actual boards instead of split wood when one considers that each plank would have been manually sawn from a heavy trunk. Nevertheless, where wood was scarce or when straight, slender roundwood was not at hand, planking was the most economical use of the raw material.

With the exception of Roman fortification towers and perhaps some signal towers, the buildings depicted on Trajan's column that are built of planked walls belong to Dacian contexts. The idea of a planked, freestanding building appears to be alien to the Romans, at least in the context of the Dacian wars. In one typical scene already cited, for example, Roman soldiers torch Dacian houses, depicted as built upon piles with horizontally planked walls and planked, pitched roofs (see fig. 5.1). The sculptor has carefully rendered the nailheads on the planking.

When evidence other than that from Trajan's column is considered, it is clear that Roman carpenters themselves built structures both large and small from sawn planks. The popularity of the practice was undoubtedly affected by availability of materials, relative cost, local tradition, and even building regulations. Sections of walls built from sawn alder planks nailed to oaken uprights and dated to the first half of the second century have been observed at Vindolanda. As Birley notes, once the planks were at the building site, construction was even faster and sturdier than wattle and daub, but the construction itself "required many weeks of preparation in the carpenters' sheds" (1994, 130). An outstanding example of planking from Italy itself is a painted depiction from Pompeii of a simple stall from which bread is being distributed; the vertical framing elements, horizontal planking, and even the nailheads are meticulously rendered, leaving no doubt as to the manner of construction (fig. 6.9).[3] The survival of a rebated corner post from Roman London shows how planks could be jointed at the corners of such timber-framed structures (see fig. 4.2, D).

Wooden Columns

Vertical columns and the horizontal beams they carry (*architraves*, an English term derived from the Latin *trabs*, or beam) are derived from the simple arrangement of upright roundwood and horizontal crosspieces that characterized the earliest wooden structures. The presence of decorative bases, fluting, paint, terracotta revetments, and capitals, either separately or in combination, can be used to distinguish a wooden "column" from a simple post or pile created by a roughly finished log.

Architectural historians have long been fascinated by the compelling evidence that the stone columns of Greek and Roman buildings are, in fact, physical derivatives of a long-vanished tradition of wooden architecture. The use of wooden columns and entablatures for public buildings had largely disappeared in Rome by the third century B.C. When we consider the constituent forms of the Greco-Roman stone column, however, the vestiges of ancient woodworking skills are striking. The vertical channels, or flutes, of the Doric, Ionic, and Corinthian orders, for example, are the only logical

Fig. 6.9. Distribution of bread from a stall in Pompeii. The plain planked counter is clearly rendered; nail heads or dowels are represented by dark dots. The painting was found in a private house (Pompeii VII, 3, 30); now in the Naples Museum, inv. 9071. H: 58 cm; W: 49 cm. First century. (Photo: Naples Museum neg. 2595)

decorations for a columnar shaft when one must use a chisel or gouge and work with the grain of the wood. The convex and concave profiles of the Attic base, so popular with Roman builders, are forms we associate with the lathe and a skillful turner. Pliny's comments on the subject indicate that wooden columns were still being made for some buildings, including temples, in his day (HN 16.188, first century). He writes that trees intended for such purposes were best felled in the springtime when it was easiest to remove their bark and thus preserve the natural cylindrical form of the shaft (most other cutting, as we shall see, was done late in the year or when the tree was dormant).

The upward flare of the classical capital, regardless of order, provided an ample bed for the butt-jointed architrave beams. A broad capital helped reduce the flexing or sagging of the wooden architrave. The manner in which the horizontal architrave was firmly seated upon the column capital is considered in more detail in the discussion of Tuscan temples in chapter 8.

As in later periods, when stone was routinely used for each component of the column, wooden column capitals and bases (when used) were usually made separately from the shafts. The broader diameters of the crowning and base elements would have been most efficiently cut from stock of greater dimension than that from which the shaft was fashioned.

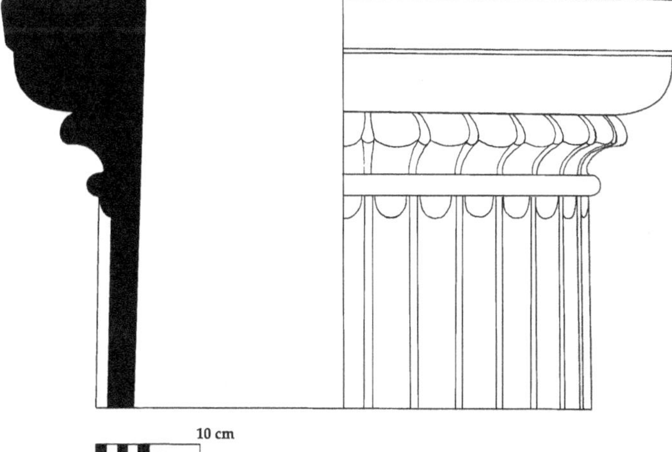

Fig. 6.10. Painted terra-cotta column capital found at the site of S. Omobono, Rome. The slight interior taper assured a tight fit on the wooden shaft. The top surface has been shaped with a subtle bevel to assist in shedding moisture. H: 27.5 cm; max. diameter: 68 cm. Capitoline Museums, inv. 16154. Sixth century B.C. (Author, after Sommella Mura 1977: 67)

Wooden columns used by builders today are often installed with bases and capitals molded from dense foam or plastic resins; the practice not only is less costly but also can extend the life of the column, since capitals and bases are most susceptible to rot, and it is important to protect the end-grain of the shaft itself from moisture penetration. Evidence from Rome reveals an interesting antique parallel. The solution was to revet the top and bottom of the wooden column with a decorative terra-cotta sleeve, molded and painted to resemble a standard base or capital. In addition, the lowest element, or plinth, of the wooden column could be a simple stone slab that acted as a footing to retard rot caused by ground moisture, insect penetration, or splash-back from the eaves.

The occasional finds of stone bases and terra-cotta capitals, without traces of the column shafts themselves, attest to the use of wooden columns in archaic contexts (particularly in the sixth century B.C.). A capital from the excavations of an archaic temple precinct discovered near S. Omobono in Rome provides an excellent example; it now resides in the collections of the Capitoline Museums (fig. 6.10).[4] The S. Omobono capital, an ornate example of the Tuscan order, was a fired clay collar for the wooden shaft of the column. The aperture in the capital indicates that the top of the column shaft itself measured ca. 40 cm in diameter. Finds like this one are very rare. Although the load would have been carried by the wooden shaft that extended fully through the collar, the capital itself would have remained fragile and easily broken. The practice was in all likelihood soon abandoned. The fact that such terra-cotta revetments can be used to estimate the size of structural timbers is of great use, as we shall see, for the reconstruction of roofing beams and door frames, all of which were commonly sheathed with similar protective plaques.

The Framing of Large Structures: Theaters, Amphitheaters, and Bridges

At certain times, and in diverse places, the Romans constructed enormous buildings entirely of wood. These included theaters, amphitheaters (or similar seating areas intended for temporary or military use), bridges (early examples such as the Pons Sublicius have already been considered), and roofs. Such projects involved heavy beams, lifting equipment, and organized work crews. Nevertheless, no matter how large the structure, basic principles of rough carpentry could be scaled up to meet the appropriate load-bearing requirements. Vertically placed timbers sustained loads, horizontal beams spanned open spaces, and diagonal struts provided reinforcement.

As is true of other forms of Roman construction, there is no evidence of engineering science, in the modern sense of the phrase, at work in the planning or construction of these buildings. Even in private contexts, Roman carpenters tended to overbuild. We can imagine that in the public realm builders were paid to be even more concerned with structural stability; public benefactors donated structures to garner public acclaim, not to attract opprobrium because of a disaster.

But disasters did happen, injuring and killing scores of people. The terrible collapse of the amphitheater at Fidenae (just north of Rome) in A.D. 27, during the reign of the emperor Tiberius, is a case in point. Here, during a gladiatorial contest, the timber framing of the seating area collapsed, reportedly resulting in thousands of casualties. Tacitus charged that the patron, a freedman by the name of Atilius, "neither placed the foundations on solid ground nor raised the wooden superstructure with 'solid joints' [*firmis nexibus, Ann.* 4.62]." Both Tacitus and Suetonius (*Tib.* 40) may have exaggerated the numbers who perished: Tacitus reported fifty thousand dead or injured; Suetonius claims twenty thousand lives were lost. Nevertheless, the amphitheater's collapse tragically illustrates the consequences of faulty timber construction in a large public building. Atilius was banished.

Stories of construction disasters spread, and builders and their patrons learned from such errors. It is notable, in fact, that the literary record does not mention more calamities than it does. There is no ancient account, for example, of an accidental bridge collapse, even though wooden bridges were built over rivers that challenge modern engineers.

TRAJAN'S BRIDGE OVER THE DANUBE, CA. A.D. 105

Trajan's stone and wooden bridge over the Danube River is notable not only because of its sheer size, but also for the revealing nature of the evidence that gives us an idea of what the bridge looked like and how it was built. This evidence consists of a literary description by Cassius Dio, composed a century after the bridge was erected; physical remains first recorded at the site of Turnu Severin (Romania) in the nineteenth century; and, most importantly, a depiction on the Column of Trajan that is universally identified with the unprecedented bridging feat of Trajan's celebrated engineer, Apollodorus

Fig. 6.11. Trajan's bridge over the river Danube, as depicted on the Column of Trajan, Rome. A.D. *113. (Anger, D.A.I. neg. 89.573)*

of Damascus (fig. 6.11). It is disappointing, to say the least, that a treatise on the bridge, supposedly written by Apollodorus himself, has been lost (Procop. *Aed.* 4.6.13; Gazzola 1963, 137). The same bridge, represented by a single span, is perhaps pictured on the reverse of a coin minted by Trajan to commemorate the project (Lepper and Frere 1988, 149).[5]

Dio's description (68.13.1) of Trajan's bridge cites some impressive figures: the wooden spans of the bridge were supported by no fewer than twenty stone piers, each 150 feet high by 60 feet wide and spaced at intervals of 170 feet. If the intervals of 170 Roman feet were measured from the center of each pier, the clear span would have been about 110 feet (32.50 m), resulting in a total length of 3,510 feet (ca. 1,032 m; O'Connor 1993, 142). Dio himself never saw the bridge. By the time he wrote his account, in the early years of the third century, the wooden spans that linked the bridge's piers had been intentionally dismantled. Hadrian had been worried, Dio tells us, that the bridge could be used for enemy counterattacks.

The image of this bridge on Trajan's column still offers the clearest idea of how the wooden superstructure was built. The relief shows five of the twenty piers, not including the two masonry flood arches built into the embankment on the left side of the composition (Lepper and Frere 1988, 149). Employing a skewed perspective that simultaneously presents bird's-eye and profile views, the sculptor has rendered the top

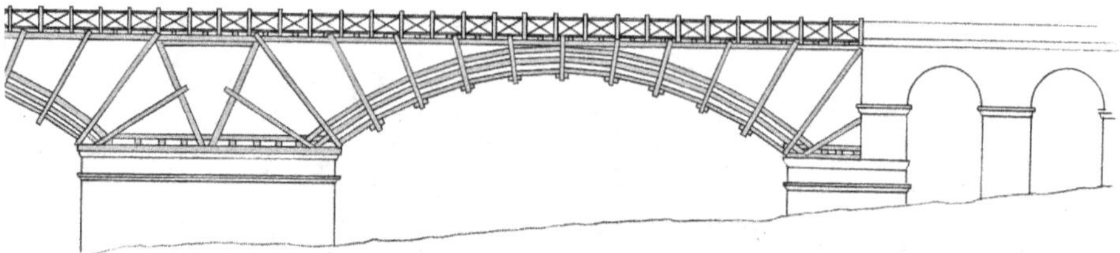

Fig. 6.12. Reconstruction of the bridge over the Danube based upon the representation on the Column of Trajan. Note the increased number of radial beams. Built in the early second century. (Author, modified from Galliazzo 1994, 323)

and both sides of the balustraded roadbed, the balustrades themselves built of uprights and crisscrossed braces. The roadbed of the bridge is depicted as being formed from a deck of squared timbers lying crosswise—these would have carried planks or wattle work covered with a weatherproof paving of clay or perhaps even concrete. The roadbed was supported between the stone piers by a wooden frame built as a shallow segmental arch, formed from three timbers held together and strengthened by radial beams. The shoulders of this arcuated framework are buttressed against paired triangular braces; each pair is affixed to a wooden deck of horizontal beams and planking that caps the top of the corresponding stone pier. I. A. Richmond has noted that the diagonal brace depicted within each triangular frame appears to have been rendered inaccurately; those used on the actual bridge would have been more reasonably angled in the opposite direction from that depicted in order to oppose the lateral thrusts of the wooden arches (1982, 35). V. Galliazzo, however, interprets these same braces as an extension of the radial beams used in the main span of the bridge (1994, 323) (fig. 6.12). C. O'Connor's reconstruction suggests that the upper beams of the segmental vault would have worked better if footed directly upon the center of the pier itself; he also suggests that the railings along the roadbed could have added further strength to the superstructure (1993, 144).

The construction of the bridge in this way eliminated the need for immense horizontal timbers to span the open water. As we will see, Roman builders never tried to span much more than 100 feet with horizontal beams. In building roofs, no builder, as far as can be ascertained, ever tried to cover a broader space using the type of segmental timber arch we see in evidence on Trajan's bridge. A shallow vault spanning something like 110 feet of open water would require four straight timbers around 8.5 m long (ca. 29 feet). The dimensional requirements are well within the range of timbers usually available to Roman builders. Even at this modest length, oaken timbers 1.5 Roman feet square would have weighed several hundred pounds each (even more if the wood was unseasoned); these would have to have been winched into position from cranes built atop the stone piers and then fastened to one another with iron pins, or thick wooden trunnels, the same squared-peg fasteners preferred by builders of New England's covered bridges eighteen centuries later.

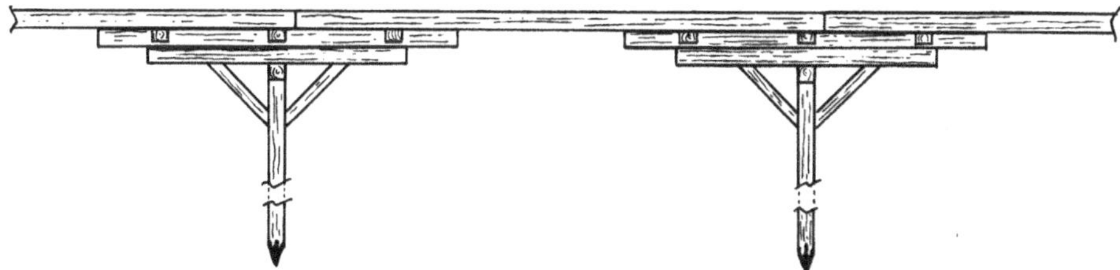

Fig. 6.13. Hypothetical reconstruction of a simple cantilevered bridge. (Author, *after* Milne 1985: 51, *fig.* 30)

The depiction of the Danube bridge on Trajan's monumental column is certainly a simplified rendition of the framing actually put into place. Just as the artist depicted only five of the twenty piers, so, too, he may well have condensed the individual elements represented in the wooden superstructure. Galliazzo's reconstruction of the bridge, for example, divides the shallow timber arches into ten segments (the relief shows only four), which would reduce the length of each bridging timber to a mere 11 or 12 Roman feet. Nevertheless, there is no reason to doubt the fundamental layout of the ancient rendition. Details such as the segmented wooden vaults make sense when the dimensional and load-bearing requirements are taken into consideration. We can only guess how the ends of the beams forming the arcades were attached to one another; some form of half-lapped joint pinned to the radial timbers seems a reasonable solution, and one that would resist flexing from loads passing by overhead. As we shall see, the principle would have been similar to the construction of a wooden wheel, the rim of which was built up of a series of curved wooden segments, or felloes.

CANTILEVERS

Another method of spanning a stretch of open water between the piers of a bridge was by employing a series of superimposed cantilevers (fig. 6.13). The wooden bridge spanning the Thames in Roman London, for example, appears to have been supported by heavy timbered piers and piled trestles—the latter in deeper water. The span of the roadbed over open water was perhaps bridged by cantilevering lap-jointed squared timbers successively outward to create a configuration representing an inverted pyramid. The basic principle used is of great antiquity, recalling the use of stone corbelling that can still be seen at Italian sites dating from the seventh century B.C. While a series of cantilevers would be satisfactory for wooden bridge construction, the technique exhibits no attempt to economize on building material and is suitable only for use in areas with abundant resources of timber. The method relies less upon advanced woodworking skills than upon strong backs to position the many heavy timbers into place.

Cantilevers of wooden beams were common in private construction as well. In Roman houses, the joists of interior upper floors often continued beyond the exterior wall surface to support an upper-floor overhang. A fine painted example can be seen in the well-known fresco recovered from the *cubiculum* (bedroom) found at Boscoreale,

Fig. 6.14. Detail of a fresco from a cubiculum (bedroom) found at Boscoreale. At the center of the painting a projecting room is supported by cantilevered joists. The foreground is dominated by a door similar to that from the Villa of the Mysteries illustrated in fig. 9.17. Metropolitan Museum of Art, New York. Mid-first century B.C. (Metropolitan Museum of Art, Rogers Fund, 1903 [03.14.13])

part of a villa owned by P. Fannius Synistor, and now on display in the Metropolitan Museum of Art in New York (fig. 6.14). The painting, dated to the first century B.C., depicts an elegant porch, protected by a high balustrade, supported by three heavy beams. The painter, no Roman builder, has rendered the beams so that the broader cross-dimension lies horizontally. In fact, as in modern practice, Roman builders used larger beams rectangular in section with the longer side placed vertically. Classical Greeks favored beams square in section. Like most painters of his era, the artist of the Boscoreale fresco was probably a Greek, perhaps from southern Italy, and more familiar with Hellenized architectural forms.

WOODEN THEATERS AND AMPHITHEATERS

Wooden theaters and amphitheaters, like bridges, were complex structures constructed by the repetition of a relatively simple component: the trestle. In its most simple form, this unit requires a minimum of three parts: two uprights plus one crosspiece. A pair of diagonal knee-braces can add significant strength. The trestle provided the support required for a roadbed or a bank of seating (see fig. 5.8). For the latter, the paired trestles would be constructed at different heights to support the slope of the seating area.

It is another matter to prove that all wooden theaters and amphitheaters were built on the principle of the trestle. There is no literary description equivalent to Caesar's proud commentary on his bridge over the Rhine. Literary sources prove only that wooden structures for mass entertainment were in fact built. Indeed, before Pompey's theater of stone, brick, and concrete was dedicated to great acclaim in Rome in 55 B.C., theaters in the capital were built only as temporary wooden structures and dismantled after the scheduled performances. Even after 55 B.C., the timbered theater did not disappear. Notable examples include the twin theaters (Theatra Curionis) built by C. Scribonius Curio in 53/2 B.C. that could host separate events in a back-to-back configuration or be rotated to face one another and serve as a unified amphitheater (Plin. HN 36.117). Several decades later Vitruvius (5.5.7) writes of "public wooden theaters" (*publica lignea theatra*) as a matter of fact. Nero, Rome's fifth emperor, constructed a wooden amphitheater in A.D. 57, the last such structure to be built in Rome before the construction of the Colosseum began a dozen years later. The impressive arena may be the building described in the verse of a Neronian contemporary, Calpurnius Siculus: "I gazed upon an amphitheater that rose to heaven on interwoven beams [so tall] it almost looked down upon the Tarpeian summit [that is, the Capitoline Hill in Rome]."[6] Pliny also mentions Nero's arena during his discussion of remarkable trees; apparently a larch of record size, salvaged from the construction of another public showplace (a *naumachia* built by Augustus for fake naval battles), was put to use by Nero's builders for the amphitheater (HN 16.200). Nero's arena may have stood for only a half-dozen years or so, another victim of the infamous fire of A.D. 64 (Tac. *Ann.* 13.31).

A few comments about the appearance of these wooden theaters and amphitheaters are possible despite the lack of physical remains. The decorative engaged columns and entablatures of well-known extant amphitheaters and theaters, such as the Flavian (or Colosseum, A.D. 80) and the Theater of Marcellus (13 B.C.), both in Rome, surely reflect the superimposed tiers of the vertical piles and horizontal timbers of wooden prototypes. Indeed, if we imagine the diagonal braces that must have been employed in the upper corners of each bay, we are presented with an elevation that approximates the later arcades of masonry structures, but conceived in terms of a linear geometry.

The relationship between the arcuated, curvilinear form of the masonry vault and the triangular brace is vividly depicted in a schematic representation of an amphitheater carved as a backdrop to a military scene on Trajan's column (fig. 6.15). The amphitheater is shown as a multitiered structure; five bays are visible along the side facing the viewer. Those of the ground level are barrel-vaulted with masonry; those directly above are framed with posts, architraves, and knee-braces, the last fixed at a point about halfway up each post. Double braces are indicated over the main entrance

Fig. 6.15. A hybrid stone and wooden amphitheater depicted on the Column of Trajan in Rome. The ground floor arcade of masonry carried a second level of wooden trestles, which in turn supported a third timber-framed story. A.D. *113. (Anger, D.A.I. neg. 89.576)*

to the arena. A third tier of open timberwork rings the top of the structure. By linking the vertical timbers with pairs of horizontal rails, an upper balustrade is created.

Here on Trajan's column the sculptor appears to have portrayed a hybrid structure. We can imagine the ground-level story of the amphitheater, bearing the most weight, being constructed of stone facing, rubble fill, and concrete vaulting. The upper story was constructed of wood, with vertical posts footed on the masonry piers below. The portrayal on the column is similar to another rendition of a partially timbered amphitheater found in Rome that may date to the period of Julius Caesar (mid–first century B.C.). The rendition includes less detail, but the use of timbered uprights and crosspieces footed on a ring of masonry is clear. Of greatest interest is a representation of a wooden roof built entirely over the structure, a feature that will be examined in more detail below (see fig. 8.18). Most of the great stone amphitheaters and theaters built during the Roman Empire, including the Colosseum itself, constructed the highest levels of seating with wood.

VII Wooden Flooring

Wooden flooring was found everywhere in the Roman world; its construction required substantial expenditures of both manual labor and materials. Wooden floors were especially important in buildings like warehouses, in which dry storage space, particularly for grain, was desired, and, most commonly, in multistoried structures like apartment buildings (*insulae*), living or storage areas on the upper levels of Roman houses, galleries and porticoes, and the mezzanines of small shops. Buildings constructed with flat roofs, a form of open-air flooring (Plin. HN 36.186), were covered with wooden structures essentially identical to those made for interior floors.

The ground floors of the earliest buildings of central Italy were formed simply from packed earth, clay, or even bedrock. The Iron Age huts discussed in chapter 6, for example, had floors of bedrock. Earthen floors remained the norm for utilitarian spaces well into historical times. Yet the first wooden floors in Italy appeared at surprisingly early dates. Recent discoveries of ground-level wooden flooring at sites such as Mezzacorona and Sanzeno-fondo Gremes in northern Italy indicate that in these thickly timbered regions planked floors were in use at least from the sixth century B.C. (Cavada 1994, 209). In these cases the floor was constructed by laying joists directly upon the soil. Once these were shimmed level with small stones, planks were laid crosswise to form the floor itself. Six centuries later this same method was apparently used to floor a modest outbuilding at the Roman town of Silchester in southern England. Here W. St. John Hope noted in 1897 a "series of dark parallel, or nearly parallel, bands, six inches wide and about eighteen inches apart. Examination showed that these were actually square-cut trenches, which, though now blocked up with earth, were such as might have been formed by filling with gravel the interspaces of a series of floor joists that had afterwards decayed" (St. John Hope 1897, 420). The modest room under discussion measured only 3.35×1.80 m. Even better preserved are the wooden ground-level floors discovered at the Roman fort of Vindolanda, where planks of alder five centimeters thick provided dry quarters for the officers of the late first-century *praetorium* (Birley 1994, 90). The use of wooden floors in larger public and commercial buildings corresponds to the development of granaries (*horrea*) and basilicas (gallery-level flooring).

Fig. 7.1. *Roman* contignatio *according to* Vitruvius: *joists* (tigna), *planking* (tabulatum, *here shown as a double layer*), *ferns, the* rudus, nucleus, *and finish surface of mosaic. (Author, after Howe in Rowland 1999, 269)*

Unlike modern planked floors of hardwood, polished and left exposed for their beauty, the wooden floors of the Romans were usually covered with a layer of concrete surfaced with tile. Only in the most utilitarian of applications, such as the loft of a storage building or the floor of a warehouse, was the wooden floor of a Roman building left exposed. The practice of covering wooden floors with masonry was, if nothing else, an effective method of fire prevention.

The Roman wooden floor of an upper story consisted of two or three elements (fig. 7.1). A few large bearer beams (*trabes* or *tigna transversa*) could support lighter transverse joists (*tigna, trabeculae,* or *trabes*) upon which the planks were nailed. The actual planked surface was referred to as a *tabulatum* (which also referred to a ship's deck), for this was a structure made from *tabulae*, a general term for boards. An alternate term was *coaxatio*, that is, an assemblage of *axes*, or planks. Rooms with short spans (*impetus* or *spatium*) might employ only floor joists, without the heavy underlying bearer beams.

The complete flooring system of joists and floorboards was called *contignatio*, that is, an assemblage of *tigna*.[1] The underside of the joists of a *contignatio*, as viewed from the room below, could be left exposed, forming a beam ceiling, incorporated into a coffering system, or covered with boards nailed to the soffits to form a flat wooden ceiling. Construction of such ceilings will be considered shortly.

Wood flooring was nailed (lit. "fastened," cf. *religo* in Caes. *BCiv.* 2.9.2) to the floor joists. Special care had to be taken so that the warping or shrinkage of the flooring would neither crack the walls into which the floor joists were set nor ruin a masonry pavement carried by a wooden subfloor. Oak, for example, while widely used by Roman woodworkers, was considered "unsuitable" (*inutile*) for wooden floors because of its tendency to warp (Plin. *HN* 36.187). Where extra strength was needed, the planking could be applied in two layers, the superimposed planks laid at right angles to those below and nailed at the ends to prevent deformation (Plin. *HN* 36.186; Vitr. 7.1.5). In New England, tradition holds that flat-sawn planks should be laid bark side up to reduce warpage.

Roman builders had to build upper-level floors with the greatest of care, for the wooden deck, while lightweight itself, was normally designed to carry a thick concrete subfloor upon which mosaics or tiles were laid. Vitruvius offers the most complete instructions for floor construction in the opening chapters of book 7 of *De Architectura*. It is clear from the details he supplies that wooden floors were designed to support masonry (6.3.9).

Vitruvius (7.1.3) prescribed that the masonry floor carried by the wooden deck was to consist of a layer of stones and concrete, the latter made with an aggregate of rubble (*rudus*) which, once packed, was to measure no less than three-quarters of a Roman foot (ca. 22 cm) in depth. A layer of ferns or straw was first placed over the floorboards to protect them from the caustic effects of the lime (Vitr. 7.1.2; Plin. *HN* 36.187). This first layer of rubble was then covered with a thick coating (*nucleus*) of *opus signinum*[2] six digits (ca. 11 cm) thick. Upon this layered subfloor—over one Roman foot thick—a final pavement of stone mosaic, terra-cotta tiles, or, in rare cases, closely fitted slabs of fine marble (*opus sectile*) was installed.

Revealing evidence for the construction of such floors is observable at many Roman sites. Often the fallen and broken slabs from a second or third story are the only physical traces left of a building's upper floor. Large numbers of such fragments, for example, have been unearthed at the Roman town of Pollentia, built in the early imperial period on the island of Mallorca off the south coast of Spain.[3]

The most fruitful places for studying such floors in Italy are the sites of Pompeii and Herculaneum, especially in the remains of the mezzanine floors of the numerous shops still standing on the periphery of nearly every city block. These floors, once accessible from the ground floor by a staircase, were used for storage and for the living quarters of the shopkeeper. In a partially excavated shop which lies on the north side of the Decumanus Maximus at Herculaneum, for example, one can still see a cross section of the entire mezzanine floor embedded in the volcanic mud which fills the back half of the room. Above a band of carbonized wood—evidently the remains of the floorboards—is a stratum of concrete ca. 12 cm thick composed of a fine, ash-gray aggregate (fig. 7.2). The floor of this Herculanean shop was less thick and employed a lighter aggregate (no stones can be seen) than that described by Vitruvius. Even with the employment of a light aggregate, such as the locally abundant pumice, an upper subfloor of Vitruvian dimension (ca. 33 cm thick), not including the finished surface of mosaic or tile, would have created a heavy load for joists and walls to bear.

Despite the potential for such heavy loads, wooden floors of the type seen at Pompeii and Herculaneum were able to bear great weight, even if the heavier Vitruvian standard was employed. This strength was afforded by the size of the joists, their

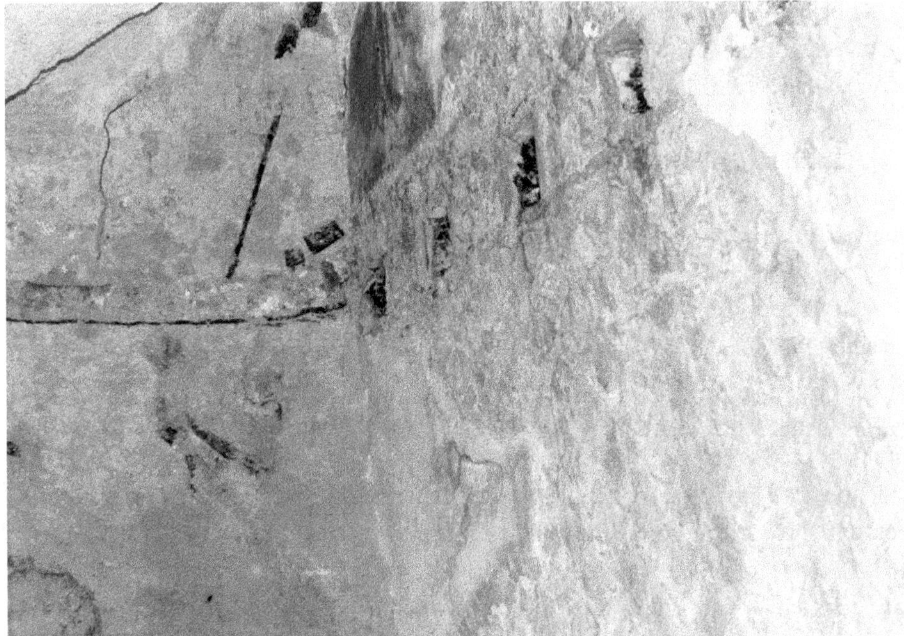

Fig. 7.2. The remains of a masonry floor supported by wooden planking still visible as a black, carbonized strip in the unexcavated strata along the north side of the decumanus *of Herculaneum. The remains of carbonized joists are visible in the beam holes on the right side of the photograph. First century. (Author)*

close spacing, and the limited clear spans they crossed. It would appear that these Campanian builders erred on the side of safety when constructing a *contignatio*, to the point where one could argue that good structural wood was actually wasted. As noted, the tendency to overbuild has been documented at other sites in the Roman world, especially where timber was plentiful. Such is the case, for example, in the timber-framed military buildings at the fortress excavated at Inchtuthil (late first century). Elizabeth Shirley, who has studied the construction techniques used at the fort, notes that the use of oversized timbers may be attributable to several factors: local supply, ease of cutting joints, and labor saved by not having to resaw timbers (2000, 13).

For the most part, traces of upper floors in Pompeian shops consist of a row of rectangular (or, more rarely, circular) socket holes that were built into the walls to receive the butts of the floor joists—the *capita tignorum* (Caes. *BCiv*. 2.9.1). One of the first questions to be answered is whether the socket holes in the upper walls of the *tabernae* are in fact accurate indicators of the dimensions of the wooden joists which were fitted into them. Fortunately, because of the preservation of carbonized wood from Herculaneum and the imprints left by wooden beams in some Pompeian walls, this question can be answered. In buildings where the construction is of good quality, where walls are well built and square, floor joists were shaped to fit snugly into the rectangular beam holes. Several of the eleven beam holes visible in a shop along Cardo V at Herculaneum

(*Insula Orientalis* II, 5), for example, contain the carbonized remains of well-squared and -fitted beams 34 cm high by 19 cm wide.

Most of the shops on the main streets of Pompeii were solidly built and kept in good repair. Squaring floor joists—at least on the top and bottom surfaces—had the obvious advantage of presenting a flat, stable nailing surface for the *coaxatio* and a sturdy soffit to rest on the floor of the wall socket. Nevertheless, it is also clear from the archaeological evidence that rectangular soffits did not always receive sawn rectangular beams. Carbonized wood and imprints in the mortar of walls indicate, for example, that a log could be split lengthwise and shoved into a socket so that the flat side was jammed up against one side of the receptacle. Gaps were filled with shims of stone, tile, and concrete. A good example can be found in an otherwise unremarkable shop (VI 14, 6) located two blocks to the east of the well-known House of the Faun. Here imprints in the concrete of the joist holes reveal timbers which were half- or quarter-rounds and perhaps only roughly hewn (that is, pointed) on their ends. The Pompeian evidence for such timber use is confirmed by comparable carbonized remains at Herculaneum.[4] Interestingly, the cheapest form of wooden joists, which consist of roundwood 10–13 cm in diameter, is often found in houses with the finest of stucco decoration. This is because the roundwood supports were not built to support the heavy weight of an upper floor, but instead were designed as an armature from which a lightweight ceiling of laths and plaster could be hung; the latter would be painted to complement the wall decoration of the room.

At Pompeii and Herculaneum, well-built upper stories were anchored in place by seating rectangular timbers into sockets which were, with few exceptions, constructed concurrently with the walls. The wall sockets themselves were commonly framed at the sides and the tops with stones or fired brick or tile. The result looks like a miniature rendition of a post-and-lintel doorway. Only rarely does one observe evidence of cutting back into a completed wall to create sockets. At Pompeii, for example, a new floor was built into the shop located at V 1, 29, creating a higher ceiling (3.17 m) than that of the old configuration (2.50 m).

The bottoms of the sockets were not framed with tiles because ancient masons tended to interrupt the construction of walls at the level of the first set of floor joists. The top surface of the unfinished wall was carefully leveled and then received a skim coat of concrete (or a leveling course of brick); then the entire structure was allowed to dry and thus at least partially cure. Later, the socket frames and upper walls were built—perhaps after the joists had been laid in place upon the top of the wall. This method would prevent both the bonding of wet concrete with wooden beams and the buckling of the uncured lower wall from the weight of the heavy joists. Thus the work of the masons could continue independently of that of the *fabri tignarii*, the carpenters.

A good example of this method of construction can be seen in the wall built to support the gallery enclosed within the large quadriporticus—the Ludus Gladiatorius, the training grounds for Pompeii's gladiators—which occupied the level ground south

Fig. 7.3. Contignatio *evidence in the west wall of the Ludus Gladiatorius, Pompeii.* Opus incertum *walls have been built upon a lower wall of* opus vittatum. *The joist holes for the gallery of the portico are clearly visible. First century. (Author)*

of Pompeii's Theater (fig. 7.3). The roof of the south side of this portico has been reconstructed in modern times, as has the gallery (now with joists of reinforced concrete) in the southeast corner. The original socket holes for this gallery are still well preserved at the southwest corner and along the west side of the porticus. Here the sides of the sockets are framed with small squared stones; the capping elements are of tile. The walls themselves are of rubblework; corners and doorways were quoined with *opus vittatum* (alternating courses of brick and squared stone).

The wall from which the gallery projected was constructed in the following manner (fig. 7.4). First, the back wall of the portico (45 cm = 1.5 ft thick) was raised to a height of 2.60 m. This was high enough to frame the doors that led to the square cells that opened off the back of the porticus. The top of the wall was leveled and finished with a thin skim coat of mortar. Then, while this topcoat was still wet, a worker scored lines alternately placed at roughly 43 and 23 cm intervals. The interstices 23 cm wide were marked with an "X," which indicated the placement of a joist (traces of such markings can still be seen). At this stage the whole wall was left to dry. In the next stage of construction, the workers built up the posts for each joist socket by aligning small squared stones with the incised lines. These holes were bridged by tiles, and the upper wall was built up of rubble and concrete. The wooden joists themselves could have been added by a team of carpenters in concert with—or subsequent to—the work of the masons.

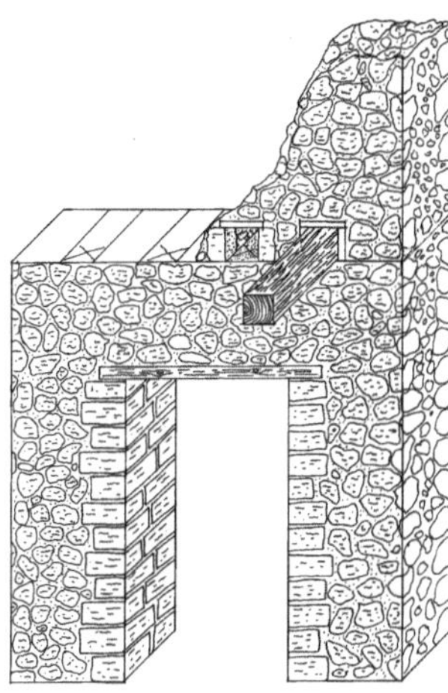

Fig. 7.4. Wall construction of the Ludus Gladiatorius, Pompeii. (from left) Step 1: The wall was built up to the level of the wooden joists; step 2: the spacing of the joists was scribed; step 3: the sides and tops of the cavities were formed with squared stones and tiles; upper wall completed. (Author)

At Pompeii one is struck by the overall consistency in the appearance of walls which were constructed to support a *contignatio*. The consistency suggests, as might be expected, that builders conformed at least to rough standards of practice, determined by tradition or even code. The spacing of floor joists, for example, appears to have been set at one Roman foot, although builders were not vigilant for exact precision. Joist spacing of 1.5 feet can also be observed, as was the case with the Ludus Gladiatorius described above. In this context, interestingly, the most common type of ruler found at Pompeii consists of two *semipes* (half-foot) rods of bronze hinged at the middle, identical in form to the example from London illustrated in figure 3.46. Such instruments could be used to quickly mark the spacing of timbers or to check the thickness of walls, especially in one-foot or half-foot increments.

Where the spacing of joists is 1.5 feet, the timber size tends to be more robust in terms of height and thickness. Thus in the Pompeian shop located at V 1,30, where the average spacing of the joists was 45.60 cm (1.54 ft), the socket opening measured, on average, 31.5 cm high and 25 cm wide (1.06×0.85 ft). These figures can be compared to those from the large Pompeian *pistrinum* (bakery) at V 4, 1, where the average joist spacing is exactly 1 foot (29.6 cm). Here the socket holes are only, on average, 18.5 cm high and 14.5 cm wide (0.63×0.49 feet). The fact that the combination of closer spacing and lighter timbering appears to be more common suggests that smaller beams were either less expensive to purchase, more easily procured, easier to handle, or possibly all three.

Since joists supporting both the *coaxatio* and a concrete subfloor were subjected to great loads, the clear spans were clearly limited. The size of Roman shops was determined by economic as well as technological factors, but it nevertheless is probable that the uniformity of commercial spaces around Pompeii and Herculaneum was at least in part determined by the ability of the builder to construct a sound upper story free of the danger of collapse. At both Campanian cities a common size for shops was about

Fig. 7.5. Shop from Ostia with holes preserved for floor joists in a wall of opus mixtum. *Via di Diana. Second century. (Author)*

15 feet across or slightly less (4.50 m). Roomier industrial spaces, such as the bakery at Pompeii V.5.1., employed interior piers to lessen the clear span of the floor joists.

The rectangular layouts and clear spans of the shops seen at Rome's port town of Ostia do not exhibit any remarkable differences from those which have been discussed at Pompeii. Although the walls at Ostia are often less well preserved than those of the Campanian towns, a surprising amount of evidence for *contignatio* still exists.

Wooden joists were still used in the second-century shops of Ostia, but in terms of spacing and application, the construction of a *contignatio* differed quite dramatically from the Campanian examples discussed above. Even within the site itself, there is no evidence of consensus among Ostian builders as to how best to build an upper floor. Joists were supported by sockets much like those seen at Pompeii but were also carried on shelves constructed from fired brick and on projecting brackets of stone.

An Ostian shop located at the east end of the Via di Diana provides a close analogy to the type of construction seen at Pompeii and Herculaneum (fig. 7.5). Here the walls are of *opus mixtum*, in this case reticulate work skillfully interlocked to a brick frame. A second-century date seems likely; Hadrianic work in this quarter abounds (Blake 1973, 160). Three pairs of beam holes, which begin 2.71 m (9.1 ft) above the floor, are built right across a juncture of brick and reticulate facing, so that the bottom half of the socket is defined by the flat faces of the brickwork, while the top half is framed with squared stone blocks and a tile cap. The span is short; the room measures only 4.37 wide×4.60 m.

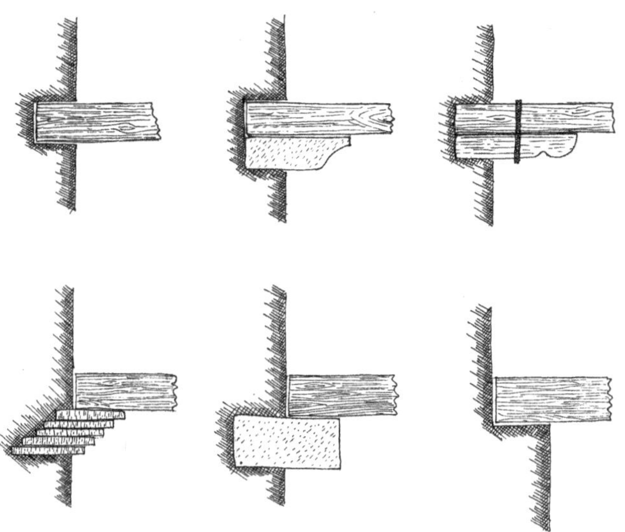

Fig. 7.6. Methods of anchoring joists to walls. (top row from left) *Joist in wall socket, joist paired with stone corbel, joist paired with wooden corbel;* (bottom row from left) *joist supported on cantilevered brick tiles, joist supported by continuous shelf of stone blocks, joist supported by offset in the wall. (Author)*

The method employed to build an upper floor framed by wood in this Ostian shop differs from the Campanian examples earlier described in two fundamental ways: first, the spacing of the beams is much greater than that observed in the older Campanian structures. At Ostia, the inside distances between the beams are 1.60 and 1.67 m, respectively (5.4 and 5.6 ft). This is three to five times greater than the spacing observed in the Campanian *tabernae*. Second, the three beams that were used to support the floor were of smaller dimensions (17×17 cm), particularly in height, than their Campanian predecessors. Thus, even if the three joists used in this Ostian shop were employed to support a lighter deck of joists and floorboards, it is doubtful that this Ostian form of *contignatio* would have been able to support the same weight as that of the earlier Campanian floors.[5] It seems that the Vitruvian formula for the construction of an upper floor, which includes accommodation for a thick masonry subfloor, had been abandoned or significantly altered by the Ostian builder.

Another apparently common form of seating the beams of a *contignatio* in an Ostian shop of the high empire was not with joist holes but upon a shelf formed by cantilevering tiles out from the wall (fig. 7.6). One well-preserved example of this can be seen in the apartment block immediately to the west of the insula of Diana. Here five courses of *bipedales* are corbeled to create a shelf approximately one foot deep, creating a ceiling just over 10 ft high (3.03 m). This method is found at Rome as well; prominent examples include the ground-floor shops of the multistoried apartment building still visible at the base of the Capitoline Hill, usually dated to the early second century (Blake 1973, 82). The support of a floor deck by a continuous shelf or offset is not limited to upper stories. Some of the smaller granaries that have been found in Britain, which were built with wooden ground floors slightly above grade, employed offsets; if the spans were modest, no other support was needed for the floor joists (Morris 1979, 35).

Fig. 7.7. Joists supported by stone wall brackets. Tabernae located at the southwest corner of the Horrea Epagathiana in Ostia (I, 8, 3). Second century. (Author)

The use of a cantilevered tile cornice to support a wooden floor makes discovering joist spacing and size most difficult, for usually there are no socket holes. In a few cases, however, the cornice is paired with socket holes, perhaps to prevent lateral shifting or twisting of the *tigna*. Good examples can be seen in the deep (4.80 wide×8.70 m) shop and its immediate neighbors located at the corner of the Via delle Volte Dipinte and the Cardo Aurighi at Ostia. Here, too, four heavy primary beams (ca. 20×40 cm) were employed to span the room, each spaced about 1.90 m from its neighbor. Construction is of brick and reticulate work similar in application and dating (second century) to the examples discussed above.

The third method of supporting a *contignatio* in an Ostian shop was by the use of projecting travertine corbels (fig. 7.6). Well-preserved examples can be found in the small shops incorporated into the Caseggiato del Larario (I, 9, 3) and in the deep *tabernae* located at the southwest corner of the Horrea Epagathiana (I, 8, 3) (fig. 7.7). In these applications wooden joists were supported by individual brackets. The shops of the Caseggiato del Larario employed only three pairs of corbels which were spaced between 1.94 and 1.98 m from one another. Here the span was only 2.90 m (10 feet). For the broader span (4.40 m) at the Horrea Epagathiana, brackets were located between just over a meter and about 1.20 m apart, still widely spaced, however, when one recalls the evidence from Campania.

Despite the variations seen in the actual method of support of mezzanine floors in Ostian shops, all appear to have employed beams that were spaced at greater intervals than those which frame the floors found at Pompeii and Herculaneum. In some cases these beams are of smaller dimensions than those of their Campanian predecessors. In others they are of comparable or larger sizes. Thus, while we can assert that the close spacing and robust dimensions of support joists used in Pompeian shops indicate that these supported heavy floors following (more or less) the formula described by Vitruvius, the same cannot be said of the Ostian floors. Since the timber framing of the Ostian version of *contignatio* exhibits more variation in terms of its load-bearing capacities, the thickness and composition of masonry subfloors must also have exhibited a broader variation.

It is difficult to explain why the method of construction for a *contignatio* used by the Campanian builder was not used by his Ostian counterpart. The difference does not seem to be a regional variation—so many other aspects of design and construction are similar. Neither is the change from simple joists to a beam and joist configuration due to a technical innovation which took place over the period that elapsed between the construction of the Campanian tabernae and those of Ostia, since the beam and joist method was understood and used in republican contexts. Factors of economic or even aesthetics might play greater roles in understanding this shift in practice.

Ostian builders may have realized, for example, that a configuration of fewer stress-grade timbers paired with a light framework of secondary joists was cheaper to build than the old Campanian model. Certainly in other contexts the use of large timbers at Ostia was far less than at Pompeii and Herculaneum. The broad doorways of Pompeian and Herculanean shops, for example, were invariably spanned by heavy beams, while at Ostia we find straight lintel arches of tile or slightly curved arches also of tile, the latter framed with light wooden boards acting as the soffits to create an apparently trabeated frame. Certainly the reduction of wood in such applications decreased the amount of flammable material used in a building, although the type of material used for the lintel of a doorway would seem to make little difference in this regard. It may simply have been cheaper to bridge a door (or window) with fired tiles rather than with a heavy solid beam.

If more can be learned about the pricing structure of raw materials at different periods in the Roman world, it may be possible to explain the differences observed between Campanian and Ostian construction techniques with greater certainty. Although we have specific evidence that the longest structural beams were becoming more difficult to obtain by the close of the first century (the roofing of the Diribitorium in Rome is a case in point),[6] the production figures for timbers of modest dimensions in the high empire are not known. Russell Meiggs has argued that deforestation was not a serious problem in Italy during the Roman period (1980, 192).

Judging from the sampling of the traces of *contignatio* examined at Pompeii and Herculaneum, it seems likely that standards of construction existed within the bounds of practicality. To support upper floors, builders laid out walls either 1.5 or 2 Roman feet in thickness, the heavier walls being used in cases of wider joist placement or to support multiple stories. Joist spacing for the *contignatio* itself ranged between 1 and 1.5 feet; the soffits of the joists were placed a little over 10 feet above the ground floor. At Ostia the placement of beams was more variable, with spacing ranging from 3.5 to 6.5 feet. The Ostian arrangement suggests that a beam and joist system of floor support was preferred to the simpler Campanian design. The nature of the construction of Campanian floors indicates that the Vitruvian description of the components of masonry subfloors reflected actual practice; in Ostia of the second century, where many of the floors may have been built to bear lighter loads, Vitruvius's description is a less reliable indicator of actual practice.

VIII Roofing and Ceilings

The visible part of a Roman roof was its external protective sheathing, originally of thatch or wooden shingles (*scandulae*) or even packed clay (Vitr. 2.1.3). Pan (*tegula*) and cover (*imbrex*) tiles of fired clay were introduced in the seventh century B.C. (Cifani 1994). By the imperial period some important public buildings were protected with roofing fashioned from stone or bronze. Wooden shingles were undoubtedly phased out to lessen the danger from lightning strikes and flying sparks.

Cornelius Nepos, the historian and biographer of the first century B.C., claimed that shingles of wood protected buildings in Rome down to the third century B.C. (Plin. *HN* 16.36). For such applications Pliny recommended oak and beech, species that could provide up to fifty years of protection. Shingles of cypress and cedar last up to one hundred years, but they are not mentioned for this purpose in the ancient sources, probably because they were not native to Italy. Shingles were split by hand; an experienced woodsman can still produce a thousand in a day (Bealer 1972, 18). Wooden shingles were nailed to underlying planks or light beams; they can be seen on houses in northern Italy. Air should circulate both above and below the wooden shingle to retard decay. Among the remarkable finds of wooden objects at the Roman fort of Vindolanda were oaken shingles still piled on the floor of a demolished building dated to ca. A.D. 100 (Birley 1994, 90). Such discoveries verify that in the northern provinces of the empire, wooden shingles never fell out of use.

Barrel- and cross-vaulted buildings were commonly covered with an exterior frame of timber sheathed with tiles so that the outer curvature of the vaults was concealed under a pitched silhouette. The roof (*tectum*, *fastigium*) of a building was generally built as a structural element independent of the ceiling and is therefore considered separately here.

The pitched roof (*tectum proclinatum*, Vitr. 2.1.3) was normally framed with wood. Exposed beams were protected by a sheathing of fired clay plaques or in exceptional cases (Hadrian's Pantheon is a likely example) sheets of bronze. Pitched roofs had been the standard for the large public buildings of the Greeks, structures like stoas, temples, and meeting halls (*bouleuteria*). From the outside a pitched Greek or Roman roof would have looked essentially the same, but, as we will see, the method of framing such roofs (prop-and-lintel vs. the timber truss) was fundamentally different.

The diagonal and horizontal elements of the pitched roof create its single most distinguishing architectural characteristic: the triangular form (*fastigium*) at the end of each long axis. On large buildings with entrances on the long axis (such as temples), the triangular frame of the roof (or pediment) dominated the facade of the structure. Not surprisingly, the form of the pediment itself came to be associated with buildings of high status. Apparently Julius Caesar's house was fitted with a pediment—perhaps in the form of a porch added to the door; Cicero considered this one of the highest honors accorded to the dictator (Cic. *Phil.* 2.110). The fact that facade pediments carried such meaning was one factor that would ensure the longevity of the traditionally pitched roof.

The degree to which a roof was pitched changed over time; generally, steeper slopes appear later. There is no specific term in Latin to indicate "pitch," although *stillicidium* may have carried this meaning, and *fastigium* indicates the presence of a pitched roof. Vitruvius (4.7.5) mentions that the *stillicidium tecti* of a Tuscan roof should correspond to "one-third of the whole." Just what he meant by this formula has never been satisfactorily explained; the proportional relationship may be between roof height and the distance (one-third of height) which the eaves (*stillicidia*) projected over the cella walls. Archaeology can supplement the sparse literary record on the topic, despite the loss of roofing timbers. From the terra-cotta revetments used on early Italic temples (6th–4th centuries B.C.) and the marble cornice moldings of later republican and imperial shrines, it is possible to compute pitch. Temples, for which we have the most information, were generally roofed with pitches from twelve to fifteen degrees in the archaic period (6th–5th centuries B.C.), then in the later republican and imperial periods (from the second century B.C. onward) between eighteen and twenty-three degrees.

It may well be that a shift in the framing of roofs from the older prop-and-lintel method to the triangular timber truss that took place over the course of the second century B.C. can explain the development of the steeper roof. In the parts of the Roman world where tiles were not the norm, degree of pitch could have been influenced by method of sheathing. In Roman Britain, for example, steeper pitches may have been used for roofs protected by wooden shingles or thatch, the latter best applied to steep pitches up to forty-five degrees (Shirley 2000, 17). (See below for a discussion of prop-and-lintel versus the timber truss.)

There existed as well a type of flat roof framed with wooden beams and covered with a thick layer of masonry and durable sheathing. Pliny attributes the invention of this type of roof to the Greeks (HN 36.186). The framing and finishing of such roofs is virtually identical to the construction of upper floors in Roman buildings (see chapter 7).

Some of the thickest and longest timbers used by Roman builders must have been reserved for the construction of roofs. The wooden framework had to bear its own considerable weight without sagging as well as the dead weight of the sheathing and, in some grand structures, an attached coffered ceiling. This total weight is known as permanent load. In addition, the roof had to be built to withstand the added stress of short-lived accidental loads. These could be imposed in the form of high winds, heavy rain, or even snow. A worst-case combination of wind and wet snow can easily add a temporary load of up to 100 kilograms per square meter on a roof. The archaeologist and engineer Cairoli Giuliani has estimated that the roofing of the Romans was characteristically built to withstand total loads (permanent plus accidental) of up to 240 kg per square meter (1990, 63). This means that a building like the senate house (Curia) in Rome would have had a roof structure capable of temporarily supporting something like 127,000 kg (70 tons) without collapsing.

The Prop-and-Lintel Roof

Most pitched Greek roofs covering rectangular spaces, unlike many of their Roman counterparts, were supported by attic-level, hidden vertical supports rising from cross-beams or interior walls (fig. 8.1). The lower ends of the principal rafters usually rested directly upon the tops of the side walls (or lateral colonnades) and were not physically tied to the cross-beams that formed the ceiling. This type of roofing is most commonly known as a prop-and-lintel. As A. T. Hodge (1960) demonstrated in his fundamental study on this subject, the mainland Greek architect either did not know of or did not choose to employ the stronger and simpler triangular timber truss. The prop-and-lintel system did not permit long, unencumbered (that is, clear) spans, and its employment is the main reason Greek temples and other large covered buildings, such as meeting halls, needed to include interior colonnades. The prop-and-lintel's one clear advantage, however, is the ease with which cantilevered beams can be incorporated to support a deep pediment on the front or back of a temple, a feature that would be maintained in later Greek shrines built entirely of stone.

In central Italy we know that Etruscan builders imitated this Greek practice in principle and consequently, since the first monumental construction in Rome (sixth century B.C.) owed much to Etruscan methods, the first great rooftops of Rome were also supported by the prop-and-lintel. The best physical evidence for the implementation of this roofing method by the Etruscans is from the archaic Etruscan rock-cut tombs that have been excavated at sites such as Cerveteri, San Giuliano, Tuscania, and Blera, all located in northern Latium and thus in close physical proximity to Rome. The exterior stone pediments of some of the wealthier burial chambers preserve features, most notably vertical props and representations of purlins, that mimic the mud brick and timbered houses of the living (figs. 8.2, 8.3). From these imitations and from ancient miniature clay models of Etruscan temples that have been found (here the sanctuary at Nemi provides important examples) it is apparent that the Etruscan builder

Fig. 8.1. The Greek prop-and-lintel system. Cross section of the Temple of Zeus at Olympia (470–456 B.C.) The rafters are independent of the ceiling joists. Each rafter is supported at four points: (1) a vertical timber placed at the center of the ceiling beam, (2) a prop with purlin supported by the interior colonnade, (3) a prop with purlin supported by the side wall of the cella, (4) the top of the entablature. (Author, after Whitley 2001, 281)

Fig. 8.2. An Etruscan rock-cut tomb from Tuscania. The triangular gable includes a representation of five "props" that carry the stone "rafters." Sixth century B.C. (Author, after Stopponi 1985, 28)

eventually created a distinctly local variation of the Greek version of the prop-and-lintel. The Etruscan realization placed great emphasis on a massive timber ridgepole and two or more lateral purlins, the ends of which, in an otherwise deep open pediment, were protected with terra-cotta plaques richly modeled with animals and human figures. Such plaques rank among the masterpieces of Etruscan art.

Etruscan (Tuscan) Temples and Prop-and-Lintel Roof Structure

Although no wooden beams from Etruscan temples have survived, the terra-cotta revetment plaques used to sheath the timbers provide important evidence for the way in which these early roofs were constructed, and even give us some idea as to the dimensions of the beams used. Three studies in particular, published over the past seventy years, stand out for helping us to understand the design of the "Tuscan" prop-and-lintel roof. In 1940, Arvid Andrén, who studied closely the architectural revetment plaques from Tuscan temples, offered a detailed analysis of Vitruvius's description of the ideal

Fig. 8.3. Etruscan tomb facade from San Giuliano (Latium), Italy. The simple facade is crowned by a rendition of a pediment supported by three heavy beams (seen from on end). (Author)

Tuscan layout. The observations of the ancient architect, which I shall turn to shortly, include an important (and singular) commentary on roof construction.

To Andrén's study can be added the evidence of Tuscan-style temples excavated at the site of Cosa, published by Frank Brown of the American Academy in Rome in 1960. Brown was able to predict timber dimensions from the architectural revetments excavated and even the degree to which wooden rafters, long disappeared, projected over the sides of cella walls, a calculation that was cleverly deduced by studying the traces of the drippings from the eaves (see *stillicidium* in the glossary) faintly visible on the surfaces of the paving stones around the temple.

Giovanni Colonna's exhibition and accompanying catalog of Etruscan sanctuaries offers an important digest of the most important Etruscan temples; analysis includes reconstructions of roofing schemes (1985). Among the most impressive of these shrines is a pair of temples discovered only a few meters from the edge of the sea at Pyrgi, a coastal site that served as the port of the wealthy city of Cerveteri (Caere). Here, too, the revetment plaques that have been recovered and painstakingly restored have been used to reconstruct the ancient appearance of the shrines. A reconstruction in model form of so-called Temple A at Pyrgi, along with its beautifully restored *columen* (ridgepole) revetment, has been added to the permanent collections on display in the Villa Giulia Museum of Rome.

The study of the revetment plaques from Temples A and B at Pyrgi, along with careful study of wall and column placement, provides an excellent idea of how the main beams of the Tuscan prop-and-lintel roof were arranged: the architraves, main rafters, the purlins, and the ridgepole. Those timbers hidden from view and thus not covered with revetments, such as interior vertical props and planked sheathing, are not possible to reconstruct with certainty, even with the help of the comments of Vitruvius. This means that the framework for the interior of the roof must remain hypothetical. Two schematic renderings of the facade of a tetrastyle Tuscan temple roofed in the manner of the shrines at Pyrgi are illustrated in (fig. 8.4) to complement the following discussion.

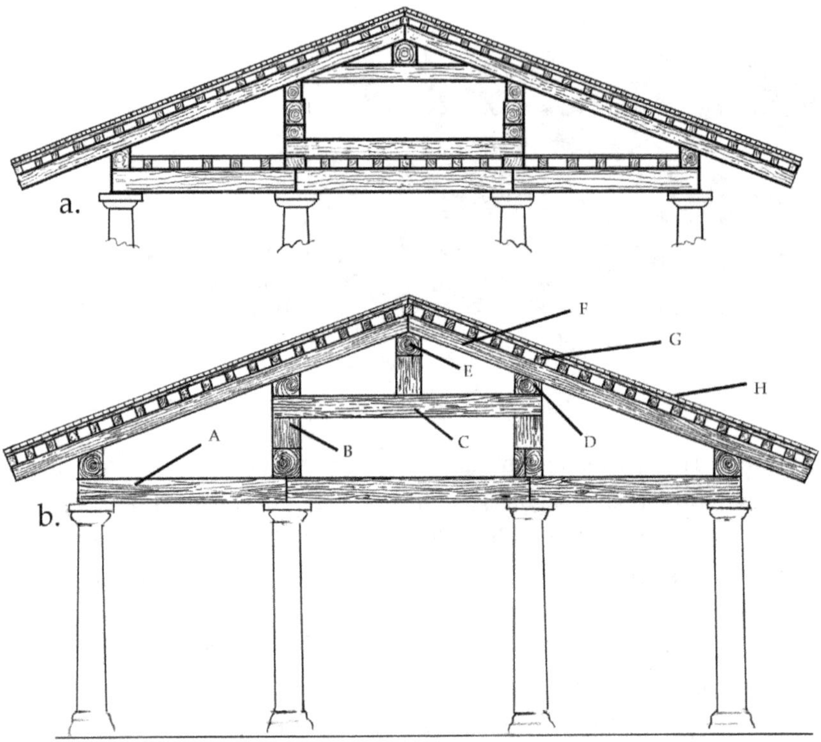

Fig. 8.4. Two hypothetical roofing systems for the covering of a Tuscan style prop-and-lintel temple. A: architrave beam, B: prop, C: lintel, D: main purlin, E: ridgepole (columen) F: primary rafter, G: secondary purlin, H: board sheathing. Terra-cotta tiles would be used to protect the wooden elements. (Author)

The Pyrgi temples have been selected as our exemplars because of their size, complexity, and period of construction (fig. 8.5). The dates of construction, between 500 B.C. (Temple B) and ca. 460 B.C. (Temple A) correspond to southern Etruria's most flourishing period, a time when its influence upon Rome, only just liberated from Etruscan rule, was still strong. The younger of the two temples, Temple A, excavated between 1959 and 1967, was erected at least a century before the tie-beam truss was introduced to Italy. It is now believed that Temple A was dedicated to the Etruscan deity Thesan, apparently similar to the Roman Mater Matuta, a sea goddess sacred to mariners.

Temple A was a large shrine (24 by 34 m) with foundations of cut stone and upper walls of mud brick. Columns of stone drums, widely spaced (there were only four placed across the facade), supported a wooden entablature and roof frame. Three wooden beams placed end to end were required for the architrave of the facade. The longest of these, spanning the central intercolumniation, was about 9.3 m long (31 Roman feet); the beams spanning the flanking columns, spaced more closely, were each 6 m long (20.2 Roman feet). It is possible to arrive at a good estimate of the height of these beams: the decorative terra-cotta plaques restored by Colonna are between 63 and 69 cm high (1970, 216).

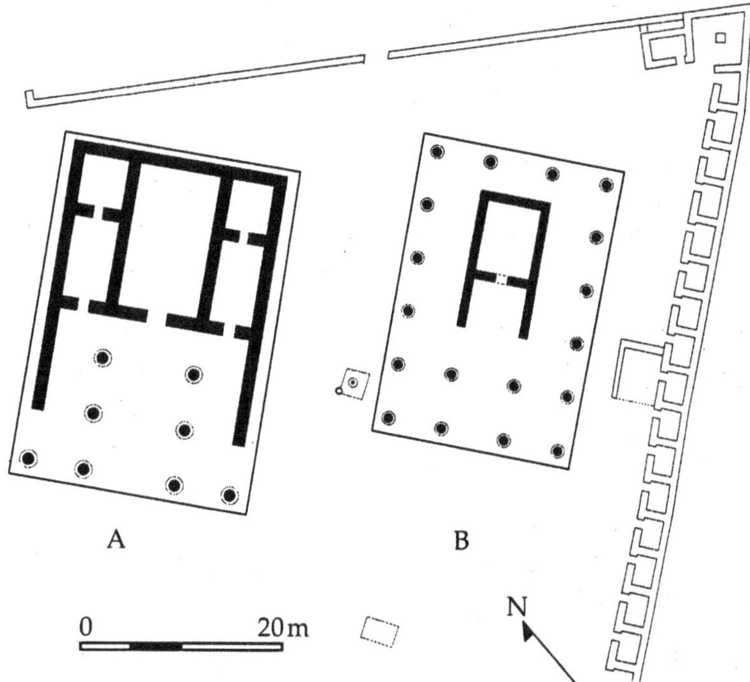

Fig. 8.5. Pyrgi: plans of Temple A (left), ca. 470–460 B.C., and Temple B, ca. 500 B.C. (Author, after Colonna 1985, 129)

It is possible, of course, that one or more of the intercolumnar spaces was spanned by a single timber. A beam ca. 15.5 m long (52.50 Roman feet), for example, could have run across three of the four facade columns at Pyrgi; yet since it was not necessary to procure, saw, and lift into place such immense beams, it is unlikely that such a timber would be used. Temple A was a large temple by Etruscan standards, and it is not hard to imagine that smaller shrines may have used single beams to span more than one intercolumniation.

Vitruvius (4.7.4) indicates that two beams could be placed side by side and lengthwise (*trabes conpactiles*) over columns to satisfy the thickness requirements for the architrave. Greek builders used this same technique with stone (Andrén 1940, lvi). Composite timbers, according to Vitruvius, were to be attached with *subscudes* and *securiculae* (wooden dowels and dovetail joints). Andrén, following T. Wiegand, suggests that beams were butted to one another with dovetails and attached laterally with wooden dowels or heavy nails (1940, lvi). Vitruvius directs that gaps two finger spaces wide were to be left between each beam so that the wood would not rot. This makes good sense, for any water that wicked into a tight-fitting horizontal joint would create ideal conditions for fungus growth; modern carpenters leave gaps between horizontal bearer beams in exterior construction. Attaching two or more beams tightly to one another, instead of leaving gaps, would make no difference to their load-bearing capacity in this application. As we will see, composite beams were also used to form the

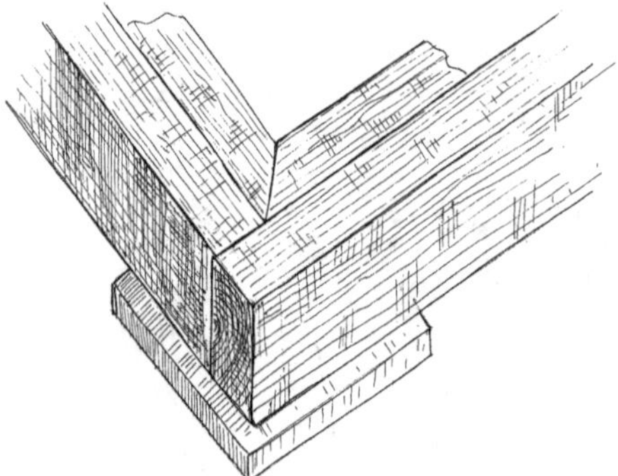

Fig. 8.6. Method of joining compound wooden architraves at corners. This method was also adapted for use with stone architraves. (Author)

architraves that Vitruvius employed in the construction of his basilica at Fanum, and composite beams made of vertically stacked timbers formed the lintels of shop doors at Herculaneum.

How were wooden architrave beams fitted when worked around corners? This is a critical point at the corners of a columnar temple facade. Mitering the ends of the beams would not have been acceptable: the beam would be susceptible to splitting at this point of the end-grain; even the presence of terra-cotta revetments would leave a vulnerable corner joint exposed to damp. Later Roman builders using stone architrave blocks faced a similar problem, and perhaps their solution is analogous to—if not directly derived from—that of the carpenter fitting a wooden architrave; the joint combined a miter on the protected inside corner with a butt on the exterior (fig. 8.6). This solution, if used with a wooden composite architrave, would allow both beams to be securely seated on the corner column shaft and provide a vertical exterior joint easily covered by revetment. Analogies exist in Greek buildings where architraves were created from adjacent blocks of marble; a good example can be found at the Temple of Aphaia on Aegina (ca. 510 B.C.; Orlandos 1966, 154).

Another difficult joint at the level of the architrave can occur along the facade colonnade where beams crossing over the porch from the side walls of the cella, parallel to the main axis of the temple, joined the facade architraves. The challenge was to create a strong right-angle joint that was supported by a single column. An example in stone from another Greek context, the Temple of Poseidon at Sounion, gives us one solution to the problem. By means of a complex beveled mortise (fig. 8.7), two heavy beams can be connected without seriously compromising the strength of either timber.

To return to the facade elevation of Temple A at Pyrgi: above the architrave, directly over each of the columns, four thick beams ran straight back; the two on the outside corners served as the bed for the lower ends of the principal rafters (*cantherii*). All four projected beyond the line of the facade columns, creating a deep overhang. Their ends were connected and concealed by boards, creating a horizontal fascia (*antepagmentum*). The two outer beams, resting directly upon the timbers of the architrave, are called the *traiecturae mutulorum* by Vitruvius (4.7.5). The two main timbers in the center,

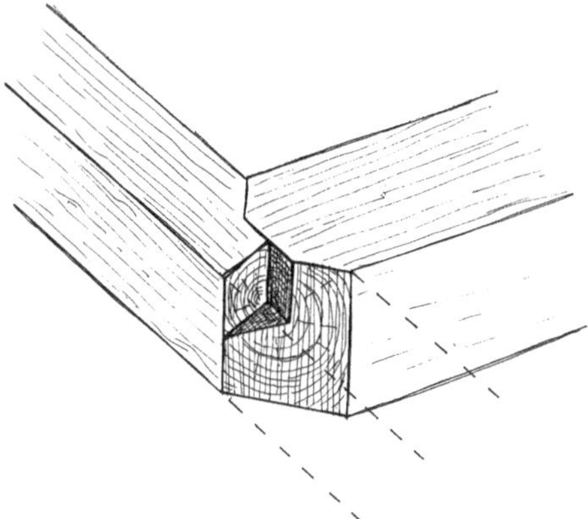

Fig. 8.7. Right-angle T join of heavy timbers over a vertical support point. The bevels and mortises can be worked in either wood or stone. The schematic here is based upon that observed at the Temple of Poseidon at Sounion, Greece (ca. 440 B.C.). (Author, after Orlandos 1966, 156)

supported by the facade columns and the cella walls, formed the lowest element for the prop-and-lintel that carried the main roof beam (*columen*). To make this critical support it was necessary to create a bridge of sorts with a series of transverse timbers (lintels) upon which the weight of the ridgepole would rest. The number of required support points along the length of the temple is conjectural. At Pyrgi the plans indicate that at least six were used. Reconstruction *b* in figure 8.4 provides a solution that requires the placement of paired vertical posts that directly support the transverse lintels. In this scenario, two additional heavy beams were carried lengthwise at the outside corner (and on top) of each of the lintels, thus above and parallel to the two beams carrying the props. These upper beams were the main purlins (sing: *laterarium* or *templum*) of the roof frame. Finally, from the center span of the interior lintel another set of vertical props supported the ridgepole, or *columen*. A slightly different supporting system is suggested in reconstruction *a*. Here a stack of longitudinal beams is placed in line with each of the two interior facade columns. The top timber, beveled to accept the main rafters, forms the uppermost course. Upon these purlins are placed the transverse lintels. In this solution, the *columen* rested directly upon the lintel. In both scenarios the ridgepole itself was beveled to receive the main rafters of the temple. These were now supported at three places: the ridgepole, the purlins, and finally at the *traiecturae mutulorum*.

Among the most notable of the terra-cotta revetments recovered from the excavations of Pyrgi's Temple A is the immense plaque that covered the end of the ridgepole from the rear elevation of the temple. The plaque, modeled with mythological scenes from the Greek story of the Seven Against Thebes, measures no less than 1.40 m across, 0.95 m on the vertical sides, and 1.20 m at its pitched apex (fig. 8.8).

The original placement of such a plaque is clear from the evidence of miniature votive models of Etruscan temples excavated at sanctuaries such as those of Nemi and Velletri; these clearly show the plaques at the apex of the otherwise open pedimental area of the roof (Blagg 2000, 83) (fig. 8.9). Interspersed among the modeled figures of the plaque from Pyrgi are holes for the nails used to attach the revetment to the roof

Fig. 8.8. Terra-cotta columen plaque from Temple A, Pyrgi. Note how the fired clay revetments appear to correspond to the wooden framing elements of the roof. Holes for nails are indicated in black. Villa Giulia Museum, Rome. Late archaic (ca. 470 B.C.) (Author, after Colonna 1996)

Fig. 8.9. Roof of a votive model temple from the sanctuary of Diana, Nemi, of terra-cotta with traces of red paint. H: 19 cm; W: 45 cm. Villa Giulia Museum, Rome, inv. 12642. Fourth–third century B.C. *(Author, after Delbrück 1912, 295)*

beam. Since some of the holes are fairly close (within 10–15 cm) to the edges of the plaque, the beam behind the plaque must have been of similar dimension (too close to the edge and the massive bronze nails used would have split the wood). I share, however, Colonna's uncertainty as to whether such a massive beam, 1.40×1.20 m in cross section, could be found for the ridgepole. Two solutions to this problem immediately come to mind. The first is that the revetment plaque was attached to some kind of wooden backing that was bigger than the ridgepole itself, thus creating the impression of a ridgepole of truly Herculean proportion. The second possibility, as Colonna sug-

Table 2. Ridgepole size as a function of overall dimensions of Etruscan temples

Temple	*Date*	*Overall Dimensions*	*Columen Size (height × width × length)*
Veii	ca. 500 B.C.	18.50 × 18.50 m	ca. 0.80 × 1.0 × 18.50 m
Orvieto	ca. 450 B.C.	16.90 × 21.90 m	ca. 0.80 × 1.0 × 21.90 m
Pyrgi, Temple A	ca. 470–60 B.C.	24 × 34 m	1.20 × 1.40 × 34 m
Tarquinia (Ara della Regina)	ca. 300 B.C.	30 × 40 m	1.24 × 1.80 × 40 m

gests, imagines the main roof beam as a composite of smaller beams, perhaps four, measuring 60×70 cm and bound together with straps of iron. Indeed, the excavations produced iron straps and nails for which no certain purpose has been assigned. The possibility that such *trabes conpactiles* may have been used for the architraves of the temple has already been considered.

The Etruscan builder's use of massive roofing beams would have added to the grandeur of the largest of the Etruscan temples. Even the rafters used at Pyrgi, if the revetment slabs provide an accurate indication, were 50 cm in thickness. If the conventional reconstructions of Etruscan Veii are correct, archaic temple roofs were also designed to carry large-scale sculpture along the ridge, placing additional stress on the main beam. At the sanctuary at Veii, the so-called Portonaccio temple, a contemporary of Pyrgi's Temple B (ca. 500 B.C.), has produced well-known examples of this kind of ridgetop decoration. Even so, the ridgepole used at the Portonaccio temple was not as massive as that used at Pyrgi (table 2). Colonna has suggested that the size of the *columen* of such temples was proportional to overall dimension: the larger the temple, the larger the ridgepole. Where *columen* plaques survive in good enough condition to provide maximum dimensions, ridgepole sizes can be estimated. Note that these dimensions indicate that ridgepoles were placed so that the greater dimension of the stock was placed horizontally. This was probably done to provide as broad a bedding as possible for the rafters, which were carried by the ridgepole and met above it (a method reminiscent of the framing of the roof of the Iron Age hut). It was therefore necessary to bevel the top corners of the ridgepole or to notch the main rafters; the former would have been stronger and is in fact suggested by the shape of the *columen* plaque found at Pyrgi.

As we have already seen, the heavy ridgepoles of the Temples at Pyrgi supported a series of main rafters. These were spaced too widely to be covered with tiles. It is likely that there were as many as three additional levels of lighter beams that formed the bedding for the final outer sheathing of fired clay. A model of the roofing of Temple B built by a former student helps one understand the proposed system (fig. 8.10). First, resting on the main rafters were a series of longitudinal beams that paralleled the heavy purlins below. These supported a layer of light rafters. The light rafters in turn supported a decking of sawn boards, to which the terra-cotta roofing tiles were finally applied.

Smaller roofs may not have needed to employ a similar multilayered grid of tim-

Fig. 8.10. Model by F. De Simone of the roofing beams used at Temple B, Pyrgi (ca. 500 B.C.): (A) main purlin; (B) primary rafter; (C) secondary purlin; (D) secondary rafter; (E) board sheathing. (Photo: Author)

bers. In the system described by Vitruvius, 4.2.1 (below), the main rafters carry only the set of longitudinal beams and a board sheathing, whereas secondary rafters are used in the Pyrgi layout just described.

Certainly in its original form (dedicated in 509 B.C.), and perhaps throughout its history, the Temple of Jupiter Optimus Maximus on the Capitoline Hill of Rome was framed with a prop-and-lintel roof; the ridgepole of this shrine had to be strong enough to carry not only the roof tiles, but also the weight of terra-cotta revetments and the terra-cotta sculpture of the enormous *quadriga* (a chariot drawn by four horses) that crowned the pediment (fig. 8.11).

From the supposed placement of the columns of the temple it is possible to estimate the longest spans bridged by the timbers of the architrave. The broadest clear spans required in this enormous shrine were, if E. Gjerstad's figures are correct, ca. 12 m (40 Roman feet) across the center pair of columns, requiring immense timbers still only a fraction of the total breadth of the roof (about 50 m, or 168 Roman feet), and a far cry from the clear spans that builders would be able to achieve four hundred years later with the knowledge of the tie-beam truss and access to the silver firs of the high Apennines (Gjerstad 1960, 181). A new reconstruction of the temple by John Stamper, however, proposes that the superstructure of the temple was in fact smaller than the podium and suggests a maximum span of 7.40 m (25 Roman feet) over the central bay, a figure more in line with contemporary large-scale Etruscan shrines.[1]

Since it was not possible to enclose large, unencumbered interiors using the prop-and-lintel, covered spaces of the early Roman republic differed fundamentally from their imperial counterparts. The tripartite cellae that characterize so many Tuscan-style temple plans (like that of the Capitoline Jupiter), regardless of the number of deities worshiped within, reflect the need to provide interior supports for the pitched

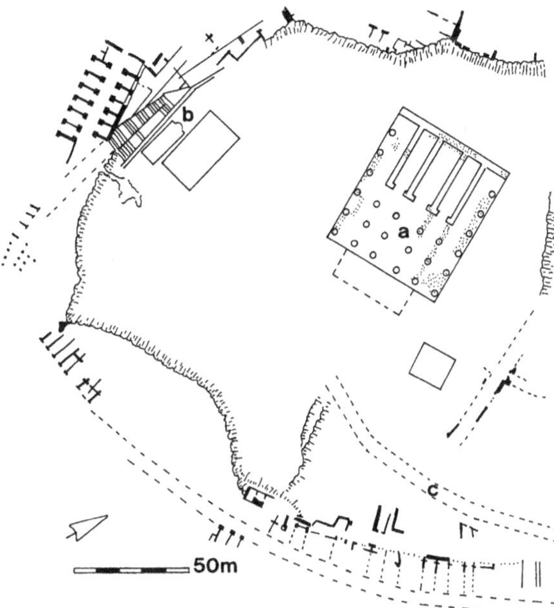

Fig. 8.11. Temple of Jupiter Optimus Maximus in Rome. Restored plan (top, a) and elevation. The roof could be carried on seven longitudinal supports (the ridgepole, four purlins, and the two lateral architraves) held up by vertical props. First dedicated in 509 B.C. *([top] Author; [bottom] Author, after Gjerstad 1960, 182, fig. 117)*

roofs that covered these shrines. The volumetric spaces we associate with the great monuments of imperial Rome were alien to the republican builder because of the prevalence of the prop-and-lintel roof. Ancient sources tell us, for example, that the original senate house of Rome, the Curia Hostilia, was still in use at the beginning of the first century B.C. This hall was considered one of Rome's earliest and most venerated monuments, its foundation attributed to the king who gave his name to the structure, Tullus Hostilius (traditional date of reign: 673–642 B.C.). Just what this first senate house looked like we cannot say, for it was rebuilt and presumably enlarged by Sulla in 80 B.C., burned in 52 B.C., and subsequently demolished. Its site remains unexcavated; there exists no ancient description. Yet even if its late republican replacement, the Curia Julia, saluted its predecessor by duplicating its basic floor plan, and the late imperial (ca. A.D. 300) rebuilding we can still visit today imitated the Curia Julia, the broad, shadow-filled box of space we associate with the senate house of Rome could not have characterized the original Curia Hostilia because there would have been no way to support the roof (now with a clear span of 60 Roman feet) using the prop-and-lintel.

The prop-and-lintel system persisted, even with the invention of the tie-beam truss; indeed, roofs large and small that exhibit the characteristic vertical supports for the ridgebeam can still be seen in Italy. The prop-and-lintel system remained a method of roofing temples in the first century B.C. The Doric temple of ancient Cora (now Cori) 63 km (38 miles) to the south of Rome preserved what Hodge has called a translation into stone of the traditional wooden form of the prop (1960, 59). The present remains of the temple were built shortly after 89 B.C. On the inside face of the facade pediment, five vertical blocks of stone have been placed in positions that would have been occupied by wooden supports of an analogous building framed with wood; a socket hole above the central stone post shows how the ridgepole of this temple was supported by a vertical prop (fig. 8.12).

Despite the traditional appeal of the prop-and-lintel roof, however, builders were restricted by the open spaces they could span. Just what this exact limit was is hard to say. The cella of the Parthenon in Athens was covered by a prop-and-lintel roof that, despite the help of an interior colonnade, had a free span of just over 11 m, about 37.3 Roman feet (Hodge 1960, 39). Spans much over this would have been more efficiently covered with a tie-beam truss (Klein 1998, 338). If longer beams are not held in tension, especially when loaded with additional weight, they tend to sag. This fault, as we will see, could be overcome with the triangular tie-beam truss.

Vitruvius on Roofing

The prop-and-lintel would have been thoroughly familiar to Vitruvius; it may have been the predominant form of roofing he saw in the Italy of his day. In his frustratingly brief discussion of the components of timber roofs of temples, Vitruvius alludes both to the

Fig. 8.12. The Doric temple at Cori (Lazio) Italy, stone framing of the pediment. The socket hole for the columen (ridgepole) is clearly visible, as are the blocks of stone used to support the two purlins on either side of the columen. Late second century B.C. Compare this to the pediment of the Etruscan tomb from Tuscania in fig. 8.2. (Photo: Author)

traditional prop-and-lintel and to the stronger, more versatile tie-beam truss. In the fourth book of *De Architectura* (Vitr. 4.2.1), he writes, "Under roofs, if the spans [*spatia*] are greater, both *transtra* [tie-beams?] and *capreoli* [are used], if a moderate size, a ridge piece [*columen*] and rafters [*cantherii*] projecting to the edge of the eaves [*ad extremam suggrundationem*]. Above the rafters [*cantherii*] there are placed purlins [*templa*], then above these, but under the tiles, planks [*asseres*], which extend beyond the walls so that the walls are covered [*tegantur*] with eaves [*protecta*]."[2]

Here Vitruvius describes two ways of framing roofs. Two key words in this passage, following Andrén's interpretation, are *transtra* and *capreoli*, and a central question is how they should be distinguished from *columen* and *cantherii* (Andrén 1940, lxii). Vitruvius says that the former pair are to be used if the span of the roof is "greater" (*maiora*), that is, if the space cannot be bridged with the traditional prop-and-lintel. The *transtra* and the *capreoli* thus appear to be the specific terms Vitruvius uses for the tie-beams and the diagonal supports (that is, main rafters), respectively (which in turn can carry purlins and secondary rafters), of the triangular tie-beam truss. Vitruvius makes no mention of what we now describe as a king post, collar-beam, or similar braces used to reinforce the simple triangular truss, unless the *capreolus* is in fact a reference to an interior diagonal brace and *columen* can also refer to a king post, as Rowland's recently published translation of Vitruvius suggests (1999, 219). Similarly, Vitruvius appears to omit the terms for such supplemental bracing in the description of the basilica he built at Fanum (to be considered more fully below). The essential tripartite division of the roof described by Vitruvius, namely, main rafter (*cantherius*) capped by the *templa* and finally the *asseres*, is reiterated later in the same book (4.2.5).

In considering this passage, one must remember that Vitruvius had in mind the prop-and-lintel roof of a Greek temple, yet the fact that specific Latin terminology existed to describe the various components of this conservative method attests to its recognized role in Roman republican architecture. Furthermore, many of the terms we learn from this passage are important for understanding the names of components used in the roofing frames built with the tie-beam truss. Both systems used horizontal beams and diagonal rafters. Both could employ a lighter decking of planks over the rafters, which provided a smooth and stable surface for the final sheathing of tiles. The fundamental difference between the two systems, as shall become apparent, is the way in which the cross-beams were tied (or not) to the rafters.

The Tie-Beam Truss Roof

Compared to the prop-and-lintel, the tie-beam truss employs smaller and therefore lighter timbers, bears greater weight, is less subject to sagging, and allows greater spaces to be spanned. In its simplest manifestation, the truss is made up of three timbers—two diagonal pieces, or rafters, and a horizontal joist, or tie-beam—forming an equilateral triangle. The tie-beam, acting in tension, counteracts the lateral thrust of the rafters. Built in this way, the trusses are self-supporting. Heavy ridgepoles are no longer needed to carry the load of rafters. Indeed, it is technically possible to build a trussed roof without a ridgepole at all (a feature of much modern construction).

Triangular trusses spanning broad spaces were presumably fortified in much the same way modern trusses are reinforced (fig. 8.13). The Roman truss, however, was undoubtedly built of thicker timbers than its modern counterpart, and therefore its internal bracing was somewhat less complicated. A vertical king post could be placed at the center of the truss, running from the apex of the triangle to the top surface (or just shy of the top surface) of the horizontal tie-beam. The element would have evolved naturally from the traditional positioning of the center prop of Tuscan roofs, but instead of being employed to support the ridgepole it was actually suspended from the top to prevent the tie-beam from sagging. In modern buildings king posts are used when spans exceed 6 to 8 m (19–26 feet). "Queen posts" positioned to either side of the king post could provide additional vertical reinforcing. For broader spans (over 12 m; 39 feet), the king post could be eliminated in favor of a pair of queen posts carrying a collar-beam, so that the triangular truss appears to have a rectangular insert (a configuration observable in the wooden beams, not ancient, now supporting the roof over the main space of the porch of the Pantheon). The rafters, forming two sides of the truss, could also be strengthened by a collar-beam attached to the king post, by knee braces, or by a combination of both.

The lower ends of the two main rafters were jointed to the tie-beam at a calculated distance from the ends; Giuliani has formulated that the thickness of the rafter can be used to gauge the point of juncture between rafter and tie-beam (1990, 64). Ac-

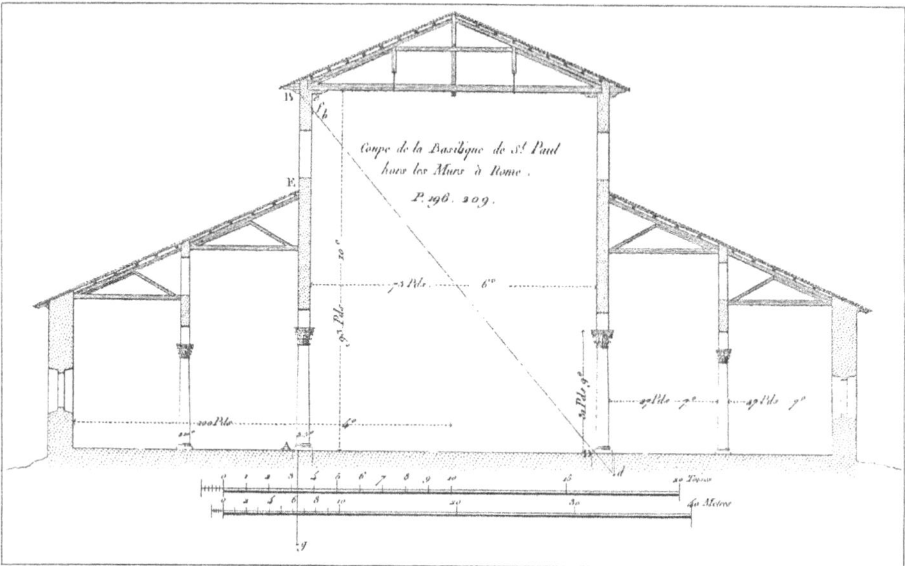

Fig. 8.13. Section of St. Paul's basilica, Rome (begun A.D. *385), drawn before the trussed roof burned in 1823. The three heavy timbers of the triangular truss support a framework of purlins and light rafters which in turn carry the tiles of the roofing. The clear span of the nave was 82 Roman feet.* (Rondelet 1814 (III), pl. 76)

cording to this simple formula, a rafter 50 cm thick should join the tie-beam at least 50 cm from the end (fig. 8.14). The actual joint used for the critical connection between the rafter and tie-beam needed to be strong enough to resist any slippage laterally by the diagonal timber. More than one solution would have worked. In figure 8.14 a bedding is shown cut into the top surface of the tie-beam. A mortise and tenon joint, recently proposed for the timbered roofs of the fortress at Inchtuthil, would also have been effective and have minimized the weakening of the tie-beam that would result from a broader cut (Shirley 2000, 25) (fig. 8.15).

The butt ends of the horizontal tie-beam were under particular stress at the point they rested upon, or were inserted into, the supporting wall. Giuliani proposes that the suitable wall thickness for support can be expressed in terms of the dimensional timber used for the truss: the thickness of the rafter plus the length of the tie-beam's projection beyond the juncture-point. From the example given above, with a hypothetical rafter measuring 50 cm thick, the supporting wall would be at least 1 m thick. Giuliani's calculations are necessarily based upon later examples of trussed roofs that still survive, so it is hard to know to what degree Roman builders formulated their construction. We have already seen that for floor construction builders tended to use heavier joists than were necessary and that there is little or no evidence of any engineering theory at work.

Tie-beams may have been further reinforced at the point of support by brackets of wood or stone, or a combination of stone brackets and knee braces (fig. 8.16). The

Fig. 8.14. Hypothetical attachment of main rafter to horizontal tie-beam of a triangular truss. The space from the end of the tie-beam to the juncture point of the rafter should be equal to or greater than the thickness of the rafter. The thickness of the supporting wall should be no less than the combined thicknesses of rafter and tie-beam. (Author, based upon Giuliani, 1990)

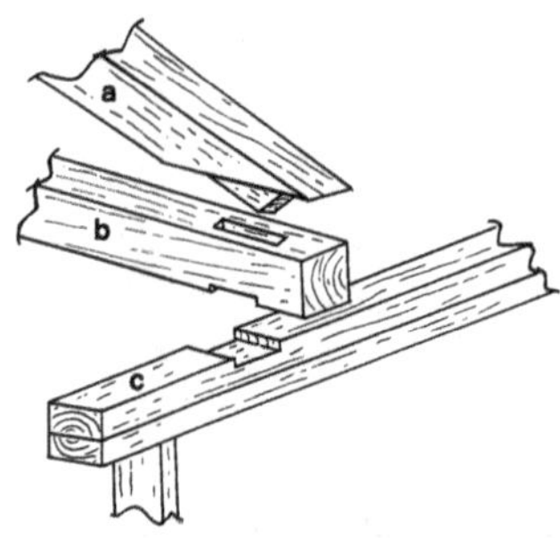

Fig. 8.15. Hypothetical method of joining a rafter (a) to a tie-beam (b) using a mortise and tenon. In this example the tie-beam is joined to the wall plate (c) by a saddle joint. If a masonry wall was employed, the tie-beam could rest upon the wall or be inserted into a socket built into the wall. (Author, after Shirley 2000, 25, *fig.* 3.6)

top surfaces of crowning cornices used for the interior engaged orders in public buildings such as temples and covered theaters would have served a similar purpose. In this context it is interesting to note that the use of interior decorative orders in temples became increasingly popular over the course of the first century B.C., just as some of the larger trussed roofs were installed in large temples. Augustus, Rome's first emperor, made good use of the trussed roof to create roomy temple interiors. The Temple of Apollo *in Circo* (ca. 28 B.C.), the Temple of Mars Ultor (2 B.C.), and the rebuilt Temple of Castor (A.D. 6) are but a few of the outstanding examples in Rome.

Specific solutions for the framing and bracing of large tie-beam trusses must have differed from site to site; the evidence available does not permit an exact reconstruction for a single Roman timber-roofed building built before the era of the first Christian churches. There is no question, however, that the timber truss was extremely strong and for this reason is still a fundamental component of modern timber framing. The Roman truss could span spaces at least as broad as those covered with barrel or cross vaults constructed of concrete, and, as we will see, the majority of the largest rectangular spaces ever covered in the Roman world were roofed with these timber trusses.

The answer as to exactly when and where the tie-beam truss was invented and first utilized has eluded modern scholars. The answer may lie, as Hodge (1960) and Klein (1998) have suggested, in the ruins of some of Sicily's larger Greek shrines, where

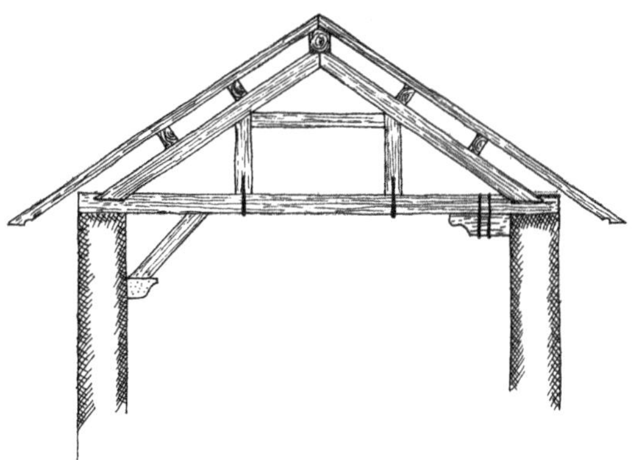

Fig. 8.16. Reinforcement of the timber truss. The composite drawing shows two methods of strengthening the timber truss: on the left a knee brace is used with a stone bracket; on the right a simple bracket is added. The main truss is also reinforced with two queen posts and a collar-beam. (Author)

the free spans surpass those of the Greek mainland and point to the introduction of the tie-beam truss. Hodge's and Klein's systematic studies of Greek roof construction have raised the possibility that the truss was known from as early as the sixth century B.C., but the evidence for this early a date remains problematic (Hodge 1960, 41; Klein 1998, 370). W. Dinsmoor thought a point much before ca. 300 B.C. was unlikely for the Greek world, at least on the mainland (1950, 242). J. Coulton has noted that the rectangular council chamber at Miletus in Asia Minor may have used a truss to cover its span of 16 m; it has been restored accordingly on paper by George Izenour (1992, 52).[3] By the time this structure was built, however (ca. 170 B.C.), the first trussed basilicas were already standing in Rome (see below; Coulton 1977, 157).[4] The fact remains that while credit for the invention of the truss may be due to an unknown Greek builder of the archaic period, its full potential for covering broad, unencumbered spaces was never recognized in Classical Greece.

Renaissance drawings of the original bronze trusses (removed by Pope Urban VIII in the seventeenth century) used to support the porch roof of the Pantheon (finished by A.D. 128) provide the earliest absolute dated evidence of a Roman building covered with a tie-beam truss in Italy, but it is clear that by this time this type of roof was well known to Roman builders (fig. 8.17). Bronze seems an unusual choice for load-bearing beams. The trusses and purlins shown in the early drawings are rendered as three sides of an open rectangle. While architectural historians have argued that no wood was used (for example, De Fine Licht 1966, 48), William MacDonald has recently doubted that open-work bronze beams alone could have supported the weight of the roof without buckling; he suggests it is more likely the bronze was used to sheathe heavy timbers (1982, 98).

A relief exhibited for the first time in 2001 in Rome offers evidence as graphic as it is perplexing concerning the existence and employment of the timber truss (fig. 8.18). That the relief was carved to commemorate a public spectacle seems clear. It was found in the area of the Campus Martius, the zone of the city associated with structures catering to public entertainment. The center of the composition is dominated by what

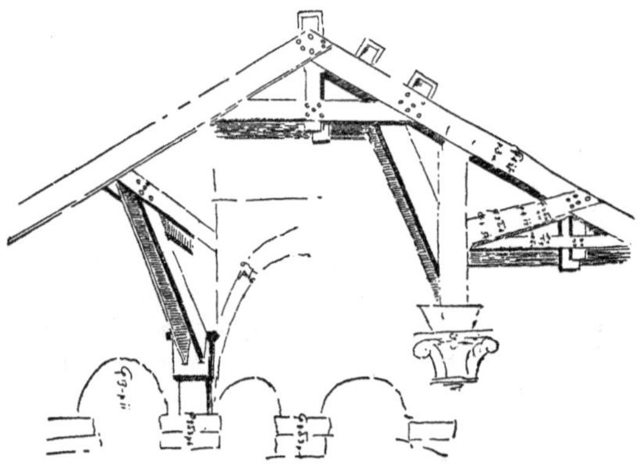

Fig. 8.17. Sketch by G. A. Dosio of the original bronze trusses that supported the roof of the porch of the Pantheon, Rome. Knee braces are used to reinforce one of three high trusses; the ends of bronze purlins are also visible. Second century. (Lanciani 1897, 483)

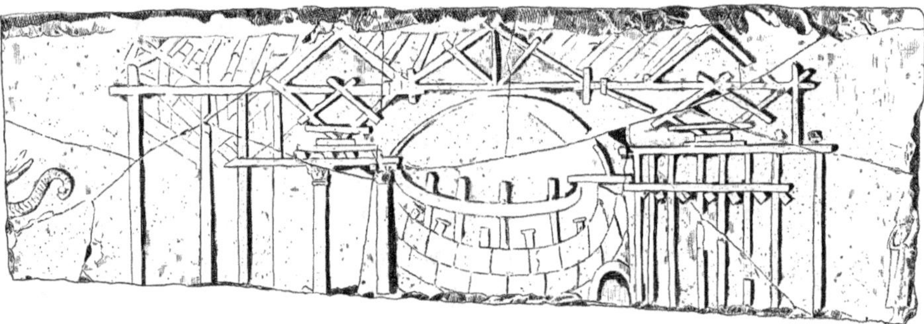

Fig. 8.18. Relief on stone of an amphitheater covered by a timber structure built of trusses. Date uncertain; possibly first century B.C. *(Author, after Coarelli 2002, 46)*

appears to be an amphitheater, similar to that already described from the Column of Trajan in that its walls look as if they have been built both of stone (at the ground level) and of wood (the upper tier).[5] The left edge of the carving preserves only the trunk and a tusk of an elephant, but this is enough to strengthen the view that the subject portrayed makes reference to a popular public spectacle.

The artist has focused his attention on the rambling timbered structure that extends over and well beyond the side walls of the arena. Three timbered trusses have been carefully rendered, as if by someone who has first sketched a structure that he personally observed. There are too many idiosyncrasies for this to be the product of the artistic imagination. The center truss is shown to be reinforced with a king post and flanking diagonal braces. The trusses to either side, of the same height as the central element but lesser in span, exhibit no internal reinforcement but are themselves supported by diagonal braces that descend and converge on what appears to be a narrow timbered platform, itself supported by tall uprights. On the left side two columns are shown carrying the load. On the right side the truss is supported by a closely spaced series of upright timbers, stabilized near their top ends with a beam or beams that tie them to the walls of the amphitheater itself.

Wooden roofs built over venues for mass entertainment were generally fixed and permanent structures in the Roman world; the best known were the music halls, or *odea* (sing. *odeum*), which represented some of the largest covered spaces ever built.[6] The timbered covering shown on the relief, however, is for the most part independent of the oval arena itself. Furthermore, the ends of the timbers have not been trimmed, suggesting hasty construction and perhaps the desire not to cut valuable beams destined for reuse. All this points to a temporary installation, which also explains the apparent scaffolding that has been thrown up to support the right side of the roof. It is also likely, given the conventions of Roman popular art, that what was once in reality a three-dimensional boxlike structure has been flattened or unfolded to result in a composition that extends so dramatically to either side of the depicted arena. In other words, the trusses we see flanking the main truss may represent the two short sides of the oval building (its "front" and "back"), while the center truss indicates one long side of the arena. Thus in plan the roof would be rectangular in layout, covering the oval footprint of the arena with a trussing system that intersected along the lines of the main and cross axes of the amphitheater. This would also explain why the two trusses shown extending over the sides of the arena are shown with lesser spans. The columns to the left may represent a columnar entryway standing at one end of the amphitheater that now helps support the timbered roof; perhaps the unadorned back of the building required the help of a forest of scaffolding. This interpretation does not, unfortunately, explain the additional projection of the roof's structure beyond the left truss.

Could a trussed wooden roof large enough to cover an arena have ever existed? Perhaps, if the arena in question was of modest size, housing a show we might associate more with a modern circus's big top, which holds a thousand spectators or so. The odeum (*theatrum tectum*) at Pompeii (ca. 80 B.C.) was covered with a trussed roof that spanned 85 Roman feet (ca. 25 m) and could hold around fifteen hundred spectators (Izenour 1992, 71). Such roofs, however, were firmly seated on four walls; the relief from Rome presents a rather wobbly looking form of support. No one would have tried such a feat with the full-sized amphitheaters of the high empire. These, we know, were rigged with cloth awnings to provide some protection from the hot Italian sun. Without knowing the date of the relief or what building it is meant to represent, we cannot arrive at a more detailed conclusion. F. Coarelli suggests the relief can be dated to the period of Caesar (mid–first century B.C.). If he is correct, then this clear representation of a timbered truss is certainly the oldest we have for the Roman period and demonstrates the confidence builders placed in the form to shelter with safety spectators below.

As far as Italy is concerned we can perhaps arrive at a date for the first use of the tie-beam truss by considering at what point it *had* to have been developed. Since tie-beam trusses are not required for small spaces; the enthusiastic adoption of the truss must have occurred out of the need (or the desire) to build large, unencumbered interiors. From literary and archaeological evidence it is clear that a new and highly popular architectural form was born in Italy in the second century B.C.: the civic basilica. Despite their differences in overall size, orientation, and siting, all Roman basilicas were distinguished by a spacious open area, the nave, normally roofed with a timbered roof carried on high walls which, in turn, were supported by a ground-level colonnade.

In Rome alone four of these new buildings were constructed over the course of the second century B.C., all bordering on or in the immediate vicinity of the Roman forum: the Basilica Porcia (184), the Aemilia (179), the Sempronia (170), and the Opimia (121). At Cosa, a modest colony north of Rome, a new basilica was built on the forum during the middle of the second century B.C. By the close of this century a basilica (perhaps only partially roofed) stood on the forum of Pompeii. None of these buildings could have been covered had the builders not known of the tie-beam truss. The proliferation of the basilican plan in the second century thus points to this period as the one in which the Roman builder first understood the tremendous possibilities this form of roofing offered. Architectural historians have long wondered if, because of its Greek-derived name (*basilica* from *basileus*, the Greek word for "king"), there were prototypes of the Roman basilica to be found in the Greek world. The absence of the basilican plan from classical Greek architecture, however, is of little surprise; whatever other reasons there may be, basilicas did not exist in Greece because native builders did not employ the tie-beam truss. Indeed, the covered meeting halls with appreciable clear spans that do eventually appear in Greece and Asia Minor may have been inspired in part by roofing developments in the west.

Thus it is fitting that in his description of the basilica he designed for the town of Fanum (Colonia Julia Fanestris), on Italy's Adriatic coast, Vitruvius offers several important and revealing details about roof construction (fig. 8.19). Here is Vitruvius's account of his basilica (5.1.6–10):

> (6) At the Julian colony of Fano, I contracted and supervised the construction of a basilica not inferior to these in dignity and grace, whose proportions and harmonies [*symmetriae*] are as follows: There is a timber-trussed nave [*mediana testudo*] between the columns 120 feet long by 60 feet across. The aisle around the nave between the columns and the exterior wall is 20 feet wide. The columns with their capitals rise with an unbroken height [*altitudinibus perpetuis*] to 50 feet and are 5 feet in diameter. Standing behind them are pilasters 20 feet high, 2.5 feet wide and 1.5 feet thick; these carry the joists [*trabes*] which support the flooring [*contignationes*] of the gallery.[7] Above these are pilasters 18 feet high, 2 feet wide and 1 foot thick which carry [*excipiunt*] the beams [*trabes*] supporting the main rafters [*cantherii*] as well as the roofs [*tecta*] of the portico, which are lower than the main trusses [*infra testudinem*]. (7) The remaining spaces in the intercolumniations between the pilasters and the beams of the columns are reserved for the windows. Across the width of the trussed area [*in latitudine testudinis*], counting the corner columns on the right and left, there are four columns at each end. On the long axis, the side facing the forum, there are eight, counting the corner columns. On the other [long] side there are six columns, including the corner columns. The two

> in the middle were left out so they would not obstruct the view of the pronaos of the shrine to Augustus, which is located in the middle of the side wall of the basilica and looks on the middle of the forum and the Temple of Jupiter. (8) [Here I omit Vitruvius's description of the tribunal in the basilica.] Above the [main] columns [of the nave] beams [*trabes*] are placed all around, made of three two foot timbers joined together [*ex tribus tignis bipedalis conpactis*] . . . (9) above these beams, over each capital and column shaft, piers 3 feet high and 4 feet square are placed. Above these [piers] are placed outwardly sloping beams [*trabes everganeae*] made from two timbers each 2 feet square around [the perimeter of the basilica]. Finally, above these, tie-beams with main rafters [*transtra cum capreolis*] placed above the shafts [*corpora*: lit: "bodies"] of the columns and the *antae* and walls of the pronaos support the single ridgepole [*unum culmen*] of the whole basilica, and a second [ridgepole] running from the middle [of the main ridgepole] over the pronaos of the shrine. (10) And so created from the roof is a double [that is, intersecting] arrangement of triangular pitches [*fastigiorum duplex*], providing a charming effect both to the exterior of the roof and to the lofty trusses [*altae testudinis*] of the interior. In addition, since the ornamental entablatures and the upper columns and parapets are eliminated, great expense and laborious details are avoided. Indeed, the reach of the columns to the beams of the trusses [*sub trabe testudinis*] seems to make all the greater the magnificence of the outlay and the authority of the result.

Covering a cavernous space like the nave of a basilica was no small feat, and it is revealing that Vitruvius devotes the majority of his description to the roofing system of his structure. This passage strengthens the argument that a *testudo* in such contexts refers to a timber-trussed roof and that *transtra* and *capreoli* indicate, respectively, the tie-beams and diagonal structural rafters of the tie-beam truss, definitions already tentatively reached from an examination of Vitruvius's earlier comments in book 4 (see above). Scholarly opinion, however, is not uniform on this point. *Capreoli* have been translated, as mentioned earlier, to refer to the diagonal braces within the trussing (Rowland 1999, 64).[8]

Of additional interest is the evidence for the use of composite beams for the architraves over the columns; the individual beams, each fully 2 feet thick (60 cm), were massive but had to span at the least only one intercolumniation (20–25 Roman feet). The nave itself, as Vitruvius tells us, was 60 Roman feet across, slightly broader than the basilicas Aemilia and Julia in the Forum Romanum and about the same span as the senate house of Rome, the Curia (at least in its last building phase of the late third century). Nevertheless, a 60-foot span for a timber truss was relatively modest when compared to the most ambitious timbered roofs built by Roman architects.

The most challenging element of the framing scheme was the intersection of two pitched roofs directly over the center of the nave. The resulting T-shaped configuration was designed to accommodate an interior shrine to the emperor Augustus built on the short axis of the basilica. This type of plan was achieved in later vaulted build-

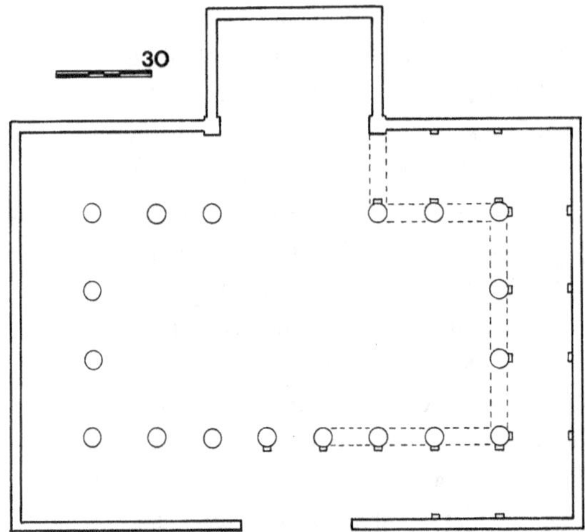

Fig. 8.19. The basilica designed by Vitruvius at Fanum. Plan (with scale in Roman feet) and interior reconstruction. First century B.C. ([plan] Author, after Alzinger 1989, 213, and Ohr 1975, 117; [reconstruction] Author, modified from Warren in Morgan 1914, 135)

ings by the use of intersecting barrel vaults. We can only speculate how this was done with wooden beams at Fanum. The disposition of the trusses here in the center of the nave is additionally complicated by the elimination of two interior structural columns so that the view to the shrine of Augustus was not impaired. One possibility is that two trusses were carried by a beam over the span of 46.5 Roman feet (13.76 m) that was not supported in the middle because of the eliminated columns (Smith 2003, 152; Morgan 1914, 135). Thomas Howe's reconstruction proposes that Vitruvius may have employed two intersecting trusses, each placed diagonally across the center of the nave so as to support the intersecting ridgepoles (Rowland 1999, 242, fig. 80). For this solution, the spans of the triangular trusses would have been slightly over 80 Roman feet (23.70 m), an impressive—but not impossible—achievement. Vitruvius himself does not provide enough information to solve the matter.

Vitruvius makes no mention of a coffered ceiling (*lacunaria*) at Fanum, even though coffering might be considered a standard feature of a prestigious public building such as this. Normally the presence of fine coffering was, as we shall see, something to boast about. Not only are coffers not mentioned, but from the description of the basilica at Fanum it appears that the trusses were left fully visible to the ancient visitor. Vitruvius claims that the view of the "lofty trusses" from the interior gave the building a "charming" effect (*speciem venustam*). The sense of the upward expansion of space and the reference to the entire truss, the *testudo*, and not just the cross-beam, the *trabs*, indicates that the architect thought the framing of the roof was both technically and aesthetically pleasing. At the same time, the elimination of the coffering would have saved a great deal of money, evidently of some concern to the architect.

The builders of the first large Christian churches of fourth-century Rome presumably studied the exposed trusses of the city's earlier civic basilicas. Trajan's great Basilica Ulpia, dedicated in A.D. 113, provided a noble model. The old senate house, the Curia Julia, had been rebuilt with trusses after a fire in the year 283. The revival of the basilican plan with its trussed roof in the fourth century can be seen in part as an attempt by the first Christian emperors to install the new state religion within a physical space associated with the glory days of the high empire of two centuries before. Although churches like Constantine's original St. Peter's no longer stand in their original form, documentation in the form of written descriptions and drawings provides important information about their general appearance and method of construction.

Rome's basilica of St. Paul's Outside the Walls is an excellent case in point. Built in the late fourth century and subsequently repaired several times, the configuration of the timber-trussed roof survived more or less in its original form until the church was severely damaged by fire in 1823 (see fig. 8.13). According to an etching of Giambattista Piranesi published in 1778 and a section of the basilica drawn by J. Rondelet (1814), the broad nave (82 Roman feet across) was spanned by paired timber trusses reinforced by a king post and flanking queen posts (fig. 8.20). We do not know if the idea of pairing main trusses was an invention of the fourth century or an imitation of earlier Roman practice. The tie-beams were carried by the walls with the aid of brackets. Such illustrations inevitably reveal some discrepancies. Piranesi shows a kind of boarded catwalk crossing the center of the tie-beams along the length of the basilica; the king posts cannot be seen. Rondelet's rendition indicates that the king posts projected between and below the soffits of the tie-beams. Both drawings indicate that the lower half of the main rafters was a composite, being reinforced by a beam running from the wall to the top of the queen post. The same method was used over the double side aisles of

Fig. 8.20. A view of the interior of the basilica of St. Paul's Outside the Walls, Rome. Etching from Piranesi, Vedute di Roma, *published in 1778. (Hood Museum of Art, Dartmouth College, Hanover, New Hampshire)*

the church, where, in combination with a diagonal strut, the configuration was that of a small triangular truss inserted into the shedlike profile of the aisles' roof. This solution may be an invention of the fourth-century architect, but it may also be a feature copied from one of the city's earlier great timber-roofed halls.

While it seems clear that exposed timber trusses were a feature of some of Rome's great covered halls, it can be argued with equal certainty that the ceiling beams of many traditionally roofed buildings were enclosed by elaborate coffering systems. These will be considered in more detail below in the section on ceilings.

Significant Uses of the Tie-Beam Truss in Roman Architecture

To appreciate the role of the tie-beam truss in monumental Roman building it is worthwhile to consider a list of the ten largest covered rectangular spaces (that have been found and described) from the Roman world. In table 3 they are listed chronologically. If buildings with spans between 80 and 84 Roman feet built as late as A.D. 400 are included, two monuments can be added to the list (see table 3A). Finally, spans between 70 and 80 feet would include most of the largest of the cross-vaulted halls of the great imperial bathing complexes from Rome and other major provincial cities.

Looking at the list of the ten largest rectangular spaces (only a few domes span

Table 3. Ten largest covered rectangular spaces (in chronological order from oldest)

Monument, location	*Date*	*Span (Roman Feet)*	*Type of Roof*
Teatrum Tectum, Pompeii	80 B.C.	87	Truss
Odeum of Agrippa, Athens	15 B.C.	87	Truss
Diribitorium, Rome	7 B.C.	100?	Truss
Odeum, Augusta Praetoria	Augustan	103	Truss
Aula Regia, Rome	92 A.D.	107	Truss
Coenatio Iovis, Rome	92 A.D.	99	Truss
Basilica Ulpia, Rome	113 A.D.	85	Truss
Temple of Venus and Roma	135 A.D.	87	Truss
"Basilica," Trier	300 A.D.	88	Truss
Basilica Nova, Rome	312 A.D.	84.5	Vault

Table 3.A

Monument, location	*Date*	*Span (Roman Feet)*	*Type of Roof*
Old St. Peters, Rome	330 A.D.	80	Truss
St. Paul's, Rome	385 A.D.	82	Truss

wider spaces), one is immediately struck by the complete domination of the timber roof. Only one vaulted structure, the late-period Basilica Nova of Maxentius, belongs in the group; it is interesting to note that Maxentius also restored Hadrian's timber-trussed Temple of Venus and Roma with a barrel vault, the largest one ever built, between 307 and 312 A.D. All the others on the list were covered with trusses. The one special case, the imperial audience hall of Domitian's palace, perhaps the largest rectilinear space ever covered in the Roman world, will require some special comment.

It is also evident that the largest covered structures were built about one hundred years after the truss was introduced for monumental public buildings. The Basilica Aemilia, for example, of the second century B.C., never required trusses over 55 Roman feet wide. By the first century B.C. spaces of dizzying dimensions were being spanned, even in smaller towns like Pompeii and military colonies such as that of Augusta Praetoria in northern Italy. A span of about 100 Roman feet appeared to be the limit a builder dared to cover, although two or three architects did surpass that, if only slightly. Most of these structures were built in the 150-year period that spanned the reigns of Augustus, the subsequent emperors of the first century, and Hadrian (d. A.D. 138).

We can perhaps account for these boundaries of both size and date by taking into account the essential role the raw materials—the trees themselves—played in realizing such massive buildings. Four species of trees were capable of producing the unjointed timbers required for the largest spans: fir, primarily *Abies alba*, some mountain pines, larch, and cedar. Suitable fir and pine, the native Italian trees, grew only in the mountains and would have not been exploited for construction in Rome's early history or even during the years of the early Republic. Even by the mid-third century B.C. central Italy had been barely tamed by Rome. It is hard to imagine that the forests of the high Apennines were logged much earlier than the late third or early second century. Other factors, such as the cessation of warfare with Carthage, created the political and social will for the great building programs of the second century B.C. in Rome, but it was access to long, sound timber that was essential for realizing the spacious, unobstructed layouts of this new generation of timber-roofed buildings.

Larch, from the Alps, and cedar, from the lands of the Levant, were known to the Romans from at least the first century B.C. as superior roofing materials with marvelous qualities. The wood of the larch was slow to catch fire, a property that prompted Vitruvius to wish for its adoption in Rome. Fragrant cedar was the wood of gods and kings: its timbers roofed the Temple of Artemis at Ephesus and Solomon's Temple in Jerusalem, and builders believed it would last virtually forever. In exceptional cases, despite the great distances, beams of these exotic woods were imported to Rome; the largest beam recorded from ancient literary evidence was that of a larch (Plin. HN 16.200). By the period of the late empire the largest trees accessible to loggers likely had been harvested for both shipbuilding and large construction projects. It is significant that when Diocletian issued his famous price edicts in A.D. 301, the largest timbers listed were beams of fir and pine 75 feet long with a girth of 6 feet (presumably 1.5 feet square).[9] This does not mean that longer timbers no longer existed. The 80-foot spans of Rome's first great Christian basilicas prove otherwise, but it seems reasonable to conclude that by late Roman times beams of significant lengths and thicknesses were rarities.

The Halls of State of the Palatine Hill, Rome

The roofing of the imperial halls of state built into the emperor Domitian's palace, or Domus Augustana (dedicated in A.D. 92), is illustrative of the full capabilities and cultural significance of the timber-trussed roof. Here we are fortunate to have secondary forms of evidence, such as literary passages and numismatics, to document the prevalence of timber construction on a monumental scale. The Domus Augustana stood in affirmation of Domitian's will to rebuild Rome after the terrifying and destructive fire of A.D. 80. Rabirius, the architect of the emperor's project, had the commission of a lifetime. His task was to create a setting suitable for an emperor of unlimited wealth and power, an assemblage of structures that was both a residential and administrative center. His access to the best of everything—materials, site, and craftsmen—was without restraint.

The result of Rabirius's efforts would represent a blend of the novel and the traditional. Due to its status and innovative qualities, the palace has been viewed by architectural historians as a hallmark of high imperial architecture, an embodiment of the New Architecture of Rome, a phase of development characterized in large part by the

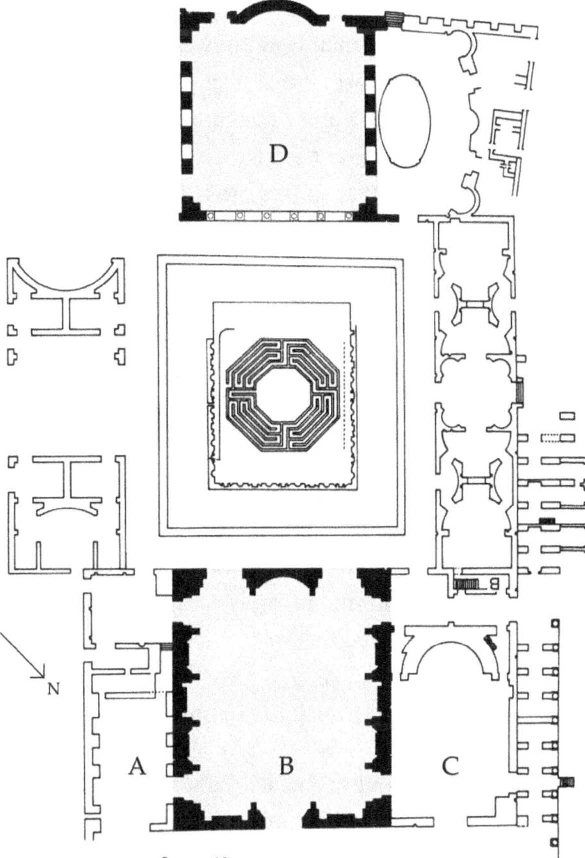

Fig. 8.21. Plan of the state halls of the Domus Augustana, Rome, dedicated in A.D. *92: (A) Lararium; (B) Aula Regia; (C) basilica; (D)* triclinium, *or* Coenatio Iovis. *(Author, after Coarelli 1974, 148)*

mature appreciation and confident application of the fluid medium of faced concrete and vaulted construction.

In light of modern scholars' fascination with the dexterity exhibited by the ancient Italian in understanding and exploiting the properties of concrete, and a conviction that the vault may have been seen in some applications as a simulacrum of heaven itself, several scholars have argued that the most important rooms of Domitian's palace were vaulted. The state rooms in question include the Lararium, so named because of an altar found there in 1725, the throne room, or Aula Regia, the basilica, identified as such because of its apsidal plan, and the dining room (*triclinium*, or Coenatio Iovis), located in the northwest wing of the palace (fig. 8.21).

While some have argued that the state rooms of the palace were all vaulted, others have proposed flat ceilings and trussed wooden roofs. Those who have championed vaults argue that the spans are too great for wooden beams or that the few contemporary literary descriptions that apparently refer to the state halls of the palace evoke images of vaulted space (Tamm 1963, 213; MacDonald 1982, 63; Meiggs 1982, 253). Wooden superstructures have been postulated mainly on the basis of the fact that the walls of the rooms, while robust, may not have been capable of handling the thousands of tons of weight and contingent lateral pressures created by the barrel vaults which would have been employed if vaulting had been used (Coarelli 1974, 150; Claridge 1998, 135; Ward-Perkins 1976, 80). No one has proposed cross vaults because the characteristic piers that support such a covering are absent from the existing footprint of the complex.

What is often overlooked in this debate is that a flat ceiling framed with wood was at least as desirable a solution for the two largest state halls of the palace, not simply because of the technical difficulties involved, but because of the fact that in certain types of buildings of status, Roman patrons and builders preferred to use flat coffered ceilings for allusive purposes. That is, in the minds of the viewer, traditional form had a greater power to impress than the innovative; in this specific case a timbered roof and a flat coffered ceiling created more prestigious associations with other great structures than could be achieved with the forms of the New Architecture. The proof that this was the case in the Domus Augustana is more than simply conjectural; a critical ancient passage referring to the palace is considered the most direct reference to the largest of the state halls in the palace of Domitian. It is also a contemporary (ca. 93–94 A.D.) eyewitness account, written by the poet Statius for the emperor himself:

> A majestic structure, huge, not with a hundred columns but rather enough to support heaven and the gods were Atlas relieved of his burden. The neighboring palace of the Thunderer views it with awe, and the gods are delighted that you [Domitian] have a house equal to theirs. Do not be in any hurry to ascend to limitless heaven. So greatly the vast expanse of the building stretches, and the reach of the far-flung hall, more unhampered than a plain, embracing beneath its shelter a vast expanse of air, and only lesser than its lord; he fills the house, and gladdens it with his mighty spirit. Libyan mountain and gleaming Ilian stone are rivals there, and much Syenite and Chian and the marble that vies with the gray-green sea; and Luna also, but only for the weight of the columns. The view travels far upward, the tired vision scarcely reaches the summit, and you would think it was the coffering of the golden sky. (*Silv.* 4.2. 18–31)[10]

Beyond the obsequious haze of the passage is a wealth of information. Just before Statius launches into the descriptive lines reproduced above, he informs us of his vantage point: "I think I'm reclining with Jupiter in the middle of heaven." He is seated close to the emperor in the midst of the wine and the food. If the passage that follows describes the setting of the banquet itself, then the columned room described is the *coenatio*—or dining hall—in the palace itself. If instead the passage is meant to evoke a more general impression of the state rooms in the palace, then Statius might have the Aula Regia in mind.

For our discussion it does not really make much difference to which of the two rooms Statius makes reference. The dining room, mentioned in the *Scriptores Historiae Augustae* (*Pertinax* 11.6) as the Coenatio Iovis, and the Aula Regia, which must have been the main audience chamber of the palace, were arguably the two most important spaces within the entire complex. They are situated within the same architectural block and exhibit nearly the same dimensions.

Were the spans of the Aula Regia and the Coenatio Iovis too great to be bridged with wooden joists? or too broad for a barrel vault? If neither solution was possible, then the spaces would have had to have been left open to the sky—or only partially covered (Giovannoni 1940, 85). These questions cannot be answered conclusively. As is clear from tables 3 and 3A, the clear span of the Aula Regia, which was a little over 100 Roman feet (ca. 30 m) if no interior columns were employed, is on a par with the largest rectangular structures of the Roman period thought to have been covered with timber-trussed roofs (Coarelli 1974, 150). Its span, however, does not exceed the greatest length recorded for timbers used for wooden ceilings (Plin. HN 16.200, 16.201, 36.102), and it is not significantly broader than other rooms known to have employed timber trusses. The banquet hall is only a little less broad (99 Roman feet; 29.05 m).

The spans of the Aula Regia and Coenatio Iovis of the Domus Augustana would have been considerably less if ranks of superimposed interior columns had been employed. The number and opulence of columns are main features of Statius's description. They bear great weight in their poetic role, which suggests that they were employed, along with the walls, to sustain the roof. This is the role of columns in trabeated, not vaulted, architecture.

If columns had been used as structural elements, as the language of Statius suggests, then the spans of the Aula Regia and the Coenatio Iovis would have been reduced considerably. In the former, for example, columns aligned with the piers on either side of the entrance doorway and with those flanking the terminal apse would have created a central nave with a span of only 9 m. While this arrangement is unlikely, even if columns had been placed close to the walls, the span would have been nearer to 26 m (88 Roman feet) because the beams would have rested on the ressauts supported by the columns. A clear span of 88 feet is quite close, if not identical, to a number of great imperial and early Christian structures that were covered with wooden roofs. There are no known structures of comparable dimension, however, that were covered with barrel vaults.

The thicknesses of the walls of the Aula Regia and the Coenatio Iovis provide little help in solving the problem in either way. Even with a wooden ceiling and roof the walls needed to be robust, particularly if the ceiling had been covered in gilt coffering (as the description of Statius suggests) and the roof tiles had been of marble or bronze. The deep niches built into the side walls of the Aula Regia define six pairs of strong piers to help support the roof.

The walls of the Coenatio Iovis appear even less capable of supporting a mas-

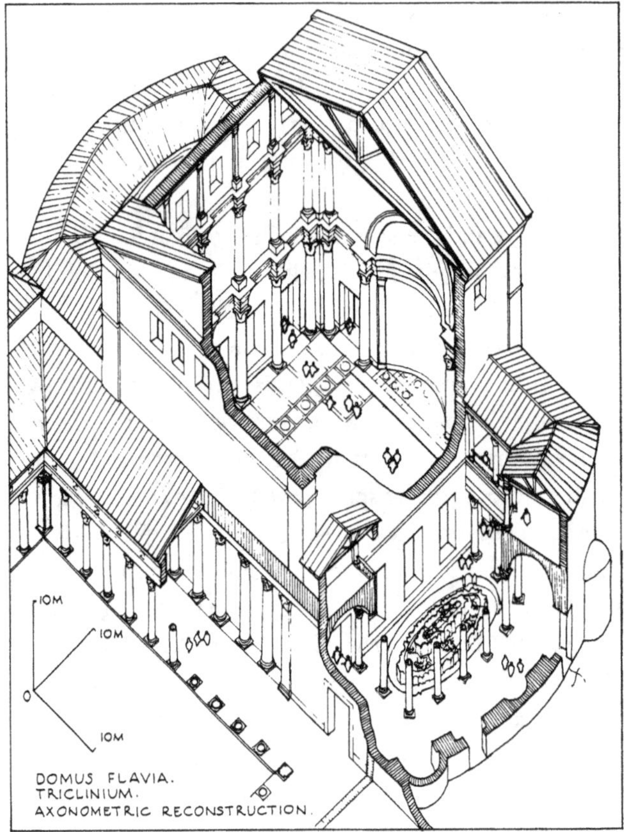

Fig. 8.22. Reconstruction of the roofing of the Coenatio Iovis (triclinium) *of the Flavian Palace, Rome, dedicated in* A.D. *92. The bracing system of the trusses is hypothetical. (Gibson, DeLaine, and Claridge 1994, 91, fig. 29, reproduced by permission)*

sive vault, and a study published by S. Gibson, J. DeLaine, and A. Claridge in 1994 has convincingly argued that this room was without doubt covered with a timber-trussed roof (fig. 8.22). Each side wall is pierced by five large apertures for doors and windows, and there is nothing in the way of exterior buttressing except at the corners. Even if the technical criteria for a barrel vault could be satisfied for either palatial hall, in these particular rooms such a solution would not have been as desirable as a flat, wooden, coffered ceiling.

One of the more interesting pieces of collateral evidence connected with the appearance of the state halls is the image of a building on a coin that was issued in A.D. 95 or 96 (fig. 8.23).[11] The structure depicted is built up in three tiers: a massive lower story with three portals, a middle story enlivened with *aediculae*, and a towering top block surrounded by tall columns bearing a triangular pediment. There is no legend to identify the building. B. Tamm, W. MacDonald, and C. Giuliani propose that the image is that of the facade of Domitian's palace, that is, the northeast side, which faced the visitor who climbed the Clivus Palatinus from the eastern zone of the forum.

If this identification is correct, the audience hall (Aula Regia) of the palace ap-

Fig. 8.23. Sestertius showing the facade of a building, perhaps Domitian's palace, A.D. 95/6. (Author)

Fig. 8.24. Reconstruction of the main facade of Domitian's palace. The trussed roof is based upon the image on the sestertius in the previous figure. The pitched roof in the reconstruction would have covered the main audience hall (or Aula Regia: see (B) in fig. 8.21). A.D. 92. (Author, after Giuliani 1977, 105)

pears to be covered with a form of clerestory or pavillion roof, the top tier decorated by, if not supported by, a columnar screen that facilitated the illumination of the hall (fig. 8.24). Even with a flat ceiling, the height of the room would have been at least 30 m (Coarelli 1974, 150). If a barrel vault had been employed here instead, the only source of light would have been from the lunettes at each end of the hall; barrel vaults were rarely punctured with windows along the shoulders of the vault. The lighting problem was probably the single most compelling reason to cover rectangular halls with cross vaults, and we have already seen that the state halls of the palace, lacking the necessary arrangement of piers, cannot have been covered with cross vaults.

The final argument for vaults in the great state halls of the palace is based upon our modern conception of ancient aesthetics and our admiration for the Roman vault. We mislead ourselves, however, if we ignore the high regard Romans held for the timber-framed roof. To consider the merits of a contrary argument based upon the aesthetics of Rabirius, Domitian, or any other influential Roman, we should consider the last line of the passage cited earlier from Statius more closely.

The Significance of *Laquearia*

The words chosen by Statius to evoke the great height of Domitian's palace hall are of particular interest: "The view travels far upward, the tired vision scarcely reaches the summit, and you would think it was the coffering [*laquearia*] of the golden sky." The key word in the passage is *laquearia*, which in English translation is often simply rendered as "ceiling." In fact the word *laquearia*, along with its variant, *lacunaria*, had a specific meaning in Roman literature. Both refer to a horizontal paneled ceiling framed in wood and often covered with gold. Gilded coffered ceilings were reserved for the most prestigious of public buildings: basilicas, temples, and meeting halls.

The importance of the Roman coffered ceiling to ancient patrons and architects should not be underestimated. It is a form connected with royal palaces in the eastern Mediterranean, the greatest of Greek temples, and the Jewish temple of Jerusalem. Coffered ceilings had been valued in Italy from Etruscan times (I shall return to the development of the form shortly).

If it is true that Domitian wished to include in his palace an element that had not only a venerable tradition full of royal and divine allusion, but even one with celestial connotation, he would have been hard-pressed to find a better solution than to cover the state rooms of his palace with a traditional gilded wooden coffered ceiling. Indeed, while the connection between the Roman vault and the heavens is in fact quite tenuous as far as the contemporary literary record goes, there is clear literary allusion between the coffered grid of a ceiling and the underlying structure of celestial bodies. It is the rational pattern of the coffered ceiling that another poet named Manilius, in his Augustan-period treatise on astrology (*Astronomica*), uses to illustrate the organization of the stars: "The fires [of the constellations] coffer heaven with various designs" (1.533).

Near the end of Manilius's poem is another reference which provides a vital clue for interpreting the language used by Statius in his description of Domitian's Domus Augustana. In the following excerpt, Manilius is considering the nature of *Spica*, an immortal who personifies an ear of wheat: "And, because the ear of wheat is inhabited by grains artfully arranged and its layout is like a building, for it provides chambers and granaries for its own seeds, [*Spica*] will make a man who carves coffered ceilings [*laquearia*] for sacred temples, providing a new heaven [*caelum*] for the roofs [*tecta*] of the Thunderer" (*Astronomica* 5.285–92).

Here again the pairing of *laquearia* with *caelum* exhibits the connection between elaborate coffering, divinity, and heaven. There is no reason to suspect, incidentally, that the ceilings evoked by Manilius are vaulted: he was writing in the early first century, when temples were never roofed with vaults; even in later times the vaulting of rectangular shrines was the exception to this rule.

Manilius's "roofs of the Thunderer" refer to none other than the Temple of Jupiter Optimus Maximus on the Capitoline Hill, built, as we have seen, with a traditional Tuscan plan, including a wooden prop-and-lintel pitched roof. The *Astronomica* attests to the fact that this great shrine was also decorated with rich coffering. In this regard it is interesting to note that Statius compares this same Temple of Jupiter to the *tectum augustum* (literally, the "great roof" or "ceiling") of Domitian's palace. Statius imagines the inhabitant of the Capitoline temple, Jupiter himself, viewing Rabirius's accomplishment with awe. The point is that the sheer achievement of covering Domitian's vast state halls and decorating them with rich coffers eclipses the glory of Jupiter's own "palace." Domitian also called himself the Thunderer. What better way to promote the connection between the emperor and Jupiter than to create an ambience that would remind a visitor of the famous shrine on the neighboring hill?

Clearly the presence of carved coffers in the Domus Augustana upholds the tradition of the most celebrated buildings of Graeco-Roman antiquity. Statius may even be making a pun. An audience who heard the words *laquearia caeli* from the passage cited above would in fact expect the last word to be *caelata*, for it is *laquearia caelata*, or carved wooden coffering, that adorns the halls of the most wealthy (Sen. *Ep.* 90.42.1).

In conclusion, Statius, in his description of the emperor's vast halls, imagines that the sky itself has become a golden coffered ceiling, so that in the emperor Domitian's realm, nature itself now draws inspiration from Roman craft, not the other way around (Ulrich 2000).

Ceilings

Ceilings in Roman houses owned by the elite and in public buildings paid for by rich patrons were afforded the same, usually lavish, attention devoted to the embellishment of walls and floors. With the exception of structural masonry vaults (themselves built upon timbered centering), Roman ceilings were permanently framed with wood. The most basic ceiling was no more than the exposed beams that carried the floor or roof above the covered room. This treatment—if it can be called such—of space would have been the norm primarily in structures built for utilitarian use: farm and storage buildings.

If the joists, or "ceiling beams," were simply topped with planks, leaving the beams themselves exposed, the result is what can be described as a beam or slot ceiling. A flat ceiling that obscured the structural timbers could be achieved by nailing planks to the bottoms, or soffits, of the joists. In such cases, the ceiling is in fact a kind of inverted *contabulatio*, the term I have used to describe planked flooring. (I shall turn to an actual example of such a ceiling from Herculaneum shortly.) Sometimes the joists of a ceiling acted merely as a vehicle for a skin of lath and plaster decoration. In other applications the wooden armature was partially or wholly concealed with carved and painted wooden panels.

Even though literary sources indicate that interior spaces were, as a rule, covered with decorative ceilings, there is evidence that in some large buildings the roofing framework was more or less entirely visible from below. We have already seen, for example, that in the basilica built at Fanum by Vitruvius, it is probable, judging from the description of the building given by the architect himself, that the framework of massive beams in the building was designed to be seen and admired from below. This particular aesthetic may have also characterized the interiors of some of the first western-style Christian basilicas, particularly those built in Rome over the course of the fourth century. Constantine's original (or "old") St. Peters is a textbook example.

There is no single word in Latin that means "ceiling." The fact that *tectum*, the most common word for a roof, can also be used to mean ceiling (for example, Sen. *Ep.* 114.9) indicates that Romans often considered the roofing of a building and the covering of a room as two elements of the same structural entity. This unity between roof and ceiling was quite manifest, in fact, for most large, prestigious buildings such as basilicas, bath buildings, temples, senate chambers, and imperial audience halls, for such chambers were, however imposing, essentially single-storied. This was also often the case in domestic settings, such as the main hall, or atrium, of the republican-period house, which will be considered separately later. For specialized ceilings, most notably flat ceilings with painted or carved wooden decorative panels, specialized terms such as *laquearia* ("coffering") were, as we have just seen, used to indicate the presence of such a feature.

Amedeo Maiuri, who published his finds from the excavations of Herculaneum in 1958, had this to say about his discovery of a perfectly preserved wooden ceiling in a wealthy dwelling known as the House of the Mosaic Atrium: "We have in this house one of the most precious examples of a ceiling and covered portico from an ancient domestic setting, specifically that discovered in the northern and eastern ambulatories of the portico, where the [volcanic] mudslide did not sweep the superstructure away" (Maiuri 1958, 291). Maiuri's praise was understandable. The chance of finding a wooden ceiling partially intact and still in place after two millennia was minuscule, even given the special circumstances of Pompeii's and Herculaneum's demise.

The wooden ceiling in the House of the Mosaic Atrium can still be observed in its original setting, now encased and largely obscured in a grid of iron and sheets of modern glass (fig. 8.25). The ceiling covers a corridor 2.85 m wide between the atrium and the garden peristyle of the house. The ceiling rises 3.52 m above the level of the pavement. The beams that supported the ceiling measure a uniform 22 cm in height, while their widths vary between 14 and 18 cm. This unevenness was concealed by the flat planks nailed from below to the beams.

Of particular interest is the fact that another set of planks was also nailed upon

Fig. 8.25. The wooden ceiling from the House of the Mosaic Atrium, Herculaneum. The carbonized planks are now encased in glass and iron frames. The top decking (A), joists (B), and ceiling planks (C) are visible in the photograph. First century. (Photo: Author)

the tops of the beams; this suggests that at some point the peristyle may have been intended to incorporate an upper-level walkway. Close examination of the surface of the ceiling itself indicates no evidence of plastering or added moldings, but the excavators did observe faint traces of red paint on the carbonized planks (Maiuri 1958, 292).

Porticoes covered with ceilings such as that found at the House of the Mosaic Atrium were apparently a characteristic of wealthy villas, particularly those found in the eastern Mediterranean. In the villas of wealthy Greeks, Vitruvius tells us, the ceilings of porticoes were decorated with fine stuccowork or wooden paneling (6.7.3). In Solomon's great temple in Jerusalem, Josephus wrote, "the porticoes, all in double rows, were supported by columns twenty-five cubits high—each made from one block of the purest white marble—and covered with panels of cedar" (Joseph. *BJ* 5.190).

It is a shame the wooden ceiling of the House of the Mosaic Atrium was not decoratively paneled, for we would finally have a unique example of an embellishment so admired by ancient patrons. In order to arrive at some notion as to what a paneled ceiling might have looked like, it is necessary to consider other evidence.

Evidence of Paneled Wooden Ceilings from Etruscan Tombs

We have already looked at Etruscan rock-cut chamber tombs (*tombe a camera*) to help understand the appearance of the post-and-lintel roof frame; I turn to them again as a helpful starting point in determining the appearance of flat paneled ceilings in central Italy from ca. 600 to 300 B.C. Caution should be exercised when using such evidence, as we will see, but the tombs in question provide an otherwise lost glimpse into the importance of wood in Italic construction for this early period.

The chamber tombs in question are those created by excavating the tomb chamber out of solid (yet relatively soft and easily worked) tufaceous bedrock. Due to the stability of the bedrock, the tomb builders were free to mimic architectural forms, in-

Fig. 8.26. A beam ceiling in the Etruscan Tomb of the Cornice, Cerveteri (Caere), Lazio, Italy. Sixth century B.C. *(Photo: Spencer Garrett)*

cluding flat ceilings, with few or no internal supports, a feat that would have been impossible had the tombs been built of quarried stone. The most important necropoleis exhibiting intricately carved flat ceilings are those of Cerveteri, S. Giuliano, and Tarquinia; the best examples date from the seventh through fifth centuries B.C. Other sites such as Chiusi and Vulci provide important corroborating clues.

The most common form of flat ceiling with carved detail represents a simple beam ceiling; the Tomb of the Cornice from Cerveteri is a good example (fig. 8.26). Parallel beams, or joists, span the shorter side of a rectangular chamber. Normally the chamber is unimpeded by any freestanding piers or columns. Both beams and interstices are unadorned by additional carving, although paint was occasionally applied to the bare surface of the bedrock. By carving smaller beams running at ninety degrees above the main ceiling joists the Etruscan tomb builder was able to create the impression of a wooden paneled ceiling (fig. 8.27). The recessed square frames formed by the crossing of the ceiling "timbers" are in fact simple coffers. The soffits of these frames were chiseled with diagonal striations, the bias of which is reversed in adjacent panels. It would appear that these are meant to represent small planks, branches, reeds, or even the "fletches" of a log cut to be employed in a ceiling.

A unique aberration of the simple paneled ceiling just described is found in the Tomba Campana of Cerveteri's Monte Abetone necropolis.[12] Here the secondary timbers run parallel to the main timbers, an arrangement that makes little structural sense, as if the ceiling has been carved more for decorative effect than as a reflection of architectural reality.

The Etruscans' interest in the coffer as a ceiling embellishment is evident in two additional tombs differing in date, manner of execution, and locale. The solitary square coffer inserted into the late sixth-century ceiling of the Tomb of the Lion in Chiusi is a carved and painted composition of fictive main and cross timbers, bordered with un-

Fig. 8.27. Coffering from a ceiling in the Tomba Cima, necropolis of San Giuliano. The main "beams" in the image are 17 cm across; the superimposed set measure 14 cm across. Ca. seventh century B.C. *(Author)*

adorned moldings that create no fewer than five superimposed recessed planes (fig. 8.28). The implied overlapping of the "planks" that formed the borders, and the illusionistic miter joints of the moldings, suggest that the carver of this ceiling was a keen observer of actual wooden coffering.

Perhaps the most elegant paneled ceiling known from central Italy is that recorded in the nineteenth-century painting by James Byres of the Tomba del Cardinale of Tarquinia (fig. 8.29). The square coffers carved into the expansive ceiling of this tomb give no indication of the wooden framework so evident in all the other examples thus far considered. The relatively late date of this Etruscan tomb (third century B.C.) is surely a factor; a coffering system that conceals the structural skeleton of the ceiling instead of being built around it is a developmental step that manifests itself most prevalently in the stone or concrete coffers used in the public buildings of the Roman period.

Fig. 8.28. Coffer carved into a ceiling of the so-called Tomba del Leone (Tomb of the Lion), Chiusi (Tuscany) Italy. Late sixth century. (Author)

Fig. 8.29. Coffering in the Tomba del Cardinale, Tarquinia. The sketch is after an eighteenth-century painting made by James Byres. Third century B.C. *(Author, after Boëthius 1978, 81)*

Wooden Coffering in the Roman Period

The importance of the coffered ceiling has already been raised in the earlier discussion of the state halls of Domitian's palace in Rome. The subject can now be considered in its broader context. In order to appreciate the importance of the Roman wooden coffered ceiling, of which no physical traces survive, it is necessary to consider how contemporary Romans viewed this type of ceiling. Despite the lack of actual wooden coffering, the actual appearance of these panels can be understood from depictions in ancient painting and from surviving examples in stone and concrete.

The term *laquearia*, used to describe the ceiling of Domitian's palatial hall, is a plural form of *laqueare*. Latin users apparently saw little difference between the words *lacunar* and *laqueare*. Both refer to coffered ceilings. Both are derived from *lacus* (a "depression," often meaning a "lake") combined with the suffix *-aris* (Serv. *Verg.* 1.726). Some writers used only one of the two terms, others used both. Vitruvius refers only to *lacunar*. From the examples cited below and in the glossary, it will be seen that there is no pattern of usage that indicates that the two variant terms carried correspondingly diverse meanings. Extant Roman literature differentiates between the flat ceiling and the curved and between flat and vaulted coffered ceilings. The ancient writer used *lacunaria* or *laqueata* alone to indicate a standard, flat, paneled ceiling.

Vitruvius (4.3.1; 4.3.5) offers the best technical definition of *lacunar*. He employs the term to describe the gaps between the ceiling joists in a Doric temple; such spaces could be concealed by metopes on the entablature (over the gaps between the ends of the joists) and by coffers on the ceiling (over the gaps between the joists themselves). The *lacunar* was thus ultimately a device to plug the regular voids formed when a row of wooden ceiling joists spanned a room.

A grid for the coffering was formed by running timbers perpendicularly to (and usually on top of) the ceiling joists. The open squares or rectangles thus formed were covered by planks to form a closed ceiling. The frames of each coffer could be carved or stuccoed and even gilded. Coffering provided practical and aesthetic advantages. The acoustics of the space were improved, and the creation of an attic provided rudimentary insulation. The framing of the roof—the joists, rafters, purlins, and assorted braces (such as king posts or collar-beams)—was hidden from view.

Coffering, however, offered no structural advantage, especially when attached directly to the roof frame. Indeed, the opposite was probably true. The main job of the tie-beams in a trussed Roman building was to contain the lateral forces created by a gabled roof laden with heavy tiles. Therefore, the main stress placed upon a tie-beam was longitudinal (tension), parallel to the grain of the wood. Coffers added dead weight along the length of the beam; king posts and queen posts could be used not only to strengthen the roof frame, as has been seen, but to support the heavy coffering.

If postclassical structures and ancient painted representations are any guide, some coffered ceilings were composed of frames and backing planks built independently of the main roofing beams, which served an essential, but otherwise hidden, role of support. The principle was the same as that used for nonstructural (or "false") vaults, which could be hung from roofing or flooring beams. Since the roofing beams were above and behind the coffers, rather than an integral framing element, such ceilings could exhibit coffers rendered in small scale, with a variety of complex geometric patterns unrelated to the ceiling beams, and even with special features such as the ability to rotate. The panels could be rendered with or without deep recesses, according to the desire of the patron. Examples of such coffering that have been carved into marble ceiling panels give some idea of what their wooden counterparts may have looked like (fig. 8.30).

From well-preserved Roman-period structures in the eastern Mediterranean it is evident that carpenters could fashion an inner ceiling built and supported independently of the main roofing beams. At the second-century Temple of Bacchus at Baalbek, for example, the stonework of the walls is in part preserved to its full height, which has permitted a plausible reconstruction of the timbered ceiling and roof (fig. 8.31).

Fig. 8.30. A marble panel carved with a geometric pattern of coffers. From the courtyard of the Antiquarium del Celio, Rome. (Author)

Fig. 8.31. Temple of Bacchus, Baalbek, Lebanon. A hypothetical reconstruction of the ceiling and roof: the entablature carried by the interior colonnade of the cella provides a broad bedding for the ceiling beams; the rafters are supported by vertical props and purlins which rest on the main beams; the rafters themselves rest on the cella walls. Second century. (Wiegand 1923, 46, fig. 92)

The cella of this imperial-period temple has a clear span of 21 m (71 Roman feet); the main cross-beams of the coffered ceiling were carried by the projecting entablatures of the interior colonnade. These also helped to support the vertical posts that carried the purlins of the roof itself. According to T. Wiegand's reconstruction, based upon cuttings in the stones of the side walls, a second, upper tier of joists was tied to the posts with saddle joints; these helped relieve the dead load on the lower joists upon which the uprights were seated. The main rafters rested on the top of the side walls of the cella, independent of both sets of underlying joists (Wiegand 1923, 46). The system reminds one of traditional Greek practice; here in the Roman east the principle of the prop-and-lintel roof still prevailed.[13]

When a Latin writer wished to convey the idea of "ceiling" in the general sense, that is, a ceiling not necessarily adorned, he used the word *tectum*. As a technical term, *tectum* means "roof," but it had a number of more general applications. To mean "ceiling" it was paired with a modifier so that the meaning would be clear. One common modifier was *laqueatum* itself. Thus Cicero (*Leg.* 2.2.93) depicts Atticus as a modest man who scorns marble pavements and *laqueata tecta*, that is, coffered ceilings. When Seneca (first century A.D.) decries the rich for embellishing their *tecta* with gold, "so that the coffers will match the floors in brightness," it makes no sense to translate *tecta* as anything but "ceilings" (*Ep.* 114.9).

Coffering in Private Houses

The comments of Cicero and Seneca just cited allude to the existence of fancy coffered ceilings in private homes. It is clear from these and other sources that coffering was a vehicle for the ostentatious demonstration of wealth. Pliny provides a similar view (*HN* 33.57): "Now, even in private houses, coffered ceilings are covered with gold . . . from coffered ceilings the practice spread to vaults and even walls, which are gilded as if they are fine vessels." Gilded coffered ceilings had once been a privilege reserved for the homes of the gods, says Manilius (*Astron.* 5.291–92). "Now," he adds, "dining rooms compete with temples."

Only the most wealthy, however, could have afforded the kinds of gilded wooden ceilings that otherwise distinguished the finest public buildings of Rome. Paneled wooden ceilings embellished with stucco or paint, however, were in all likelihood a common feature of the houses of the merely well-to-do. After all, closed ceilings did offer the aforementioned advantages of modest insulation and better acoustics. Surely the first applications of simple coffers in a domestic setting were found in the atrium, the most important room in the house. The entrance hallway, or *fauces*, of a house may have been similarly embellished, if painted imitations, such as that from the House of the Samnite at Herculaneum, are reliable indicators of the practice (Clarke 1991, 88) (fig. 8.32).

Although examples of wooden coffering from the atria of Roman houses are lost, we can imagine that in such halls the two main support beams of the impluviate roof (the framing of which is considered in more detail below) would have remained visible, while the paneling covered everything else. We will see how Vitruvius mentions the coffered ceiling (*lacunaria*) in his advice concerning the proper heights of atria. Once again we are reminded of the simple yet decorative "beam ceilings" of Etruscan "palatial style" tombs from Caere; the possession of decorative wooden ceilings is evidently a source of pride for the patron.

Coffered Ceilings in Public Buildings

Expensive embellished ceilings in Roman settings were ultimately associated with the luxury of fabled foreign places, particularly the legendary cities of Asia Minor, Egypt,

Fig. 8.32. Painted coffers from the ceiling of the entrance vestibule of the House of the Samnite, Herculaneum. First century B.C. *(Author)*

and North Africa. It seems no Roman writer could conjure the image of a foreign palace without noting the presence of its golden coffers. In Ennius's depiction of the palace of Priam (in fact, the Troy of legend had fallen a millennium before), the gold and ivory coffered ceiling provides a canopy of "barbarous elegance" for the Trojan king (Cic. *Tusc.* 1.85). In Cleopatra's palace, says Lucan, writing nearly a century after the Egyptian queen's suicide, "the coffered ceilings exhibited wealth, and thick gold covered the beams" (*Bell. Civ.* 10.112–13). Virgil imagines the queen of Carthage in such a setting (*Aen.* 1.726). In his version of the story of the sword of Damocles, Cicero depicts the palace of Dionysus I with a coffered ceiling through which the fabled sword can be lowered.[14] When a Greek client-king wished to honor a Roman god, he knew how to do it in oriental style. Thus Livy describes the promise of King Antiochus to build a temple to Jupiter Capitolinus "of which not only the coffered ceiling, but all the walls, were to be covered with revetments of gold" (41.20.9).

The wooden ceilings of the cellae of Rome's state temples were surely coffered and probably gilded, if the stone coffering used in the *pteromata* (exterior aisles) of surviving shrines is any gauge. Vitruvius (5.2.1) also indicates that senate houses were embellished with coffered ceilings. The most important of such buildings, the Curia of Rome, was undoubtedly decorated (and has been so restored) in this way. Vitruvius calculates the proportional requirements of the interiors of temples by measuring from floor to coffering: *a pavimento ad lacunaria* (4.6.1). Cicero describes Verres looking upon the beautifully coffered *tectum*—or ceiling—of the Temple of Castor, which stood at the edge of the Roman forum (*Verr.* 1.133). As we have already seen, the most important shrine in Rome, the Temple of Jupiter Optimus Maximus on the Capitoline Hill, was richly coffered.

Given the previous examples, it is no surprise to find references to coffered ceilings in the audience halls and state dining rooms of emperors such as Domitian. We have already seen how this first-century emperor used coffering in his palace to draw comparisons between himself and Jupiter. But Domitian's palace was not the first such imperial residence to be so decorated, as is clear from the literary record.

Nero, who turns out to have had something of a partiality for trick coffered ceilings, doused his dinner guests with perfume sprayed from pipes hidden behind the panels of his dining rooms (ca. A.D. 65).[15] The biographer Suetonius claims that the panels that turned were made of ivory, an added touch available only to the most wealthy (*Nero* 31.2). The coffers in Agrippina's bedroom were less benign: Nero tried to rig them so they would collapse upon his mother while she slept (*Nero* 34.2). The late imperial audience hall at Augusta Treverorum (Trier), built in the early fourth century, can hardly have been covered with anything but a coffered ceiling and has been so restored following the serious damage suffered by the building during the Second World War.

The Roofing of the Atrium in the Roman *Domus Italica*

As the most important space in the traditional Roman single-family dwelling (or *domus*), the atrium is worthy of special consideration. To cover these halls the tie-beam truss was not a solution, for the roof, as is well known, generally sloped inwardly on four sides in order to capture rainwater. The awkward physical stresses on the framing posed a special challenge to the builder. The *domus* has been best described and understood through the excavations at Pompeii and Herculaneum and through references in literary texts, of which Vitruvius is once again an important source.

The *domus italica* as such is primarily a phenomenon of the republican period, a time when, as we have seen, vaulting did not play a major role in Roman architecture (with a few notable exceptions, such as in some bath buildings). Building materials were, in central and northern Italy, primarily of mud brick and wood. Ashlars of cut stone were saved for city walls or high-status public buildings or for the facade walls of the homes of a few wealthy individuals. By the middle of the second century B.C. concrete and rubble construction was increasingly dominant in the area of the Bay of Naples, but wood was still essential for the construction of stairways, doors, door and window frames, lintels, cornices, upper floors, ceilings, and roofs. Of these applications, it was the roofing of the atrium that was most ambitious.

The atrium was the focus of the republican *domus*. Described as *cavum* ("cavern-like" or "enveloping") by Vitruvius, it occupied and described a space as long, broad, and high as the building site permitted and the patron could afford. In early Italic dwellings the atrium may have been an enclosed courtyard open to the sky, but at least by the third century B.C. *atria* were all or partially roofed over. The illumination of these spaces was always a challenge because the atrium was surrounded by other rooms: bedrooms (*cubicula*), alcoves (*alae*), the entrance corridor (*fauces*), and the reception bay (*tablinum*). Indeed, the word *atrium* itself suggests a dark space (*ater* means "black" or "gloomy").

Table 4. Dimensions of the largest atria excavated in the area of the Bay of Naples (measurements in Roman feet)

House	*Location*	*Atrium Size (L × B)*	*Type*
House of the Wooden Partition	Herculaneum	31.5 × 27	Tuscan
House of the Bicentenary	Herculaneum	36 × 31	Tuscan
House of Sallust	Pompeii	49 × 31	Tuscan
House of the Menander	Pompeii	33.7 × 24.5	Tuscan
Villa of the Mysteries	Pompeii	39 × 27	Tuscan
Villa at Oplontis	Torre Annunziata (Oplontis)	47 × 33	Tuscan

The grandest of these halls were illuminated by a centered, rectangular opening (*compluvium*) in the ceiling and roof that also admitted air and, when it rained, water. Vitruvius calls the opening of the *compluvium* simply *lumen*, a word that in its common usage means "light" (6.3.6).

In his discussion of the proportional rules of rooms that open off of the atria, Vitruvius (6.3.4) categorizes the halls by their size: smaller atria have a length of 30 to 40 feet; medium-sized spaces measure 40 to 50 feet in length; other categories include atria in the range of 50 to 60 feet, 60 to 80 feet, even 80 to 100 feet. It is the widths of the atria, however, we are particularly concerned with here. Since the width of the atrium was shorter than its length, the main timbers that spanned the atrium would run from side to side, across the main axis of the space. Vitruvian proportions include rules for atria up to 60 feet in width. If such private spaces were ever built on this scale and roofed in the Tuscan way, the timbers needed to achieve the span would have been in a league with the largest public buildings of the Roman republic.

Not one of the atria excavated at Pompeii or Herculaneum comes close to the maximum width considered by Vitruvius; nor have other sites turned up anything of this size that belongs to the realm of domestic architecture. Even most spacious private atria (table 4) fall well short of the Vitruvian maximums, and these were well-appointed houses of wealthy owners. Nevertheless, the spaces to be spanned were considerable, and their roofing may well have been the greatest challenge a local builder—or the *tignarius* working for him—ever had to face.

Vitruvius appreciated that the roofing of the atrium was a major issue; it is the very manner in which the atrium is roofed that, for Vitruvius, defines its type. The brightest and airiest of the atrium styles Vitruvius lists is the Tuscan (*cava aedium tuscanica*). The Tuscan atrium (its name perhaps derived from the Etruscan method of covering domestic, if not palatial, halls) was characterized chiefly by its lack of interior supports (fig. 8.33). While four (the tetrastyle type) or more (the Corinthian atrium) columns were employed to support the *compluvium* and the inwardly sloping roof of other domestic atria, the main beams of the Tuscan atrium bore the entire weight of the roof. The dramatic, unobstructed view afforded by the Tuscan atrium from the entrance passage (*fauces*) to the terminus of the main axis, the *tablinum*, where the *paterfamilias*

Fig. 8.33. The Tuscan atrium visualized in a nineteenth-century reconstruction of the interior with coffered ceiling. In the standard Tuscan atrium, only two main bearer beams supported the roof. From the House of Pansa, Pompeii (VI, 6, 1). First century. (Gell 1857, pl. 37, from an engraving by C. Heath)

conducted his daily business, may explain why the majority of Pompeian patrons preferred the Tuscan style above all others. Presumably the Tuscan type was also the oldest of the variants and its use evoked tradition and solemnity.

Although simpler in plan and aspect than the other types of atria, the Tuscan atrium was in fact more difficult to build, for the entire weight of the roof, its tiles of fired terra-cotta and any decorative ceiling elements such as carved wooden coffering, had to be supported by just two massive cross-beams (*trabes*) that were not held in tension (because of the inward, "impluviate" slope of the roof) by the employment of diagonal trusses. This massive load was borne by the timbers, concentrated near the middle of the span, and transferred to two opposing walls of the atrium. Builders must have looked to the thick trunks of lowland oak and beech to suit the job. These species, particularly in Latium and Campania, were locally available and relatively plentiful. Yet oak does not produce long, straight lengths, and we are tempted to speculate that the proportions of some republican atria were determined more by the lengths of stout beams available than by any other factor. Indeed, it may have been for structural reasons and availability of material that other atrium types, such as the tetrastyle, were planned or even retrofitted into place. As noted, Roman builders preferred long, straight single beams; a scarfed joint at the center of a beam in a Tuscan atrium would have invited disaster.

To construct an atrium a team of masons and *tignarii* worked closely together. The earliest *domus*, with walls of mud brick, may have had to incorporate vertical beams of wood or stone into the walls to support the localized and intense loads generated by

Fig. 8.34. The wooden roofing components of the Tuscan atrium: (left) the main cross-beams are put into place on the partially finished wall; (right) the wall is constructed to its full height, the arca is put into place, and the rafters are added. (Model: Author)

the paired *trabes*—if such a roof could be used at all. Before concrete was discovered, good building stone like blocks of tufa or limestone, if available, was also used for the main bearing walls. Construction in ashlar has always been the most expensive kind, owing to the skills involved, and therefore the earliest atrium-style houses were owned only by the very wealthy. To put this another way, we should consider the possibility that the shift from walls of mud brick to walls of cut stone and eventually concrete and rubble in domestic housing of central Italy was influenced by the desire to incorporate a specific kind of interior space, the atrium, and roofing system, the impluviate.

The increasing use of concrete over the second century B.C. to construct load-bearing walls eliminated some of these early obstacles and allowed people of modest means to enjoy the beauty and quiet that even a small atrium could afford. The vast majority of supporting walls in the atria of Pompeii and Herculaneum are of concrete and rubble construction, and from these remains it is possible to create a clear picture of how the main beams of the Tuscan atrium were fitted into place (fig. 8.34). The method is similar to that used in constructing upper floors (see chapter 7). Masons constructed the walls of the atrium to the height of the main supporting cross-beams and paused while the walls set firm. Even before construction began, the two massive cross-beams had in all likelihood been measured, ripped by saw or trimmed by adze, and brought to the site. These could now be hoisted to the top of the stable side walls, whereupon the masons could now build around the four ends of these two beams to the full height of the walls. Built into the wall in this way, the beams would have been difficult if not impossible to remove or replace once the upper wall was finished. The observant visitor can still clearly see the horizontal seam in the rubble facing of the wall of an atrium where construction was halted to position the twin bearer-beams of the atrium.

With the two main beams in place, the *arca* (or *area*), or frame, of the central opening (the *compluvium*) could be built. On two opposite sides the *arca* was framed by the paired bearing beams. Two additional timbers, probably called *interpensiva*, were then fitted into place. Just how the cross-beams and the *interpensiva* were joined is uncertain.

C. F. Mazois, who published detailed drawings of the buildings he observed at Pompeii between 1809 and 1822, depicts the *interpensiva* resting on top of the *trabes*. Raffaele Oliva, an architect who worked closely with Maiuri at Herculaneum and whose sections of the houses excavated there are featured in Maiuri's final publication, depicts the *interpensiva* as being flush with the main timbers. If Maiuri discovered carbonized main timbers from the atria of the houses he excavated, details about the joinery are lacking. It is clear from the sections drawn during the excavations that most, if not all, of the upper walls of the atria excavated had collapsed well below the level of the roofing timbers.

If Oliva is correct, then the *interpensiva* presumably would have been notched into the main bearer-beams; a carpenter would be reluctant to do this, for the main beams would be weakened at a critical juncture. Vitruvius worried about the load imposed by these main framing elements (6.3.1). He provides a hint of how the *arca* was constructed in his discussion of the ideal proportions of the atrium. The height of the atrium up to the main bearer-beams is to be three-quarters of the length of the atrium, while "the rest is to be planned as the dimension of the coffering and the *arca above* [*supra*] the main beams" (6.3.4).[16] The trick here is to figure out what Vitruvius means by "the rest" (*reliquum*); in any case he clearly indicates by the use of *supra* that the ceiling elements are carried by the beams.

Consider a house with an atrium 32 Roman feet long. The height to the bottom of the main beams should be, according to Vitruvius, 24 feet (three-quarters of the length). This leaves a full 8 feet for the thickness of the main beams and the depth of the *arca*; it seems excessive. Vitruvius must mean something else, but the fact that he regards the dimension from the bottom of the beams to the top of the *arca* (where the ceiling begins) as a significant dimension in the planning of an atrium suggests that the *arca* was built on top of the bearer-beams. This, as we have seen, makes better structural sense.

Once the *arca* had been installed, four more strong beams were needed to complete the basic frame for the impluviate roof. These were the timbers that ran down from the top corners of the atrium's masonry walls to the corners of the wooden *arca*. The inward slope of these corner beams defined the valleys, or *colliquae*, of the atrium's roof. Vitruvius does not specify these timbers by name; Mazois referred to them as *tigni colliquiarum*, evidently a coined, yet nevertheless appropriate, term. The lower ends of the corner beams would have been notched to fit tightly against the *arca*; the upper ends were embedded in the masonry walls themselves and further secured by long iron spikes (sing. *clavus trabalis*) driven through the beams into the walls. These nails, the largest used by Roman carpenters, have turned up in a number of modern excavations.

The main load-bearing beams of the atrium were now in place. The roughly trapezoidal fields defined by each side of the atrium and the framing timbers were now filled with the rafters (*cantherii*). The most complicated joins were those that connected the successively shorter rafters to the diagonal corner beams; each had to be custom fit with a compound diagonal cut. Terra-cotta tiles could be placed directly over this framework, or the rafters could be covered with planking (*asseres* or *tigna*) to form a sheathing (*operculum* or *contabulatio*) upon which the roofing tiles and decorative water-spouts could be laid.

Just how much of the roof structure was visible from the floor of the atrium is hard to say, despite the fact that the modern visitor to Pompeii and Herculaneum invariably peers up to the skeletal frame of faithfully reconstructed roofs—even if the beams are now of reinforced concrete or I-beams of steel! Vitruvius, as noted, prescribed the dimensions of the atrium up to the *lacunaria*, or coffering. There can be little doubt, if the literary passages cited in this chapter are any guide, that the *atria* of the wealthy were covered with lavish coffered ceilings. The moralists might have disapproved of this display of excess, but the fact remains that embellishments of this type were regarded by the Roman patron as one important signifier of elevated status.

Centering

There can be no doubt that some of the greatest wooden roofing armatures ever erected in the Roman world were the timbered frameworks set upon high masonry walls which served as forms for concrete vaulting. Generally known by the term *centering* (or *centring*), these skeletal structures not only had to span the same broad spaces that would be covered by masonry vaults, often without the benefit of supplementary supports other than those at the extreme ends, but also had to be built to hold the load, however temporarily, of curing concrete or partially assembled stone voussoirs, without deformation or collapse. As Rabun Taylor has recently observed, the carpenters who built these forms are truly the "unsung heroes" of Roman architecture (Taylor 2003, 178).

Despite the importance of these wooden forms to Roman architectural practice, it is a startling fact that we know virtually nothing from ancient sources about how they were constructed or what they looked like. Neither Vitruvius nor anyone else mentions even in passing—let alone discusses—the process of centering, though the practice would have been highly refined by Vitruvius's day. There are no representations of centering in Roman art either, perhaps not surprising in the public realm, in which building processes are not part of the iconographic repertoire, but the absence of any visual evidence among funerary reliefs, or of even a term that might indicate a specialized skill practiced by a carpenter, is disappointing. The sole physical evidence extant for ancient centering is the imprints of boards that can be seen in the *intradoses*, or undersides, of vaults formed from concrete.

The widespread use and mastery of wooden centering, therefore, are not in any dispute, but the manner in which it was done is necessarily speculative and the specialized vocabulary of the practice essentially lost (see below). A number of modern scholars have nevertheless addressed the problem with considerable insight, most recently Taylor (2003), L. Lancaster (2000), J.-P. Adam (1994), and J. Rasch (1991). One way of approaching the topic has been to consider better-known postclassical analogies. John Fitchen's study of the construction techniques of Gothic cathedrals (1961), for example, has been cited to offer a glimpse at the obstacles faced by the Roman carpenters in charge of such projects, even though there is only a little more specific contemporary technical detail (for example, some manuscript illustrations, plans, and depictions on stained-glass windows; 1961, 6) than the dearth of evidence from the Roman period. The result is that modern books and articles offer a number of hypothetical framing designs employed for Roman centering. Each is informative, as long as it is remembered that all are based upon educated guesses and backward extrapolation from postclassical models. In the reality of Roman building practices, probably no single solution was used. Since we know of no lost treatise on the subject, it is very possible that the exact manner of centering was transmitted from master carpenter to apprentice, solutions were based upon years of personal experience, and, to judge from the observations already made about floor construction, overbuilding was the best insurance policy.

Yet some basic observations about the manufacture and installation of Roman centering can be offered here. It seems important to consider how the better-documented woodworking practices in the Roman world might be employed to begin to address the question of centering.

The raw materials necessary for the centering itself were simple: wooden beams square or rectangular in dimension and flat wooden boards for the smooth exterior surface of the form against which the cut stone or concrete of the permanent vault was laid. The procurement, installation, and ultimate disassembly of these same raw materials, however, were the factors that required a carefully planned strategy. It goes without saying that the availability of thick, sawn boards for the outer skin, or lagging, of the centering was a crucial prerequisite. The boards themselves, while expensive to acquire, could be reused again and again and may well have been some of the same planks used for the shuttering of the foundations of a large project. Where dependable supplies of timber did not exist, the building of vaulted spaces created a special challenge. Those that survive exhibit evidence of admirable resourcefulness on the part of the builders. In Roman North Africa, for example, a most elegant solution was found: the centering was made from interlocking hollow tubes of terra-cotta that were left in place after the mortar of the vault had set (Adam 1994, 177).

The main framing timbers that defined the form of the vault and bore the heavy loads during construction would have had to be found in dimensions comparable to those used for trussed roofs, built with strong joints that resisted shifting, and probably hoisted into place from the ground as completed subassemblies. No one can say for certain whether fewer arced frames of heavy timbers spaced widely or many closely spaced frames of lighter timbers were preferred by builders (DeLaine 1997, 168). Actual practice was probably determined by what was at hand and by what the master carpenter in charge of the project was accustomed to. We have already seen that in floor construction there was a tendency, at least judging by the comparative evidence between

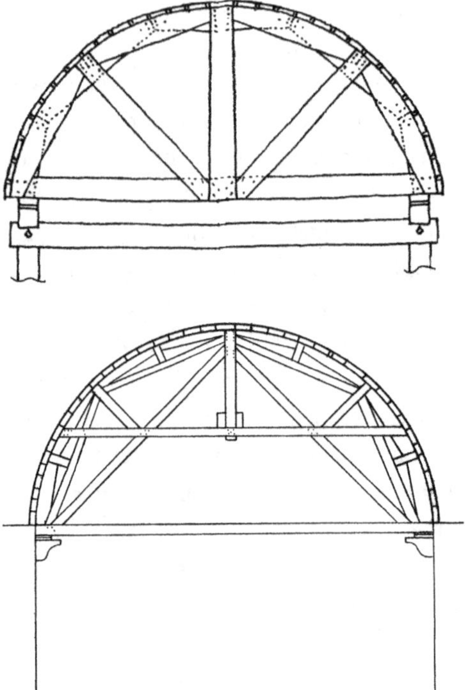

Fig. 8.35. Centering with a timber framework, hypothetical reconstructions. A simple spoked arrangement of timbers (top) can be contrasted with a more complicated framework for a broader span that incorporates the principle of the triangular truss (bottom). (Author [top after Wasmuths Lexicon der Baukunst, 1931, p. 497; *bottom after Adam 1994, 175, fig. 418])*

Pompeii, Herculaneum, and Ostia, to shift from closely spaced joists of small dimension (first century) to more widely spaced heavy timbers that supported an upper deck (second century).

My discussion of the timber truss concluded that large rooms covered by any kind of roof only rarely exceeded a clear span of 90 Roman feet; the largest barrel- or cross-vaulted structures have a span of just under 90 feet. This raises the question of whether the size of the vaults spanning rectangular spaces was to some degree determined by the same limit faced by the builder to the timber-trussed roof: that is, access to long, straight timbers of sufficient dimension that would, in the case of centering, be used to frame the lowest horizontal element of the form. If this is true then we can form some preliminary assumptions about the framing of broad concrete vaults over high spaces: first, builders preferred not to use supplementary wooden supports across the middle zones of the centering, but rather a flying configuration if at all possible; second, the shape of the centering itself was more likely to have been a semicircle than an arch of polygonal configuration (such as that used, for example, to support the roadbed of Trajan's bridge over the Danube).

The actual configuration of the component parts for each arcuated support of the centering for barrel or cross vaults is, once again, an exercise in speculation (fig. 8.35). The same is true for the centering of domes, which, because of the nature of the volume being defined, would have been framed somewhat differently.[17] It is impos-

Fig. 8.36. Modern centering in use at the site of Herculaneum for the restoration of barrel vaults. (Author)

sible to prove any reasonable solution as wrong. Some solutions, however, seem more plausible than others. It is tempting, for example, to see a parallelism between the development of the timber truss, which was perfected, as has been suggested, in the second century B.C., and the contemporary experimentation with vaulting large spaces with concrete. The load-bearing and spanning capacities of the truss inserted within a semicircular centering armature would appear to be a complementary and logical development of roof-framing systems over the course of the late second–first centuries B.C. Knee-braces of wood or stone corbels could be used to enhance the stability of a centering system, a practice probably well known, as we have seen, to the builders of timber-trussed roofs. Knee-braces and wedges, the latter placed at the outer ends of the centering, could also facilitate the removal (dropping and dismantling) of the form after the masonry vault was in place. Of course the jointing methods and dimensional requirements of the trussed frame would already be familiar to a Roman carpenter.

Another framing configuration worth consideration is one that resembles a spoked half wheel. For larger spans this might well have placed too much stress upon the middle of the main transverse beam, but for smaller spans the rigidity of the radially disposed struts would resist deformation under heavy loads of cut stone or concrete, especially if it was possible to place a central prop under the middle of the transverse beam. This basic configuration was used recently to rebuild concrete barrel vaults at Herculaneum (fig. 8.36); here modern steel scaffolding provides several points of support under the vulnerable base of each semicircular frame. The spoked half wheel configuration could have been assembled much like the turning mechanisms of the Great

Wheels used in crane construction, the jointing of which, as we will see, seems to have been rendered for disassembly from below with relative ease, a *sine qua non* in the art of centering.

The kind of framing employed for Trajan's bridge over the Danube, which describes a curvilinear (but not a semicircular) form, is another solution that could have been applied to centering schemes. There are some clear advantages to using what we may refer to as a polygonal scheme: the possibility of great spans, less reliance on single long, massive beams, and assembly and disassembly in shorter, more manageable units. Once again, we cannot assert one way or the other if this kind of segmental framing was used. It seems reasonable to suspect, however, that roofers of Roman buildings would prefer not only simple solutions but also ones for which they already understood, however intuitively, the basic dynamic principles. The elimination of the tie-beam in timbered clear spans of any appreciable dimension seems heresy, particularly when we are faced again with the fact that barrel- or cross-vaulted spaces never exceed the very limitations that were in place for the standard timbered, tie-beam-trussed hall. In the absence of the rigid tensional strength offered by the tie-beam, it is easy to imagine an increased possibility of deformation during the application of the concrete.

This speculation raises another vexing question. If Roman carpenters were so adept at creating armatures for masonry vaults with precise semicircular arcs and at spanning rivers with segmental arcs of heavy timbers, why not cover some rooms with an attractive framework of wooden beams that supported the sweep of a self-supporting curved board roof? In fact I have already mentioned that in domestic spaces like bedrooms and reception rooms in private houses, wooden semicylindrical "vaults" of lath and plaster hung from hidden, horizontal roofing beams. These, of course, were used only in small spaces and were not capable of supporting their own weight.

The question concerning the apparent absence of permanent arcuated wooden frameworks in larger spaces can be approached first in terms of tradition and simplicity. A standard triangular truss was easier to build than a semicircle closed at the bottom by a horizontal timber. Both framing devices would have defined a flat, planar ceiling, which, if boarded or coffered, would have hidden the structural framework. As part of their own building tradition, Romans preferred the exterior profile of an angled pitched roof; even barrel vaults were covered by timbered frameworks to appear pitched from the exterior.

Does this mean that timbered "vaulted" roofs did not exist? Since there are no depictions of such a structure, evidence for a curved wooden roof needs to be teased from a suitable Latin passage. A possible candidate comes from the description of a large structure Varro calls a *peristeron*, a barnlike structure that could house up to five thousand pigeons. The building, Varro says, is *testudo magna, camara tectus*, which can be translated as "a great covered structure, roofed with a vault" (Rust. 3.7.3). Of particular interest here is the juxtaposition of *testudo* with *camara*. A *testudo* has already been associated with a trussed roof. A *camara* is often used to refer to an arched or vaulted space or form, including the form of a trellis (Plin. *HN* 19.69), the inside of a furnace (*HN* 34.101), and even a broad-beamed boat (Tac. *Hist.* 3.47), not to mention a standard (concrete) vaulted bath building (Vitr. 5.10.3). Varro's description of the airy, windowed pigeon coop does not suggest he has a masonry vaulted space in mind. It seems plausible that for this rather unconventional type of building a light, trellislike, wooden framework was constructed, a kind of wooden vault not used in public buildings or houses. It is, finally, tempting to find in Varro's use of *camara* a reference not only to the type of curved wooden structure found in an agricultural building, but also to that important application for which we have no certain Latin term: centering.

IX Interior Woodwork

Opus Intestinum

Woodwork occupied a prominent place in the treatment of Roman interior spaces; it is the missing component of an environment now understood primarily through stone mosaic, marble revetment, and stuccoed walls. The contributions of the finish carpenter would have been found in both high- and low-status construction.

The Latin term *opus intestinum* ("inside work") explicitly indicates interior woodwork: door and window frames, cornice moldings, decorative paneling, balconies, screens, and built-in cabinetry. *Opus intestinum* being a distinct term that appears in both literary sources and on funerary inscriptions, there is every reason to believe its practitioner was a specialist with many years of training and a collection of specialized hand-tools such as fine-toothed saws and molding planes. Until recently, this distinction was recognized among modern woodworkers: carpenters performed rough framing work—the part of the structure that remained invisible—while finish carpenters or joiners worked exclusively on interior, nonstructural woodwork. With the advent and dominance of power tools, however, the line between specialists has blurred; a modern craftsman may be adept at every aspect of construction involving wood (Williams 1990, 2).

Portals, Doors, and Shutters

SINGLE AND FOLDING

The Romans, like the Greeks, employed the skills of the woodworker to create doors of fine joinery. The main door to a house was designed to impress the visitor with a preliminary declaration of the owner's wealth and status. Ancient literary descriptions of doors include important passages from Vitruvius (especially book 4, in which the architect focuses upon the proportional requirements of monumental doorways in the Greek tradition) and a lengthy inscription discovered at Puteoli (*Lex operum Puteolana*) that describes the components of a large public portal, detailing even the species of wood to be used in its construction. That such basic specifications were carved into stone attests to the attention, expense, and pride manifested in well-crafted entryways. Archaeological evidence consists of painted and sculptural representations of doors, the door openings and thresholds preserved in buildings, carbonized doors from Herculaneum preserved wholly or in part, plaster casts of doors and jambs made during the excavations of Pompeii and at nearby Oplontis, and even the occasional preserved door found in waterlogged conditions, as seen in examples found at Vindolanda in Britain and Lake Nemi, near Rome. The visual evidence cannot be weighted equally. Plaster casts of actual doors found at Pompeii, for example, provide functional models that are not necessarily representative of the elaborate doors commonly depicted in funerary art (Righini 1965, 399).

Doors were made from wood or wood sheathed with bronze and, in some exceptional cases (as can be seen at extant tombs), from solid marble.[1] Door frames (the jambs and lintels) were similarly embellished: at Etruscan Pyrgi (fifth century B.C.), for example, the jambs and lintel were covered with elaborately molded terra-cotta slabs. Thus the wooden framework of many of the finest of doors would have been sheathed under a skin of precious materials. By early Christian times, if not before, monumental doors were made from wood with carved surfaces. The most outstanding extant examples in Italy are the carved cypress doors of the Church of Santa Sabina on the Aventine Hill in Rome. Panels carved with floral designs and scenes from the life of Christ date from the mid–fifth century.

The tradition of masterpiece doors can be traced to the extravagant building projects sponsored by Greek tyrants and kings. We hear, for example, of a door of gold and ivory made for the archaic Temple of Athena at Syracuse (ca. 480 B.C.), the prosperous Sicilian city originally colonized by Corinth in the eighth century (Cic. *Verr.* 2.4.56). The paneled marble doors from the fourth-century B.C. royal tombs unearthed by Manoles Andronikos at Vergina in northern Greece corroborate the literary accounts. Greek mansions, Vitruvius tells us, exhibit main doorways (*ianuae*) "of appropriate dignity" (6.7.3). No doubt inspired by such famed examples of Greek luxury, Augustus graced his magnificent Temple of Apollo on the Palatine in Rome (28 B.C.) with double doors (*valvae*), a celebrated masterpiece (*nobile opus*) paneled in Libyan ivory (Prop. 2.31.12). The carved ivories were presumably mounted upon a frame of well-joined wood. One door depicted the Gauls cast down from the snowy slopes of Parnassus,[2] the other the deaths of Niobe and her children. Private Roman patrons donated ornate doors to public projects; thus Pliny the Younger praised his wife's grandfather Calpurnius Fabatus for funding not only a public colonnade, but also the expensive decoration of its doors (*Epist.* 5.11.1).

Latin inscriptions can confirm the importance, attention, and expense that could be apportioned to the creation of a fine public portal fashioned from wooden beams. We can consider the rigorous specifications recorded on stone for a building project contracted in 105 B.C. at the coastal town of Italian Puteoli, near modern Naples. This *Lex operum Puteolana* (or, more formally, the *Lex Puteolana parieti faciendo*), as it is known to modern epigraphers, includes a list of materials for a monumental door protected

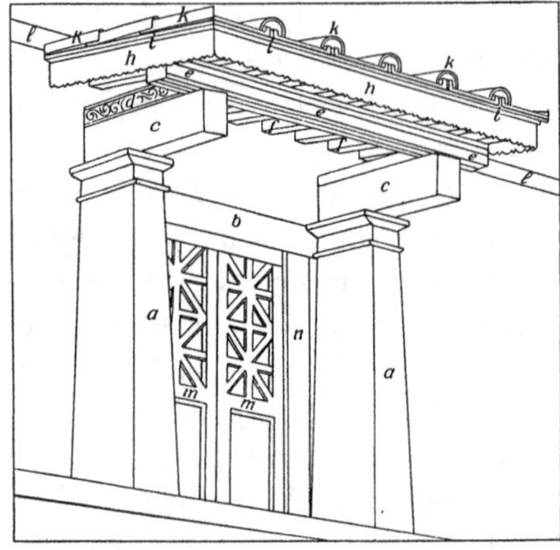

Fig. 9.1. Reconstruction of the portal at Puteoli: (*a*) anta; (*b*) limen; (*c*) mutulus; (*d*) sima picta; (*e*) trabicula; (*f*) asser; (*g*) operculum (*not visible*); (*h*) antepagmentum; (*i*) cumatium; (*m*) fores clatratae; (*n*) postis. *The portal was built in the late second century* B.C. (CIL 1.2.1 1918, 526)

by a porch that gave access to a sacred precinct located opposite an existing Temple of Serapis (Dessau [ILS] 5317; CIL 1.577; 10.1781) (fig. 9.1).

As is often the case with other Latin sources, while the inscription from Puteoli provides many important details, including the species of materials to be used for the door and its porch, the dimensions of most of the contingent wooden members, decorative features, and even the number of rows of tiles required to cover the protecting roof, it is difficult to know for certain just how the portal was assembled. While this may disappoint, it is not surprising, since what was recorded was a formal contract including a list of specifications, not a treatise on construction techniques (Anderson 1997, 74). Theodor Wiegand's thoughtful commentary on the inscription, published in 1894, is still of great use.

The relevant passage begins by specifying that the opening in the wall (*lumen*) is to be 6 feet wide by 7 high. Two piers, each 1.25 feet wide, are to flank the door; on either side, these are described as *antae.* Next in mention is an oaken lintel (*limen robustum*) 8 feet long, 1.25 feet thick, and 9 *unciae* (0.75 of a foot; see comments on Roman measurements in the glossary) high. It is not clear from the passage if the lintel rests upon the *antae* or is in fact placed over the doorway; the latter seems more likely (Wiegand 1894, 732). Perpendicular to the corners of the lintel and projecting over the *antae* are two beams or mutules (*mutuli*), also of Valonia oak; these are 1 foot high and 8 *unciae* across. They project 4 feet from the wall, which means that they continue beyond the tops of the *antae* for 2 feet, creating a deep overhang. This treatment is the same I have described for the *traiecturae mutulorum* (a term used by Vitruvius) that were cantilevered over the facade of a Tuscan temple.

From this point on the description is even less easy to picture. Two firwood beams (*trabiculae*) 6 *unciae* square are placed crosswise above the *mutuli.* At the top of (or above) the *mutules* is a painted band (*sima picta*) affixed with nails. The whole structure is then apparently roofed in with light beams, or *asseres,* 4 *unciae* square and placed not more than 9 *unciae* apart from one another. Whether these are installed to form a simple, sloping, shedlike roof or a double-pitched roof in the style of a small pediment is hard to say. The specifications now mention sheathing (*opercula*) of firwood, using planks 1 foot long (?) (*ex tigno pedario*, or perhaps "of inferior grade"). As we have seen, such a sheathing would be placed over the rafters to create a good bedding for the tiles. Finally, firwood *antepagmenta* 9 *unciae* wide and only 0.5 *uncia* in thickness are added. A cymation molding (*cumatium*) is attached. Here again we can perhaps understand the nature of the *antepagmenta* by considering their use for the Tuscan facade: the thin boards in question were used to cover the ends of the *asseres*. The roof is finally sheathed with a row of six tiles.

After all this, the description of the door is rather anticlimactic: a double-leaved arrangement that is to be grated (*fores clatratas*). The doorposts are to be of the best oak, *aesculus* (the varieties of oak will be considered in chapter 12). From the lack of additional information about the doors we can assume the term *fores clatratae* provided enough information to the contractor, a certain C. Blossius, that he could fulfill his obligations. We shall see that the practice of flanking exterior doors on Roman homes by pilasters is similar to the treatment described in this inscription.

A SIMPLE WOODEN DOOR: THE SHUTTERED ROMAN SHOP

In contrast to high-status paneled examples, the most simply constructed Roman doors were little more than a row of vertical planks inserted side by side into grooves cut into the stone threshold and the wooden lintel of the door frame (*coassamentum*). Once the individual planks were in place they could be secured from behind by a horizontal bar. This type of door, more accurately described as a large shutter fixed in place, was used to close the shops (*tabernae*) lining the sidewalks of urban streets throughout the empire. At most Roman sites, the only visible traces of such doors are stone thresholds, each with a channel into which the vertical planks were fitted. At Herculaneum, however, one can still examine *tabernae* that retain enough of their original wooden door frames—and even a door itself—to allow a more complete description.

A fine example of such a door frame, surviving in the form of carbonized wood, is located at shop number 4 on the north side of Herculaneum's *decumanus maximus* (figs. 9.2, 9.3). As in other Roman shops, the spacious dimensions of the doorway, 3.13 m wide by 2.30 m high, maximized exposure to pedestrian traffic.

In order to span the opening of this shop, two heavy beams 4 m long, 14 cm high, and 24 cm deep were stacked upon the flanking brick piers of the street facade. A third, shorter beam (3.13 m long) was placed beneath. This fit snugly between the brick piers and was directly supported by two vertical timbers 9 cm wide by 38 cm deep that formed the doorposts proper of the shop front. Vertical channels cut into the faces of the three beams forming the lintel may have once accommodated iron straps to bind them tightly. Finally, a smaller beam on each inside face of the side posts and another on the soffit of the composite lintel were chiseled with grooves 3.5 cm wide and 2 cm

Fig. 9.2. A lintel and door frame of a shop from Herculaneum, now carbonized, from shop no. 4 of the decumanus. *First century. (Author)*

deep; these once received the ends of the planked door. This arrangement indicates that the construction and fitting of doors and their frames were undertaken as a separate phase of construction. Most important, periodic repairs or replacements could be undertaken without tampering with the main structural components of the opening.

A common feature of Roman *tabernae* was a smaller, inset door that could be used for private access when the vertical shuttering of the shop had been locked into place. The indicator for such a feature is a rectangular cutting, often with its pivot point preserved, in the stone threshold block. The shops of Herculaneum are no exception. The cutting in the threshold of shop number 4 clearly shows the location and the width (72 cm) of the narrow door-insert.

Thousands of such shops were built in Roman towns throughout the empire, but it is exceedingly rare for any wooden elements to have survived. While shop number 4 at Herculaneum is not the only such structure there to preserve traces of its wooden door frame, it is the only location where a significant portion of its shuttered door, now carbonized, has survived in good enough condition to permit close examination (fig. 9.4). The remains of the door consist of seven vertical planks of wood connected by a variation of a modern tongue-and-groove joint. The opening of the shop indicates that the planks would have stood 2.34 m tall; the four longest carbonized planks, however,

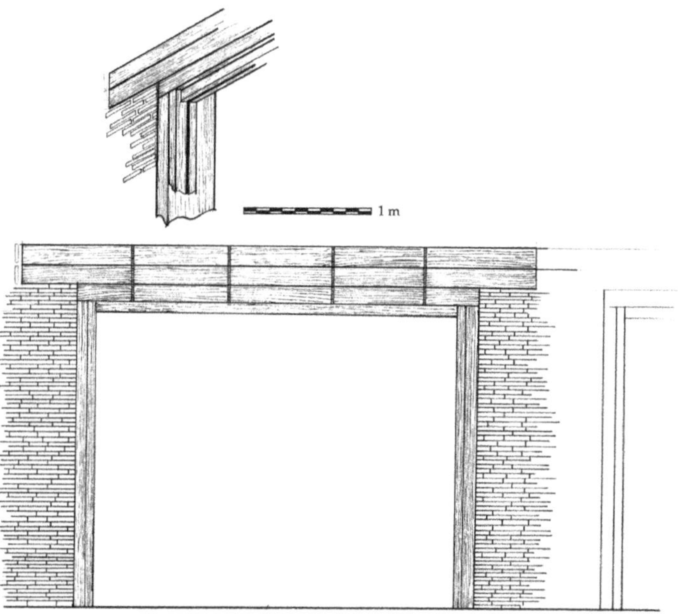

Fig. 9.3. Reconstruction of the shop entrance shown in the previous figure; (top left) *a detail of the inside upper left corner of the wooden framing elements of the shop.* (Author)

now measure approximately 2.20 m; deterioration is worst on the lower ends. Fit together, the seven planks measure 1.46 m wide, less than one-half of the 3.13 m width of the shop's doorway. The planks are of uneven widths, ranging from 16 to 24 cm across.

As the shopkeeper slid the individual planks into place one at a time, the door took its form. Each plank had a deep groove cut into one vertical edge and a bevel into the other. If one faces the exterior of the door, the grooves are in the right edge of each plank. The beveled edge would have been quicker to cut with a plane than a tongue and also less susceptible to snapping or splitting. A door fitted into place in this way had several advantages: it would have been very secure against theft; a shopkeeper could handle the single planks by himself, whereas a one-piece door of this size would be impossible for one man to lift and position; no complicated or expensive hardware was necessary; worn, split, or rotten planks could be discarded and replaced individually. The major disadvantage of such a door was the fact that individual boards tend to warp, making assembly potentially difficult. The type of cupping that bends long planks on the short axis, however, would not have affected the edge-to-edge joins. Rot was clearly a problem. Seating the long planks in a threshold groove channeled moisture into the end-grain and promoted decay. It is perhaps for this reason that this shop door from Herculaneum is so poorly preserved on the lower ends of the planks; the wood fibers may have already been rotted when the shop was destroyed. Awnings or cantilevered shed roofs (*protecta*) over shop doors would have retarded this kind of damage; their use in towns like Herculaneum was common. Old shops in the monastery town of Farfa (near Rome) still preserve heavy wooden lintels and horizontally shuttered facades with inset doors (fig. 9.5)

Fig. 9.4. Shuttered door from a shop on the north side of the decumanus *at Herculaneum. The drawing indicates the actual state of the carbonized door;* (bottom) *detail showing three ways in which planks were finished to fit tightly against one another. First century. (Author)*

The smaller inset doors of the sort one might find built into a Roman shuttered shop were probably made in a fashion still familiar to anyone who has seen a rustic door to a barn or stable. These were fashioned from several vertical planks against which horizontal timbers were bolted on the inside. A utilitarian wooden door found during the excavations of the Roman fort at Vindolanda provides a rare example of its type (fig. 9.6). Both the frame and the planks of the door, dating from the second century, were found virtually intact. According to Birley's description (1977, 115), the door survived because it had been removed from its frame and placed at ground level to form a simple platform; its horizontal position in boggy ground provided the conditions necessary for its preservation. Such an unpretentious door might still be used in a farm building of Britain today. Vertical planks were attached to a rectangular frame consisting of two uprights, or stiles, and three horizontal rails. The door was stiffened by a stout diagonal timber into which a broad bedding was cut to receive the center rail. Interestingly, even though this was a simple workaday door, the basic frame of two stiles and three rails is also, as we will see, characteristic of the finest of Roman doors.

Fig. 9.5. Shuttered shops in Farfa, Italy. The high sill protects the planks from water damage. The wooden lintel set into the masonry wall is identical to ancient Roman practice. (Photo: Author)

Fig. 9.6. Roman door from the Roman fort at Vindolanda. Only the main framing elements have survived. Second century. (Birley 1977, fig. 57, reproduced by permission)

Fig. 9.7. Exterior door from the House of D. Octavius Quartio, Pompeii. Cement cast of the wooden original. The bronze disks, or bosses, that decorated the rails were found in situ. First century. (Photo: E. Sullivan)

MAIN ENTRYWAYS TO HOUSES

At the main entrances to their spacious city houses, proud Roman patrons embellished tall wooden doors of fine joinery with decorative bronze fittings. Bronze attachments were characteristically round, shieldlike bosses attached in horizontal rows on the rails of the door, presumably derived from decorative nail heads (fig. 9.7). The most elaborate domestic entryways surely imitated the portals of great public buildings (themselves influenced by Hellenistic models). The immense doors (6 m high) of the ancient senate house (Curia Julia) in Rome were sheathed with bronze and can still be seen in the narthex of the papal Church of St. John Lateran. Virgil imagined the doorway of Juno's great temple in Carthage sheathed and hinged with such bronze embellishment (*Aen.* 1. 448). In this and other literary examples poets undoubtedly drew upon contemporary realities as inspiration for their imaginary settings.

Whether in painted representations or in well-preserved archaeological sites, the exterior doors of the elite can be distinguished from interior doors by the physical appearance of the door frame. Doors placed on the facades of houses and public buildings tend to be framed by engaged columns or pilasters (fig. 9.8). In domestic set-

Fig. 9.8. Exterior doorway to a house in Pompeii. Note the use of flanking engaged pilasters. (Photo: Author)

tings these posts are often rendered in stucco and usually carry a classical entablature with a prominent projecting cornice. We are reminded of the *antae* that framed the portal described in the inscription from Puteoli. Doors on the interior of wealthy houses and public buildings are framed by a molding that ran around the sides and top of the opening. Rules governing the proportions of this molding will be discussed shortly.

At Pompeii, the public entrance to the atrium-style house was through a narrow hall, or *fauces* ("jaws"), that was commonly accessible through two sets of doors. The outer set (*ianuae*) closed flush to the exterior facade; the second entered directly to the atrium. In wealthier homes a slave (holding the role of *ostiarius*) was positioned in the *fauces* to regulate the flow of traffic from the street to the interior hall. At night the outer doors were closed against thieves and locked from the inside.

A brief examination of the preserved stone thresholds in the *fauces* of one of Pompeii's most visited houses, the so-called House of the Faun, offers a telling example of the disposition of exterior doors in grand urban houses (fig. 9.9). The *fauces* itself was 3.10 m wide and closed by two sets of double doors. The first pair, closest to the street,

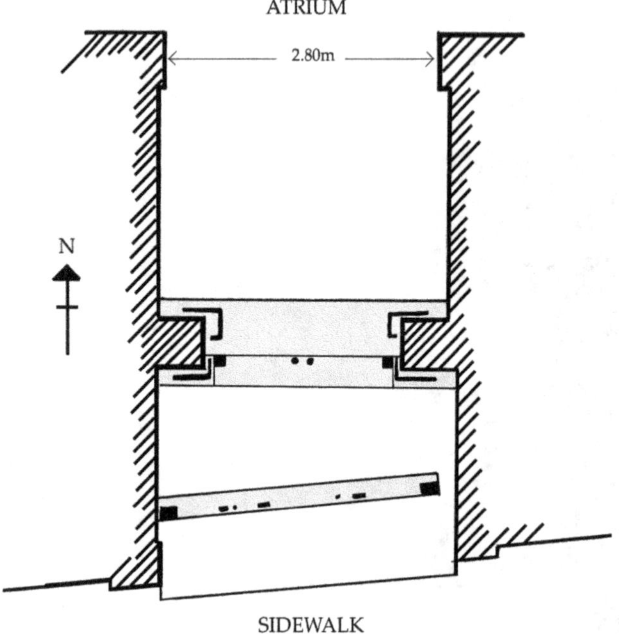

Fig. 9.9 Plan of the main (west) entrance of the House of the Faun, Pompeii. (Author)

turned in pivots 2.50 m apart (shown as black squares in fig. 9.9) on a threshold that was installed on a bias to the main grid of the house so that the leaves when closed would be parallel to the sidewalk and facade wall. The inner doorway was narrower than the first, its pivots spaced only 1.60 m apart. L-shaped grooves in the threshold were once beddings for woodwork that was carved to represent decorative doorjambs. These vertical framing elements were added only for decoration; they conceal masonry piers, and since the doors rotated on pivots set in the threshold and lintel, a strong frame to support hinges was not required. The inner doors were actually locked from the outside, which suggests that at night the *ostiarius* remained in the small space left between the two sets of doors.

The shallow L-shaped grooves cut into the threshold block around the corners of the masonry doorjambs are found as well around the bedroom doors of the House of the Faun inside the main atrium (fig. 9.10). The boards used to frame the doors were at least 3 cm thick (the width of the grooves) and probably even slightly thicker. A simple tenon could anchor the framing pieces to the grooves, and the overhang of the board itself would hide the juncture point. At Herculaneum, where the preservation of wooden remains tends to be better, we can observe carbonized fragments of such wooden doorjambs still *in situ*. The wooden jambs extant in the Collegio degli Augustali, for example, attest to the prominence of such woodwork even in public buildings otherwise decorated with fine painted stucco (fig. 9.11).

The finest wooden doors (and window shutters) were formed from stiles (*scapi*, the vertical framing members), horizontal rails (*impages*), and sunken panels (*tympana*);

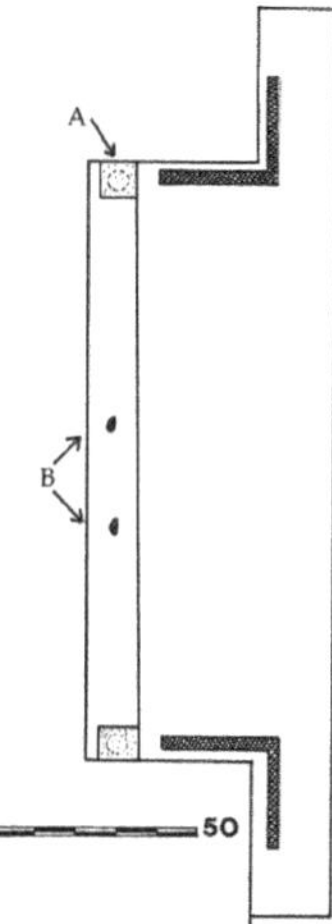

Fig. 9.10. Plan of a threshold from a bedroom (cubiculum) *in the west atrium of the House of the Faun, Pompeii. Cuttings for wooden doorjambs, sockets for pivots (A), and cuttings for the locking mechanism (B) are clearly indicated. (Author)*

Fig. 9.11. View of the fauces of the Collegio degli Augustali, Herculaneum, from the inside of the room. Traces of the wooden panels used to frame the doorjambs are visible as black (carbonized) material. (Photo: Author)

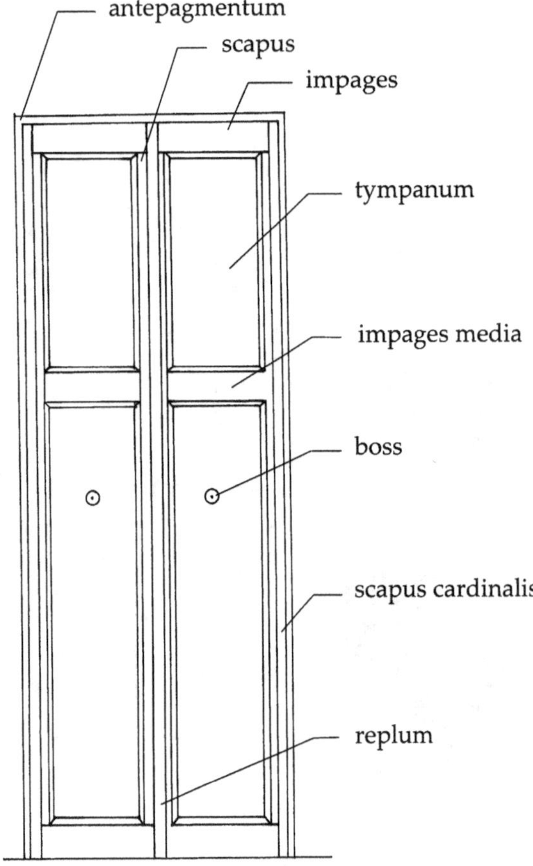

Fig. 9.12. Elements of a double-leaved Roman door. The image is drawn to scale from a plaster cast now in the atrium of the Villa of the Mysteries, Pompeii. H: 3.20 m; W: 1.20 m. Ca. 50 B.C.–A.D. 50. (Author)

at Pompeii and Herculaneum the most prevalent form exhibits two superimposed rectangular panels on each leaf; the larger of the panels occupies the lower two-thirds of the door (fig. 9.12). Doors with three sunken panels are also documented; in some instances the uppermost panel was defined by a metal (or in some cases wooden) grate. Paneled doors were most often arranged in pairs (*bifores*) or as pairs of folding leaves (perhaps *valvatae*, cf. Vitr. 4.6.6). An example of a three-leaved door, now preserved in the form of a cement cast, is visible at the Villa of the Mysteries at Pompeii (fig. 9.13).

The pivot point on a Roman door was located at the bottom of the stile closest to the doorjamb (the *scapus cardinalis*). Most doors swung on bronze or even hard wooden pins (*cnodaces*) which rotated in shallow bronze cups (*armillae*) set into a stone threshold. Excellent examples of the latter exist in the atrium of the Villa of the Mysteries at Pompeii. In the thresholds of the *cubicula* (bedrooms) of the House of the Faun in Pompeii, the cuttings for the *armillae* measure 10 by 8 centimeters. Fine bronze hinges strikingly modern in appearance have also been found in great numbers at Pompeii and Herculaneum; these were used on multileaved window shutters and doors and for the

Fig. 9.13. A triple-leaved door from the Villa of the Mysteries, Pompeii. Cement cast made by the excavators. The upper and lower sunken panels are clearly visible. The lintel, now in concrete, is modern. The original door dates from ca. 50 B.C.–A.D. 50. H: 2.38 m; W: 1.68 m. (Photo: Author)

smaller doors of cabinets. Double-doors were held shut in the middle by rods which dropped vertically into stone thresholds and probably by a corresponding fastener that slid upward into the soffit of a wooden lintel—the system is still used in Italy today. For greater security, locking mechanisms and heavy bars could be used.

Judging from the few actual examples of Roman doors that survive and the dozens of openings for doors left in masonry walls, there is little evidence to suggest that Romans favored the dramatic tapering of the vertical elements of the frame or door exhibited in earlier Etruscan tomb reliefs and paintings, a design element associated with the eastern Mediterranean (Righini 1965, 399). Moderately tapered doors do exist in some of the older, sumptuous Pompeian homes. The House of the Faun, for example, exhibits lofty bedroom (*cubiculum*) doors 1.46 m wide at their base and 6 cm narrower at their lintels, placed a good 4.08 m above the floor. Even a slight taper, with the consequential loss of right-angled relationships, makes the joinery of the wooden pieces much more complicated. Most functional Roman paneled doors can be described as elongated rectangles, each leaf roughly six times higher than it is wide.

A select group of specific examples from sites in Italy can serve to illustrate the major structural and decorative features of well-crafted Roman doors and shutters. The latter, as we will see, exhibit close similarities to doors in terms of assembly and appearance. An instructive example is a wooden shutter recovered from the bottom of Lake Nemi, which should rank as the single most important site in Italy for the history of Roman-period woodcraft.[3] The lake was partially drained between 1929 and 1931 so that two wooden pleasure barges constructed of pine, fir, and oak during the principate of Caligula (A.D. 37–41) could be extracted from the muddy bottom for ultimate display in a lakeside museum. The two enormous barges, measuring 67 and 71 m in length, stood safely in the museum from 1940 until 1944, when they were destroyed by fire. The publication of the excavations by G. Ucelli in 1950 offers a tantalizing glimpse into the wealth of the artifacts, wooden and otherwise, recovered from the ships. Surviving metal fittings are now displayed in the Museo Nazionale (Palazzo Massimo) in Rome.

Among the finds were at least two actual wooden shutters, crafted to resemble full-size paneled doors (Ucelli 1950, 162). The best-preserved one was identified as a window shutter because of its modest dimensions: 1.30 m high by 45 cm wide (fig. 9.14). It appears to have been assembled from seven separate pieces: two stiles, three rails, and two sunken panels. The sunken panels are of unequal height. As is characteristic of all paneled Roman doors, and of their Hellenistic counterparts, the lower panel, or *tympanum*, was taller than the upper. At Nemi the lower panel measured 65×24 cm, the upper only 40×24 cm. The panels were framed by broad compound moldings ca. 7 cm wide. Near the center of the lower panel was a circular bronze boss and a ring that served as a handle. The long, narrow proportions of the Nemi shutter are owing to the fact that this was one side of a pair for a window measuring 1.30 m in height and 90 cm across. A literary parallel is provided by the poet Ovid's description of moonlight filtering through "double-shuttered windows" (*bifores fenestrae*) as he lies awake in exile from Rome (*Pont.* 3.3.5). Another striking parallel is a carbonized window shutter preserved *in situ* at Herculaneum (fig. 9.15). The windows of many old houses in central and southern Italy are still fitted with pairs of heavy wooden (and louvered) shutters.

Other fragmentary doors and shutters were recovered from Lake Nemi. One hinged example, 2.25 m in height and 90 cm in width, served as the left side of a pair that framed a doorway 1.80 m (or 6 Roman feet) broad. This example was never fully published but seems in every way similar to domestic doors that were common in contemporary Campania, particularly those that have been discovered in the sites around the base of Mount Vesuvius.

While many Roman doors either opened in pairs or acted alone to close apertures, broad doorways and windows were closed with four leaves arranged in pairs of two. Latin refers to this type as *quadriforis* (perhaps *valvae bifores*, to denote each pair of two). These were usually interior doors facing on an enclosed peristyle garden, and they were extra wide so that a fine view was afforded the occupants within. Examples of *quadrifores* in doors and windows can be observed *in situ* at the imperial villa of Oplontis, located in the town of Torre Annunziata near Pompeii. The wood itself no longer survives, but the structures have been preserved in the form of cement casts made by

Fig. 9.14. Wooden shutter recovered from the excavations at Lake Nemi. The construction is identical to that of wooden doors. (left) Reconstruction; (right) original. Note the handle in the center of the lower panel and the surviving hinges on the stile. The artifact was destroyed in 1944. First century. (Ucelli 1950, 163, fig. 169, courtesy of the Istituto Nazionale di Archeologia e Storia dell'Arte, Rome)

the excavators. A large window in one bedroom, or *cubiculum*, for example, exhibits dimensions that would be amply suitable for a doorway in a less opulent home. Pliny described such windows at his seaside villa, which he claims were "no smaller than folding doors" (*Epist.* 2.17.5). At Oplontis four wooden leaves filled an opening just under 2 m high by 1.62 m in width. Each leaf was 1.95 m high by 40 cm across. The lower sunken panels measured 1.07 m high by 27 cm wide, while those above had the same width but were only 50 cm in height. The window itself was framed by an elegant wooden molding 18 cm across (fig. 9.16).

Another complementary body of evidence concerning the appearance of ancient

Fig. 9.15. Carbonized window shutters at Herculaneum along the north side of the decumanus. *The vertical planks are held in place by upper and lower crosspieces mortised into stiles. First century. (Photo: Author)*

Fig. 9.16. Four-leaved window shutter from a bedroom (cubiculum) *in the villa at Oplontis. The drawing was made from a cement cast of the wooden panels. Vertical lines on the right indicate observable remains of the window frame. Overall dimensions: H: 1.95 m; W: 1.62 m. First century. (Author)*

doors consists of the many painted representations surviving on Roman walls. These are usually part of larger compositions, most dating from the first century B.C., and belonging to a genre of painting that has come to be known as Second Style. The best examples have been recovered from the sites buried by the eruption of Vesuvius, with one outstanding exception from the Palatine Hill in Rome. The beautiful doors depicted by Campanian wall painters, however, must be regarded with certain precautions, especially if they are being used to understand the appearance of contemporary Roman doors, however refined. For these painted doors, like the superb example found on the back wall of a reception room known as *oecus* 6 in the Villa of the Mysteries at Pompeii, those found painted on the atrium walls of the nearby wealthy villa at Oplontis, and

Fig. 9.17. A painted doorway from the back wall of oecus 6, Villa of the Mysteries, Pompeii. Mid–first century B.C. (Photo: Author)

those at Boscoreale (see fig. 6.14), are usually shown in fantastic settings, commonly thought to have been inspired by the type of stage decoration seen in Roman theaters of the first century B.C. The painted doors may well represent the kind of portal that, in the artistic imagination, represented the luxury associated with a royal Hellenistic palace. Nevertheless, despite the exotic nature of the painted examples, many of the details depicted afford interesting comparisons to the type of archaeological evidence, such as plaster casts, discussed above. The painted door at the Villa of the Mysteries is a useful example.

Situated at the center of the rear wall, the painted door serves as the decorative focus for the entire room, for it dominates the view from the main entryway (figs. 9.17, 9.18). The double-leaved door is depicted about half-size, 0.57 m wide at the bottom (including the door frames), then tapering gently toward the top to a height of 1.37 meters, including the depicted lintel. The portal as a whole is even higher because the lintel is surmounted by a decorative cornice and pediment that add another 0.47 m to the door frame. The door and its framing elements are flanked by painted Corinthian columns rendered to give the impression that the door is situated behind a colonnade.

Like actual Roman-period doors found in this very villa and in other houses in the vicinity now preserved in the form of plaster casts, each leaf of the painted door in

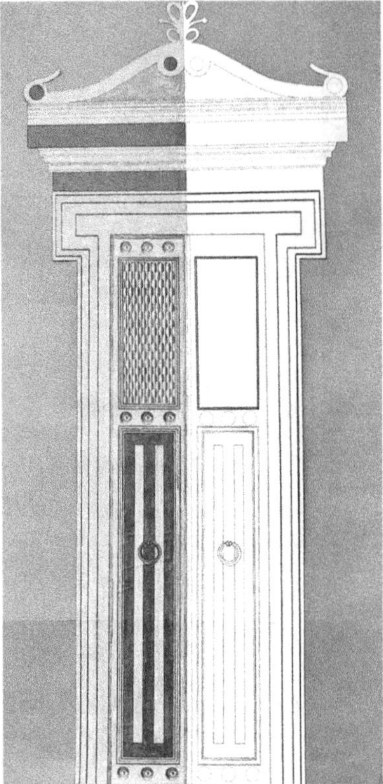

Fig. 9.18. A drawing of the painted door shown in the previous figure. Author.

oecus 6 is divided into two superimposed panels, or *tympana.* The bottom panels are a little over twice the height of the top panels (0.765 m vs. 0.33 m). From the meticulously painted details it is possible to discern that all four *tympana* were rendered as if sunken below the rails and stiles and bordered by a delicate ogee (cymation) molding. Bronze rings for opening each leaf are painted at the center of each lower tympanum. Additional prominent metal decorative attachments are shown as miniature round shields, or bosses, painted in horizontal rows of three upon each of the rails. Thus each leaf is decorated with nine bosses. Through skillful shading the artist has depicted each boss as convex in profile with a centered dimple. In addition, a row of projecting metal pins has been painted down the center molding, or *replum*, of the door. Bronze bosses have been found *in situ* at Pompeii in both public and private contexts (Righini 1965, 405) (see fig. 9.7).

A closer look at the decoration of the painted door suggests that no wood was left exposed (this is true of similar painted examples also). The lower *tympana* are decorated with vertical bands of deep red and cream; the same color scheme is used to decorate the rectangular fields of the upper *tympana.* These upper panels were filled with a scale, or squamate, pattern that resembles the overlapping laurel leaves one sees carved on

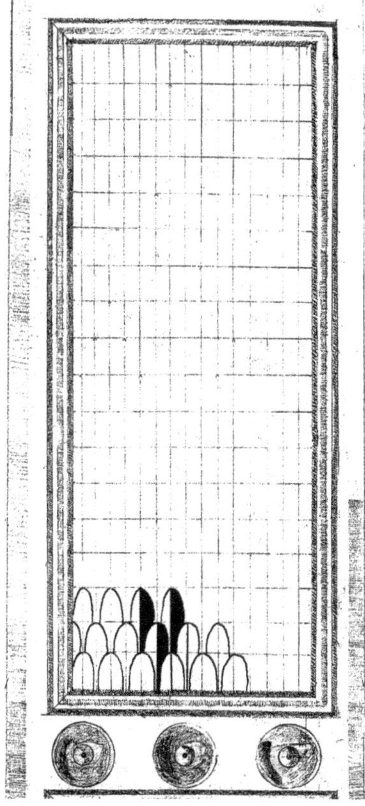

Fig. 9.19. The layout of the decorative pattern used for the painted door from the Villa of the Mysteries. (Author)

the convex moldings (*tori*) of early imperial column bases. Here the stylized leaves are painted in eighteen horizontal rows. In all, 153 painted leaves were required to fill each tympanum. The manner in which this design was laid out gives us some insight into the art of the artisan who made inlays and veneers. The painter divided the tympanum into a grid consisting of 288 rectangles of equal size: eighteen from top to bottom and sixteen across. Once this grid was laid out, the leaves could be quickly and easily painted without the need for stencils. The slight variations in the profiles of each leaf indicate that they were in fact painted freehand (fig. 9.19). The dark red and cream color scheme is standard for this type of decoration; it is rendered similarly, if not exactly the same way, on other contemporaneous painted doors and columns.[4] Since the artist added no shading to this part of the door, the decorated area is meant to be understood as flat. The impression created is that of a fine surface inlay. The colors suggest the kinds of luxury material known to have been used on door panels, in this case ivory and either marble or tortoiseshell. The individual shapes required for the pattern could be cut easily with a fine saw, as there are no difficult inside corners or tight curves to contend with.

The manufacture of a high-quality door was particularly time-consuming because once the component pieces had been fit together, they were glued and then clamped for months or longer until the wood was fully seasoned and no longer in danger of distortion. Pliny (HN 16.215), for example, reported that the doors of the Temple of Artemis at Ephesus, among the most celebrated temple doors from the ancient world, had been glued and clamped for a full four years before they were installed. At the time Pliny wrote these comments, the cypress doors of this shrine, still functioning, were over four centuries old, and Pliny's knowledge of any details of their construction is perhaps no more than a rendition of local lore. Nevertheless, Pliny's single greatest source, the treatise of Theophrastus, has this to say about the making of fine doors: "They do not complete the doors at once; but, when they have assembled them, set them up, and then finish them off the next year, or sometimes the next year but one, if they are doing especially good work. For in summer, as the wood dries, the work gapes open, but it closes in winter. The reason is that the open fleshy texture of the wood of the silver fir absorbs the air, which is full of moisture" (Theophr. *Hist. Pl.* 5.3.5).

Vitruvian Proportions of the Door

Vitruvius devotes an entire chapter (chapter 6) of book 4 to address the proper appearance of temple doors. This attention reflects the importance accorded to creating a suitably impressive entrance to the cella of a temple. Using Greek terminology, Vitruvius (4.6.1), defines three main styles of doorways: the *thyroma doricum*, the *thyroma ionicum*, and the *thyroma atticurges*, that is, Doric, Ionic, and Attic variants of the door and its door frame. Although his discussion is focused upon rules Vitruvius describes as Greek, the information is meant to be of practical use to the Roman builder, who, in terms of architectural decoration, was wont to imitate Greek style. Even though most functional Roman doors did not, as we shall see, exhibit Vitruvian ideal proportions, the terminology used for the individual components can in fact be applied to most Roman doors of good quality:

> The doors [*fores*] are so assembled [*compingantur*] so that the hinge stiles [*scapi cardinales*] are one-twelfth the width of the total opening. Between these stiles [*scapi*], the sunken panels [*tympana*] are three-twelfths wide. The rails [*impages*] are arranged so that by dividing the height of the door in five parts, two parts are given over to the upper section and three to the lower. Let the middle rail [*impages*] be placed *super medium* ["above the center," thus the upper tympanum will not be as high as the lower one]. The other [rails] are placed at the bottom and top [of the door]. The width of the *impages* is to be one-third [the height of?] the *tympanum*; the decorative border [*cymatium*] one-sixth the width of the *impages*. The breadth of the inner stiles [*scapi*] is to be one-half that of the *impages*; the center molding [*replum*] is two-thirds the width of the *impages*. The *scapi* against the door frame

[*antepagmentum*] are made one-half [the width of] the *impages*. If [the doors are] *valvatae*, the heights are the same, but let the breadth of the opening be increased. If [the doors] are *quadrifores*, add to the height. (4.6.4–5)

Vitruvius's description of the rules governing the doorways of Greek temples makes it clear that the doorway comprises three basic elements: the actual door (*foris*), the door frame (*antepagmentum*), and the crowning elements, cornices or friezes or both. The rules for the *foris* itself have just been described. The characteristics of the opening of the doorway and its framing elements are summarized below. All of these constituent parts could be constructed from joinery, although by the imperial period the largest temple doorways were framed with jambs and cornices of fine marbles. The rules, or *rationes*, of the entrances (*ianuae, ostia*) and their framing (*antepagmentum*) are as follows:

THE DORIC DOORWAY (Vitr. 4.6.1)

1. The top of the crowning cornice [resting on the *antepagmentum*] should be level with the tops of the column capitals in the pronaos.
2. The opening [*lumen*] of the double folding doors [*valvae*] should be 0.714 of the total height of the walls (five-sevenths of the distance from floor to ceiling).
3. The width of the opening [*latitudo luminis*] at the bottom should be 0.458 of its height [just under half, or 5.5-twelfths, of the height].
4. The doorway should be tapered [*contrahatur*] toward the top as follows:
   a. one-third width of door frame if less than sixteen Roman feet tall.
   b. one-quarter " " " " " height is sixteen–twenty-five Rom. ft.
   c. one-eighth " " " " " twenty-five–thirty Rom. ft.
   d. Perpendicular sides if door frames are higher than thirty feet.
5. Jambs [of the door frame] should be contracted at the top by 0.07 [one-fourteenth] of the base width.
6. Lintel [*supercilium*] height = width of the jambs [*antepagmenta*] at the top.
7. Lesbian *cymatium* on the jambs should be 1/6 of the jamb width.
8. The frieze above the lintel [*hyperthyrum*] is to be as high as the lintel, bordered on top with a Doric *cymatium* and a Lesbian astragal.
9. The projection of the unornamented cornice should equal cornice height.
10. The ends of the T projections [*proiecturae*] on either side of the lintel should be joined with a miter[? *in ungue*] with the *cymatium* of the jambs.

THE IONIC DOORWAY (Vitr. 4.6.3)

1. The opening of the door is determined as described for the Doric.
2. Bottom width of door should be 0.4x or two-fifths the height of door.
3. Contractions [taper] of jambs are to be the same as Doric.
4. Width of door frame = 0.07 [one-fourteenth] of door opening [note that the width of the jambs was not specified for the Doric doorway].

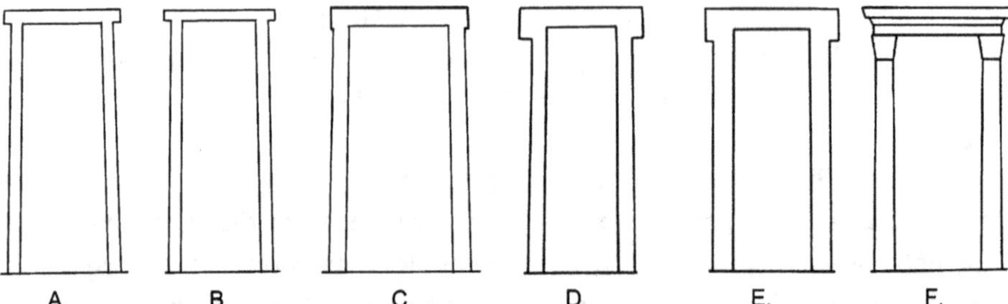

Fig. 9.20. Door frames and openings for double-leaved doors redrawn as if all are of equal height while retaining their true proportional characteristics: (A, B) actual doors from the atrium of the Villa of the Mysteries, Pompeii; (C) the Vitruvian ideal for the Doric/Attic doorway; (D, E) painted doors from the Villa of the Mysteries and the Villa at Oplontis, respectively; (F) the entry doorway from the House of the Samnite (Herculaneum). Where actual wooden framing was used (A, B) the border tends to be thinner. Generally the actual examples tend to emphasize height by increasing the width:height relationship from the Vitruvian ideal of 4.5:10. (Author)

5. *cymatium* width should be one-sixth of the width of the door frame [same as for the Doric].
6. For the division of the *fasciae* on the architrave: First, divide the architrave into 12 parts.
    a. first *fascia* and astragal = three-twelfths
    b. second *fascia* = four-twelfths
    c. third *fascia* = five-twelfths
7. Tops of doorways should follow Doric rules; consoles [*ancones* or *parotides*] should also be employed.

THE ATTIC DOORWAY (Vitr. 4.6.6)

1. Same as Doric but:
    a. The *fasciae* run all the way around the frame.
    b. Doorways constructed without latticework.
    c. "nor are they to be *bifora* but *valvata*, and open outwards" [*et aperturas habent in exteriores partes*].

The Vitruvian formulae outlined above include specifications regarding the relationship between the width and the height of a door, depending on which style the door is conceived in. For the Doric and Attic styles, the width of a given door should ideally be 45 percent of its height. The Ionic style presents a slightly narrower door, proportionally speaking; here the door width is only 40 percent of total height.

It is instructive to compare the Vitruvian ideal with actual doors excavated in the area of the Bay of Naples (fig. 9.20). Doors from the sites of Pompeii, Herculaneum, and Oplontis are particularly useful for such analysis. Because they are often preserved to their full heights, it is often possible to recover the dimensions not only of the door opening but also the door frame, and many would have been roughly contemporaneous with the period in which Vitruvius writes his treatise. Moreover, this was an area

Table 5. Proportions of Roman doors as a factor of width vs. height

Location	*Door Type*	*Width × Height*	*Width as % of Height*
1. Pompeii-1	Painted, luxury	0.40 × 1.29	31
same, including door frames	Painted, luxury	0.57 × 1.39	41
2. Pompeii-1	Atrium/peristyle	1.08 × 3.20	33
same, including door frames	Atrium/peristyle	1.42 × 3.36	42
3. Pompeii-1	*cubiculum* 3 (fine)	1.19 × 3.13	38
same, including door frames	*cubiculum* 3 (fine)	1.53 × 3.28	46
4. Pompeii-2	*cubiculum* (fine)	1.46 × 4.08	36
same, including door frames	*cubiculum* (fine)	1.85 × 4.34	42
5. Pompeii-3	Street entrance (fine)	1.60 × 4.15	38
same, including door frames	Street entrance (fine)	2.58 × 4.45	58
5. Pompeii-4	Street entrance (util.)	1.52 × 2.54	50
6. Herculaneum	Street entrance (fine)	1.52 × 3.98	38
6. same, with door frame and cornice	Street entrance (fine)	2.28 × 4.43	51
7. Oplontis (Villa)	*cubiculum* 11 (fine)	0.93 × 2.71	34
same, including door frames	*cubiculum* 11	1.13 × 2.80	40

Notes

Numbers of individual rooms (e.g., "*cubiculum* 3") are taken from the standard published plans of A. Maiuri or F. Coarelli.

Pompeii-1 = Villa of the Mysteries (door frame of example 2 estimated from scars in plaster).

Pompeii-2 = House of the Faun (door frame of *cubiculum* estimated from scars in plaster).

Pompeii-3 = House of the Menander (lintel is a modern restoration; no upper cornice factored).

Pompeii-4 = House of L. Ceius Secundus (exterior door; plaster cast).

Herculaneum = House of the Samnite (based on measured drawing by R. Oliva).

profoundly influenced by Greek culture, and the Vitruvian proportional system, as has been noted, was based upon Greek models. From the half-dozen examples reproduced in table 5, it is clear that the proportional system used in high-quality domestic settings in the Bay of Naples area closely corresponds to the dictums of Vitruvius—but only if the door frames are included in the calculations. The implication is that wealthy patrons encouraged builders to construct doorways that resembled, proportionately, those that one would see in monumental public structures such as temples.

X Wheels

The wheels of the carts, wagons, and chariots of the Roman world were of wooden parts often reinforced with metal. The craft of the wheelwright was characteristically conservative; changes in methods of construction occurred slowly over a span of time measured in centuries. When innovations were introduced, older methods of construction were not necessarily eclipsed. Techniques used by Roman wheelwrights were themselves of venerable age and, to a remarkable degree, practiced by their successors with little change for another thousand years or more—in some parts of Europe and England well into the twentieth century.

The skills mastered by the wheelwright were considerable; we can imagine that apprenticeship lasted for many years. It is indicative that the image of the wheel figures so prominently on the tombs of these specialists (fig. 10.1; see also fig. 3.15). The artisan had to be able to identify suitable wood from standing timber. His inventory of wooden blanks was best stored and air-dried for five years or longer; these would produce the most durable wheels. He assembled the essential components of the wheel without metal fasteners. Yet from an early time, and certainly throughout the Roman period, wheels were reinforced with bronze and iron; thus the wheelwright either personally understood the principles of blacksmithing or worked side by side with such a specialist.

The comments offered below can only summarize the meticulous scholarship that has been directed at the history of the wheel and its role in transportation, agriculture, and warfare. Of particular relevance to this book are the methods of joinery used to make wooden wheels and the types of trees recognized as best-suited to this application. Monographs by such scholars as S. Piggot (1983) and Woytowitsch (1978) include detailed analysis of the typology and distribution of wheels and their applications in ancient vehicles.

The most primitive of wheels could be fashioned from a single slab of wood. These had the appearance of a solid disk, pierced at the center by a round or square aperture to accept an axle. Other wheels of the solid disk type were fashioned from three or more planks. A second category is that of the crossbar wheel, which has a rec-

Fig. 10.1. A funerary marker of Q. Minicius. The craftsman poses with a wheel of eight spokes and a short-handled adze. From Fossano, in northern Italy. Now in Turin. (Koppermann, D.A.I. 74.1564)

ognizable wooden rim reinforced by struts that are placed on the chords, but not the radii, of the wheel's circumference. The third and best-known category is the spoked wheel, the construction of which requires the most demanding skills. All three types were known, built, and used in Italy; none was of Roman invention per se. While the planked, solid wheel was a precursor to the fully developed spoke-type, both forms continued to be used throughout the Roman period and beyond.

The spinning disk of a wooden wheel is subjected to rapidly changing forces that are unique to its form and function. Whereas the load upon a wooden element of architecture or furniture is predictable and generally more or less constant, the forces in play upon the wood fibers of a spinning wooden disk carrying a load shift from instant to instant, thousands of times over a short period of use. Furthermore, the forces acting upon the wooden disk may operate on two different planes at any given moment. For

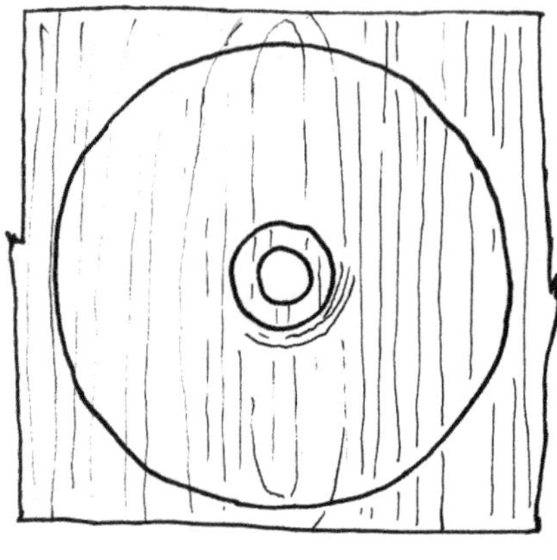

Fig. 10.2. The single-piece wheel, including nave, formed from a thick wooden blank cut longitudinally from the trunk of the tree. In the section the growth rings are indicated from a hypothetical top view. (Author, after Piggot 1983, 19, fig. 3)

example, a cart wheel moving around a sharp bend or along a slope is subjected to both vertical and lateral stresses. Potholes, rough paving stones, or curbs add momentary, intensely concentrated pressure. It is important to keep the role of such physical forces in mind, for these are the key to understanding the development of the wooden wheel.

The idea for the fixed wheel doubtless evolved from the use of the roller: easily cut roundwood was placed under the runners of heavy sledges. At some point the simple roller might have been held in place by pins; no archaeological evidence can confirm such an intermediary step (Piggot 1983, 39). With time the roller became a fixed axle, to which was connected a rotating disk, the wheel itself, at either end. The appearance of solid wooden wheels is widespread throughout Europe in the third millennium B.C.; virtually intact examples have been recovered from the Netherlands (Vogel and Waterbolk 1972).

As Piggot (1983, 17) has observed, the notion that primitive circular wheels were created by sawing disks from a thick tree trunk, as if "slicing a salami," cannot be sustained. Given that the grain of the wood runs from the outside to the inside faces of such a wooden disk, the wheel would have rapidly split and disintegrated. In any case, if an unwitting wheelwright ever tried such a stunt, it would have occurred long after the introduction of the wooden wheel, for the method requires the invention of the full-size crosscut metal saw, which did not yet exist when the first wheels were being fashioned with nothing more than wooden wedges and the most primitive of adzes.

Wheels made from a single piece of wood were therefore fashioned from a thick slab cleaved from a longitudinal section of the tree (fig. 10.2). Ideally the plank would be cut slightly off-center to avoid the pith of the trunk (and thereby reduce the chance of the wheel splitting). If the plank was of sufficient thickness, a hub, or nave, could be

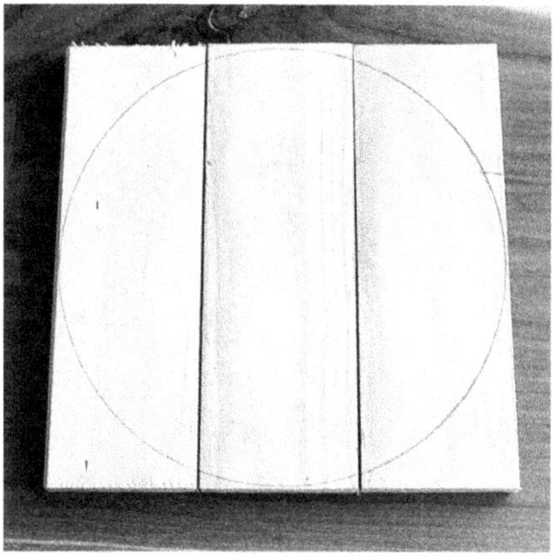

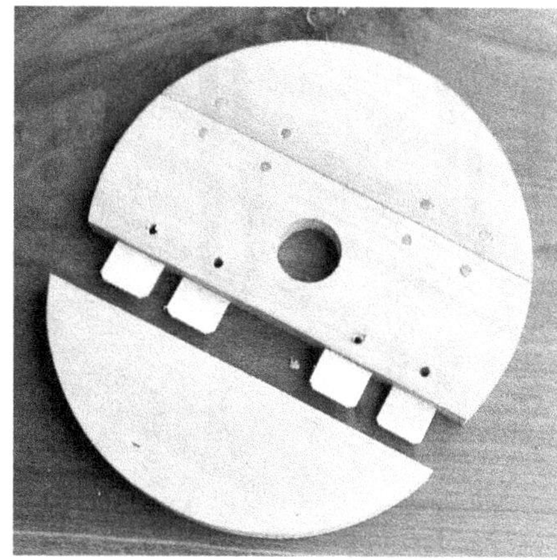

Fig. 10.3. Model of a tripartite wheel under construction: (left) the shape of the wheel is marked out on three planks; (right) the three elements are mortised to one another. Battens can be added to strengthen the wheel. The opening for the axle would normally be reinforced with a nave. (Author)

integral to the single-piece construction. This method, however, wasted wood or required thick stock that was not available. Another solution was to fashion a tubular collar, or nave, for the axle that could be inserted through a hole bored at the center of the disk. The use of a nave prevented the wheel from wobbling on the axle. Solid, one-piece wheels, then, were limited in size and rate of production to the dimensions and suitable species of available timber. Slow-growing, dense-grained woods that grew to sufficient girth provided the blanks. In northern Europe, where such early wheels have been found virtually intact, the timber was cut from oaks at least 250 years old. Although even older trees could be found, hardwood such as oak develops cracks or rotted cavities in the center of the trunk at extreme age (Piggot 1983, 19). Even when such wheels could be fashioned, splitting and cracking along the grain would have been a frequent nuisance. We can imagine that repairs were frequent in the form of planks attached to the opposite arcs of the wheel that were formed parallel to the grain of the wood.

Indeed, from such makeshift repairs and the dwindling supply of mature trees of sufficient dimension, the transition to wheels constructed of three (tripartite) or more edge-joined planks was an inevitable development. The fashioning of large-diameter wheels from planks milled from smaller-diameter trees must have resulted in even broader availability of the wheel. This serviceable form would always be cheaper to construct and require less skill than its spoked counterpart. For heavy, slow-moving vehicles the planked wheel was ideal. The primary challenge in its manufacture was to create a rock-solid edge-to-edge join. The preferred method was the use of mortises, tenons, or dowels so familiar to the shipwright for edge-fastening the planks of a hull (fig. 10.3). Just as the shipwright reinforced such joins by inserting interior ribs behind the planked hull, so too the wheelwright could reinforce his wheels with battens that spanned the exterior surface of the wheel's planked disk. These battens could further

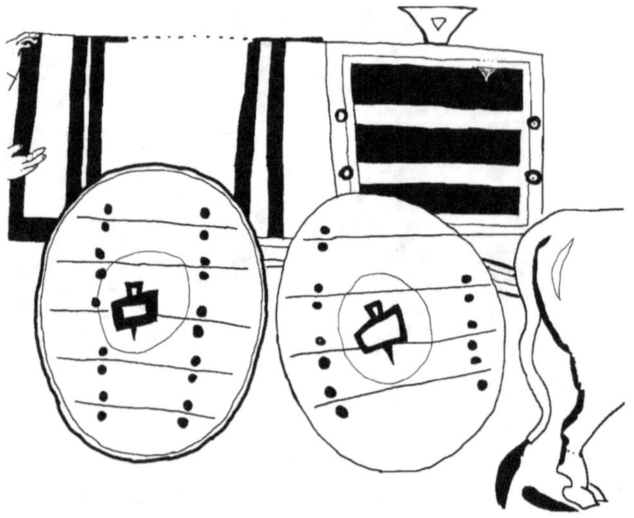

Fig. 10.4. A cart used for transporting wild animals is depicted with planked solid wheels. From the "Great Hunt" mosaic at Piazza Armerina, Sicily, ca. A.D. *300. (Author)*

stiffen the wheel and protect it from lateral stresses by being bedded in shallow channels running across the grain of the planks. Although the battens, which do not radiate from the hub of the wheel, are not spokes in the proper sense of the word, they represent the idea of reinforcing the disk where it is most vulnerable. Batten-reinforced wheels in which the batten is placed in both a straight and slightly flexed configuration have been found. From the Bronze Age on, the battens were commonly fashioned of wood, as can be ascertained from actual examples found in northern Italy, but there was no reason that as metal fittings became more common these elements could not be made of iron (Säflund 1939, pl. 98; literary evidence for this practice will be considered shortly.) A final modification of the planked wheel was the addition of apertures, usually in the form of two lunate cutouts diametrically opposed in the flanking planks, which would have moderately reduced total weight. As weight loss was minimal, there may have been other reasons to account for this development, such as providing handholds to help lift wheels bogged down in mud (Littauer and Crouwel 1977, 103). Planked wheels never fell out of use during the Roman Empire. A late representation in mosaic from ca. A.D. 300 from the villa at Piazza Armerina in Sicily clearly depicts the individual boards (five and six, respectively) that make up each of the wheels (fig. 10.4). Lines of nail heads indicate the presence of wooden battens on the inside face of each disk. Here, too, rectangular axles and oversized cotter pins suggest that the wheels shown cannot turn independently. A dark band of mosaic *tesserae* along the perimeter of the rear disk suggests the presence of an iron tire.

The crossbar wheel serves as a kind of missing link between the planked tripartite wheel with reinforcing battens and crescent cutouts and the fully radially spoked wooden wheel. It was certainly easier to manufacture than its spoked counterpart and offered significant reductions in weight from solid wooden models. M. Littauer and

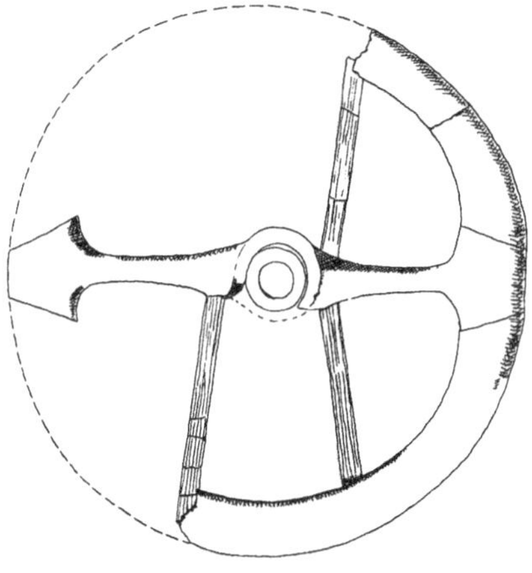

Fig. 10.5. Crossbar wheel. Late Bronze Age wheel from Mercurago, northern Italy. Walnut, diameter 88 cm. (Author, after Piggot 1983, 97, fig. 53, and Childe 1954, 214, fig. 135)

J. Crouwel (1977, 95) have defined the essential characteristics of the crossbar wheel: a single diametric crossbar sufficiently wide and thick to house the nave and two or more crosspieces that traverse the nave tangentially, the ends of which are mortised into the felloe (in some cases the crosspieces may be formed of shorter rods mortised into both the felloes and the crossbar). In Europe the oldest crossbar wheels have been found in an early Middle Bronze Age context (ca. 1800–1100 B.C.) from the site of Mercurago in the Piedmont of northern Italy (Barfield 1971, 74). The fine example illustrated in figure 10.5 was apparently of walnut, slightly less than a meter in diameter (88 cm). The spade-shaped ends of the crossbar formed part of the rim of the wheel itself (Piggot 1983, 98). Two, or perhaps four, felloes made up the rest of the rim. The crosspieces, which appear to have been slightly flexed, were fitted into rectangular openings in the crossbar and mortised into the felloes. The tubular nave had been fashioned separately and inserted into the center of the crossbar. A second wheel from the same site, of similar design but of more robust construction, was fashioned from oak. The crossbar wheel is also known through depictions on late Roman-period reliefs, the funerary stele from the National History Museum of Bucharest being a good example (fig. 10.6). The form continued in use throughout Europe long after the age of Rome's domination.

It is not clear whether the crossbar wheel did in fact represent a developmental stage toward the fully spoked wheel or is simply a parallel variant. Both types appear in Italy virtually at the same time, and examples of the more complicated spoked wheel have been identified in artistic representations and fragmentary examples from the eastern Mediterranean (Egypt and Syria) from ca. 2000 B.C. Certainly by the time of Rome's traditional foundation (753 B.C.) the spoked wheel had been known in Italy for at least several centuries.

The three elements of the familiar spoked wheel, rim, spokes, and hub, are designed so that the stresses of the carried weight and roadbed contact are best absorbed by the fibers of the wood employed. The dynamics of the wheel are such that the weight of the cart both hangs from and is carried by the spokes. Ancient wheelwrights paid

Fig. 10.6. Crossbar wheel. Funerary relief from the Romanian National History Museum, Bucharest. Late imperial period. (Singer, D.A.I. neg. 70.3605)

close attention to both the seasoning of the raw materials and the species of the wood employed. Massive timbers of the type needed to build solid-disk wheels were not required; indeed, a spoked wheel can be manufactured from relatively small constituent parts.

As noted, the nave of the wooden wheel could be fashioned as a distinct and separate element from a very early period. For the spoked wheel, elm was preferred for the nave, although Pliny also mentions ash and holm oak, the latter for its reliability when subjected to friction (HN 16. 229). The cylindrical form was first shaped on a lathe. Bored with a series of augers of increasing sizes to accept the axle and mortised with a chisel around the outside of the cylinder in order to receive the tenoned spokes, a wood that resisted cracking and splitting was needed. The nave was sometimes reinforced inside and out with metal cylinders; such a metal bushing on the inside of the nave reduced both friction and noise. A good example in the form of a well-preserved oaken nave sheathed with bronze was discovered in 1967 at Ischia di Castro in north-

Fig. 10.7. An eleven-spoked wooden wheel from Newstead, Scotland. Total diameter: 92 cm. Second century (?). (Curle 1911, pl. 59)

ern Lazio, Italy. The find has been dated to the sixth century B.C. Spoked wheels as well as many solid disk and tripartite counterparts rotated freely on a fixed axle, a desirable feature in making turns, for the wheels need to rotate at different speeds on corners to prevent skidding.

The spokes of the wheel were crafted from a strong and springy wood like ash, although other tough woods, like oak, were used as well. We have no description of how the spokes were cut from blank stock. In postclassical times artisans favored cleaving, not sawing, the individual spokes from a block of wood, for such practice revealed hidden cracks and ensured strength. The rough blanks could be shaped with drawknives or spokeshaves into their final form (or, if they were of sufficient diameter, turned on the lathe, see White 1984, 136).

The felloes of Roman wheels were rendered from multiple curved sections or even single pieces of stock steamed and bent to form a perfect circle. An example of the latter, excavated at the second-century Roman frontier fort at Trimontium (Newstead, Scotland), exhibited a nave of elm roughly 40 cm long, spokes (originally ten) of willow terminating in both round (for the felloes) and square (for the nave) tenons, and an outer rim (the felloe) created from a single piece of bent ash, secured at the seam with an iron plate; the total diameter was 92 cm (fig. 10.7) (Curle 1911, 292). Pliny regarded ash as suitable for war chariots; the best guess is that he has the wheel (and perhaps the frame) of the vehicle in mind (HN 16.228).

The wheel itself was ultimately reinforced with a "tire" consisting of an iron hoop that protected the wooden felloe. This reinforcing band offered obvious benefits: uneven wear on the wooden wheel was diminished, and the joints between the individual felloes were protected. Tires would have benefited solid and tripartite wheels as well; there is evidence from as far back as the third millennium B.C. that wheels were protected by leather straps or a combination of leather and hobnails (Littauer and Crouwel 1977, 98). Fitting the iron tire tightly enough around the rim of a wooden spoked wheel was a challenge; again, I turn to nineteenth-century analogies. Once the blacksmith had forged the iron tire to the precise dimensions of the wheel, the metal ring was heated to a high temperature, fitted red-hot to the wooden felloe, and cooled rapidly with multiple buckets of water so that it would cool and shrink to make a tight fit before charring the wooden wheel.

Wooden disks fashioned from parallel solid planks or as a combination of outer rim, spokes, and inner hub were used in a variety of applications other than that of transport. Indeed, although we do not have an ancient description of the steps required for the manufacture of a cart or wagon wheel, we can find some details of construction technique for agricultural applications. In his discussion of a solid wooden disk used in a pressing-room, for example, Cato prescribes that the edge-to-edge planking be joined with dowels of cornel wood and that the whole assembly be further reinforced with iron crossbars held in place by nails (*Rust.* 18.9). In another context we might easily believe that the passage describes the construction of a solid wooden wheel.

Large-scale wheels, perhaps several meters in diameter, were constructed of wood to serve in diverse applications, including the transportation of heavy stones from quarries, waterwheels, and heavy lifting cranes. Such a device, or "machine" (*machina*), Vitruvius tells us, can be described fundamentally as "a unified [assemblage] of joined timbers [*continens e materia coniunctio*]," with the capacity to lift heavy weights, "moved systematically [*ex arte*] by the rotation of disks [*circulorum rutundationibus*]" (Vitr. 10.1).[1]

Vitruvius credits the Cretan architect Chersiphron (mid–sixth century B.C.) and his son Metagenes with the invention of great wooden disks, 12 feet in diameter (3.5 m), built around squared architrave blocks which were fitted with pivots in their ends so that the whole assembly could be hauled by oxen (10.2.11). In short, the blocks to be transported acted as massive "axles" for the oversized wheels. The solution was ingenious, although maneuvering the assembly around corners must have been cumbersome, since the wheels could not move independently of one another. We are offered no details as to how the disks of the wheels themselves were fashioned. It is likely, based upon the enormous weight of the stones being transported, that the planked disks were built up in two or more layers, each with the grain of its planks running at right angles to those adjacent.

Heavy lifting cranes and waterwheels were built on the spoked-wheel principle (fig. 10.8). The general stresses placed upon such construction were quite different

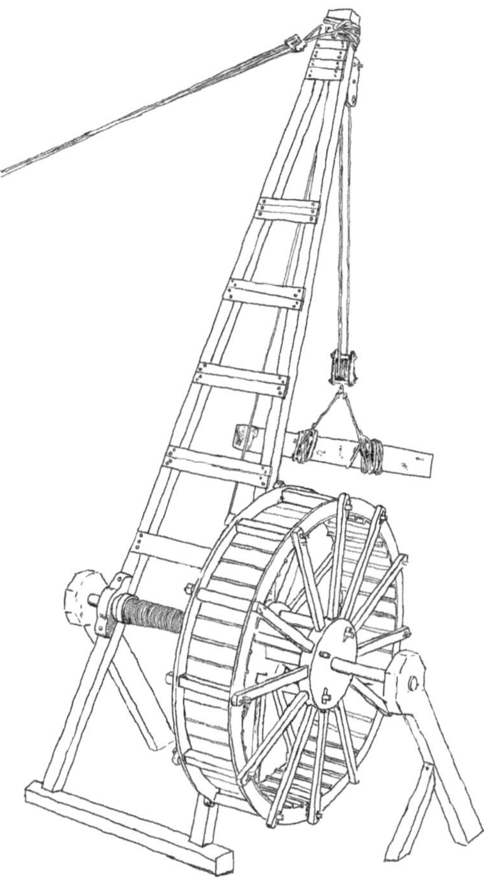

Fig. 10.8. Reconstruction of a "Great Wheel," or treadmill-powered crane. The use of wooden pegs to attach the radial timbers to the exterior rim of the wheel is based upon the depiction of the Haterii relief shown in fig. 10.9. The drawing is based upon a model built by the Istituto e Museo di Storia della Scienza, Florence. (Author)

from those upon wheels intended for road travel. Rotational speed was generally slower, and of course iron tires were unnecessary, although with their absence an important binding component for the overall assembly was lost.

The appearance of the wheel used for the heavy lifting crane is best known from actual depictions on Roman funerary monuments, including that carved onto the marble slabs of the Haterii funerary relief (late first century) (fig. 10.9) and a similar depiction from Capua (Adam 1994, 46; Zimmer 1982, 159). The cylindrical "motors" of these cranes were powered by men who caused the central axle to rotate by shifting the weight of their own bodies; this action took place inside the wooden cylinder itself, estimated by Adam to be 8 m in diameter (Adam 1994, 46). In comparing depictions of these cranes to the small spoked wheel used on carts, one is struck by the extremely thick hub and the fact that the spokes are not attached by mortises to the inside face of the outer rim but instead are lapped to the curved exterior timbers and attached by transverse timbers to their counterparts at the other end of the cylinder. On the Haterii relief the spokes appear to be attached to the rim by heavy pins or pegs. This form of construction is not only useful for creating very strong joints, but also makes it possible to re-

Fig. 10.9. Treadwheel-powered crane from the Tomb of the Haterii, Rome. Vatican Museums. Late first century. (Author)

place individual components easily and even facilitates disassembly and reassembly of the cylinder for transport to the job site (not only over distances but also for movement through narrow passages). The greater diameter of the hub creates more area for attachment of the spokes, as torsional force is transferred from the outside of the cylinder to the axle; unlike the freely rotating wheel used for transport vehicles, the machine wheel turns in unison with the axle itself. This is evident on the Haterii relief through the clear depiction of a nave that houses a squarish housing and axle at the center of the wheel.

XI Furniture and Veneers

Over the past century a number of publications for a broad range of audiences have appeared on the subject of Greco-Roman furniture. The topic was considered with special zeal in the nineteenth century, when craftsmen and homeowners consciously imitated classical examples to furnish their revival-style homes. Handbooks such as J. Pollen's *Ancient and Modern Furniture and Woodwork*, published in 1875, reached a large audience and provided models for imitation (fig. 11.1). In the last quarter of the twentieth century, studies of ancient furniture focused less on aesthetics and more upon the technical abilities of the ancient craftsman; much of recent scholarly attention has been devoted to the conservation, reconstruction, and preservation of precious examples of furniture. The remarkable finds of fine wooden furniture (eighth century B.C.) in the royal tumulus at Gordion in Anatolia, the carved sarcophagi of late Greek and Roman date found along the shores of the Black Sea in the nineteenth century, and the carbonized furniture extracted from the hardened volcanic mud of Herculaneum by Amadeo Maiuri after the Second World War provide outstanding examples.

Modern interpretation of the appearance and use of Roman furniture depends on several forms of evidence similar in nature to those used to understand the role of wood in architectural applications. Literary references supply the Latin terminology associated with furniture and give indications of usage and value. Sculpted and painted representations of furniture otherwise no longer extant compose a vital body of evidence, as do freestanding examples of functional pieces rendered in bronze or carved from stone. Finally there exist a few actual pieces of wooden furniture that have miraculously survived. From this last category, including the household furnishings of wood that have been found in Egyptian tombs, many examples extant are hardly Roman in terms of find-spot and date. Nevertheless, the trade of the furniture maker was a conservative one, and the technical problems encountered by makers of furniture who used wood were solved by solutions that must have been widespread throughout the ancient world. By the period of high empire, skilled woodworkers, even in Rome, were often freedmen of Greek or eastern Mediterranean descent. Many Greek types were adopted

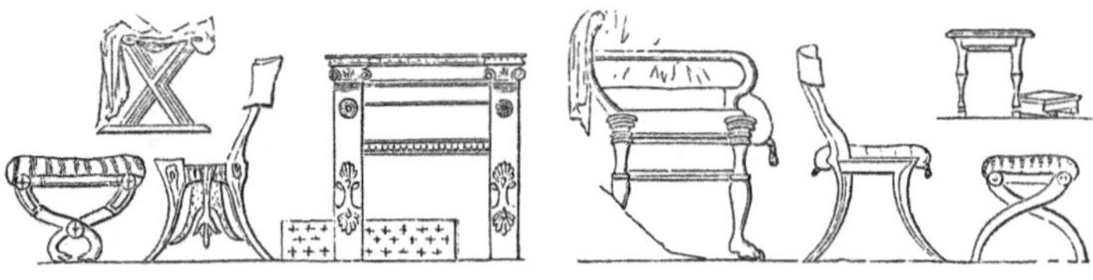

Fig. 11.1. Illustration from J. Pollen's Ancient and Modern Furniture and Woodwork *(1875). The backed chairs on the top register exhibit the form of the Greek* klismos. (Pollen 1875, 11)

by Roman patrons and throughout much of the Roman period wooden furniture both simple and elegant was manufactured by skilled Greek craftsmen. Thus some examples of furniture from provincial contexts are certainly analogous, if not identical, to the types of furniture that were used in Rome itself.

The single most important group of wooden furniture recovered from an Italian site is from Herculaneum; a detailed study has been published by Stefan Mols (1999). We have already seen that the peculiar conditions of Herculaneum's interment were conducive to the carbonization, and thus preservation, of wooden objects. Unfortunately, much of the wooden furniture, laboriously extracted from the petrified mud of Herculaneum, was found over a century ago, and many pieces have since been inadvertently destroyed or otherwise lost. The stabilization of the surviving pieces in a binder of opaque wax has obscured surface detail that would otherwise help to identify technical aspects of joinery. Mols's inventory of the extant wooden furniture from Herculaneum includes the following list of recovered objects: thirteen beds and couches (including an infant's cradle on rockers), six small tables, six table legs, one backless bench, various chairs and associated fragments, three benches, fourteen armoires (of which four functioned as household shrines, or *lararia*), and one chest. Not all the furniture from Herculaneum, it must be stressed, has been inventoried (Mols 1993, 489).

From such assemblages of furniture, the secondary evidence of literary descriptions, and artistic representations, we can consider five broad categories of Roman furniture: chairs (including benches and stools), tables, shelving, chests (including cupboards and private shrines), and beds (or couches).

Chairs, Benches, and Stools

According to Isidore, the first objects made by Daedalus, the mythical father of woodcraft, were a stool and a table (*Orig.* 20.1). Roman chairs were built to seat one or more persons; some had backs and arms, most took the form of simple benches. Only backless wooden benches have survived from Herculaneum. Chairs with backs and armrests are known from literary descriptions and artistic depictions. From what can be observed of actual examples, wooden legs were mortised into the underside of the seats. In some cases, as with modern wooden chairs, an open mortise was used. The legs themselves were sometimes strengthened by horizontal rails, also attached with mortises. Many Latin terms for chairs are derived directly from Greek. Varro (*Ling.* 5. 128) considers much of the nomenclature to be derived from the Latin *sedere*, to sit; as is clear from a survey of terms, variations often indicate the scale or capacity of otherwise similar forms.

The two most elaborate chairs in terms of complexity of construction and decoration were the *cathedra* and the *solium*. The first of these was a type of backed chair, sometimes with arms and usually with gracefully curved legs (see fig. 11.1). The canonical form is best known from Greek contexts, primarily from representations on vase painting and relief sculpture (especially from the sixth to fifth centuries B.C.). This Greek prototype, the *klismos*, has been described as "the most famous, beautiful, and successful of all chairs" (Morely 1999, 25).

The Roman version of the *klismos* was not as common as its Greek counterpart (Richter 1966, 101). The splayed legs, which provided stability and comfort, were fashioned either by bending the wood with steam or by cutting the legs from naturally curved stock. Either method would have reduced the splitting that comes from cutting an arc from a straight-grained plank. The legs were tapered to form a blunt point at the foot. The side rails of the seat were then joined with tenons to the squared tops of each leg. The back of the chair was also curved, bowlike, and supported either by the continuous extension of the back legs or by a vertical member mortised into the top of the rear legs. The wood of the back was a thin sheet of material, presumably cut from a plank with a frame saw and then bent by steaming or soaking in hot water. Sometimes this piece was further supported in the center by an additional splat of thin material.

From literary sources it is clear that the Roman *cathedra* imparted some distinction to its occupant; teachers might sit in *cathedrae* while pupils sat in simpler *subsellia* (Hor. *Sat.* 1.10.90; Isid. *Orig.* 20.11.9), and the *cathedra* was reserved for elders of the early Christian church, thus the practice of naming a bishop's official church "cathedral." In private homes it was considered a fine piece of moveable furniture, worthy of specific mention (Plin. *Epist.* 2.17.21).

The *solium* may have represented, like the *cathedra*, a type of chair associated with prestige, particularly in the domestic context. The Latin *solium* is considered in some cases an equivalent to the Greek term *thronos* and is thus commonly translated as

Fig. 11.2. Venus seated on a solium *with deeply turned legs and footstool. From a fresco found at the Villa Farnesina, Rome. Late first century* B.C. (Mon. dall Inst. 12 (1885), pl. 21)

"throne." Isidore states flatly that the two terms are equivalent (*Orig.* 20.11.9). Indeed, in certain Latin passages in which gods or kings are described as being seated on *solia* (for example, the *solium regale* of Ov. *Fast.* 6.353) or in which an emperor dreams of standing next to the *solium* of Jupiter (Suet. *Calig.* 57.3), "throne" is doubtless the best English rendering.

The correspondence between the Greek and Latin terms and actual examples of Roman-period chairs leads one to the conclusion that *solium* was used to refer to two types of backed chairs that share some basic features but in fact look fundamentally different. The throne type is a backed chair with armrests (and often an accompanying footstool) that is supported by vertical legs in the form of deeply turned cylinders (resembling, as we shall see, the legs of the couch) or four freestanding flat slats with decorative cutout profiles (fig. 11.2). Similar to the *solium* with four freestanding legs is a version with one solid slab for each side of the chair, the forward faces of which are rendered with a serpentine profile that was carved (certainly in the stone examples that have survived) with the same kinds of animal leg and claw motifs we see on benches and armoires supported with slabs (see below). This form of heavy, monumental construction seems better suited to stone than to wood.

In addition to the rectilinear throne type of *solium* there appears an important variant that, like the *sella curulis* (see below), may be derived directly from Etruscan precedents. Numerous examples and depictions of these chairs from Etruscan sites have been found in contexts dating back as far as the seventh century B.C. Prominent examples include the full-size Barberini Throne of the first half of the seventh century B.C. and additional large-scale examples carved in the Etruscan tombs of the celebrated Banditaccia necropolis of Cerveteri dating from the sixth century B.C. (MacIntosh 1974, 22). From these and many smaller-scale renditions in a variety of media and contexts the form of the Etruscan version can be sufficiently described: a chair resting upon a cylindrical or conical base, with a curved back that melds seamlessly into the arms of the chair. It may be this distinctive "bucket seat" shape (described in German as *Rundthron*) resulted in the term *solium* being used eventually (by the first century) to describe bathtubs, seats in bathtubs, or even bathtub-style sarcophagi. Since the imposing stone chairs carved in the tombs of Caere are found in settings that evoke the interiors of fine houses, it seems reasonable to assert that the type is an indicator of prestige in the homes of the elite.

It is easy to imagine this form, based upon the cylinder and the cone, being first carved from the massive trunk of an ancient tree. Indeed, such an artifact, now heavily restored, was discovered in a tomb from the Italian site of Verucchio and is now displayed in the Museo Civico of Bologna (Jurgeit 1990, pl. 7). In this regard it is interesting to note that both Virgil (*Aen.* 8.178) and Ovid (*Fast.* 3.359) describe *solia* of maple wood from Rome's distant past; whether their mention of the same wood is merely coincidence or the maple was associated with the oldest *solia* is hard to say. The only Latin reference to a contemporary wooden *solium* is a curious account of Augustus bathing in a wooden tub (or sitting on a wooden seat placed in a tub) (Suet. *Aug.* 82.2).

While Latin passages attest to the importance of this type of chair, none describe the appearance of the *Rundthron* in the Roman period. Excavations have not recovered any certain examples. The Etruscan evidence just considered suggests that Italic *solia* were built upon a base with solid sides in either a conical or roughly cylindrical form. It seems safe to assert from artistic depictions, considered below, that the *solium* of the Roman period usually included a backrest, armrests, and footstool (or *scamnum*), but the exact appearance certainly varied from region to region and over time. As in the case of other high-status furniture items, wood may have formed only the hidden frame of the chair. The curvilinear Etruscan examples suggest a bentwood frame over which a flexible material such as leather or sheets of bronze was placed. In any case, the desired profile would have required knowledge of wood bending, presumably by steaming or soaking in very hot water.

The *Rundthron* form of the *solium* is in fact depicted frequently in Roman funerary art. The deceased, male or female, may be shown seated in the chair, or else mourners, including musicians, may be seated in *solia* that have been placed next to the funerary bed (*lectus*) (figs. 11.3, 11.4). The scenes, while funerary in content, reflect, like their old Etruscan forebears, ideal domestic settings. An armless form appears to have been used only in Italy (Kampen 1981, 65). Another variant, apparently most popular in the provinces of northern Italy and Germany, was made of wickerwork; carved renditions can be seen on sarcophagi in the British Museum, the Vatican collections, at Trier, and from Desenzano, on the south shore of Lake Garda (fig. 11.5). We will see that Pliny

Fig. 11.3. The Simpelveld Sarcophagus. (top) The deceased lies on a funerary bed (lectus funebris) protected on three sides by paneled boards; a solium is carved next to the bed (at the foot of the bed a building, perhaps a villa, is depicted. (bottom) A stand (abacus, urnarium?) holding three jugs, a mensa delphica, and an armarium. Sandstone. L: 2.05 m. Ca. 175–225. (Courtesy of the Rijksmuseum van Oudheden, Leiden)

praised the qualities of the willow for "luxurious [wicker] armchairs" (HN 16.174). He uses, however, the Greek term *cathedra*, not *solium*. Late Roman sarcophagi depict the *solium* upon a four-wheeled cart pulled by a team of horses. The scene is most commonly one of a funerary procession; in such cases the *solium* is capable of accommodating two adults. Examples can be seen in the collections of Stockholm, Turin, and Rome (Himmelmann 1973, pl. 50–5). On the Arch of Galerius in Salonika (ca. A.D. 300), the emperor is shown in triumphal procession seated upon a *solium* mounted on a chariot.

The *solium* represented the authority of the patriarchal head, the *paterfamilias*, of the Roman household. Even the well-appointed farmhouse, says Cato, should have a pair of such chairs (*Rust.* 10.5). A century after Cato's day, Cicero, describing the celebrated statesmen and orators of the Roman republic, is explicit in his reference to these chairs: "In the old days individuals sought out these men . . . seated in their *solia* at home, not only to consult them on a legal issue, but also about arranging a daughter's marriage, buying a farm, tilling their land, and finally about every kind of responsibility or business (*Orat.* 3.133)." The *solium* was not associated with display in public settings, other than in the processional examples noted above.

The chairs called by the terms *cathedra* and *solium* were not representative of common wooden furniture. Most seats were, structurally speaking, little more than a hori-

Fig. 11.4. A solium *with attached footstool depicted on the sarcophagus of Sosia Juliana. An adze is depicted in relief below the seated figure. Ravenna, Museo Nazionale. Mid-third century. (Author, after Gabelmann 1973, pl. 50)*

zontal slab supported by four legs. Expense seems to have been lavished more on constituent materials: wood types, veneers, precious metals, and ivory. Of the class of backless benches, the *subsellium* was apparently ubiquitous (Mols 1999, 54). Literary sources use the term in a broad spectrum of settings. We can read about *subsellia* in private homes, in theaters, in the senate house, in the law courts, and in schools. Produced in great numbers and highly portable, the majority of such benches were made of wood. We can assert that this was so, even without having any certain wooden examples, through literary accounts. Three examples from Suetonius's biographies make the point: when Julius Caesar's body was burned in the Roman forum by an angry mob, we read that *subsellia* from the senate house and the courts were smashed to pieces to fuel the fire (*Caes.* 84.3). There is a humorous account of the benches collapsing under a fat man's weight in the theater (*Claud.* 41) and of a brawl in which the combatants swing broken benches at one another (*Nero* 26.2). In all such accounts we imagine that the *subsellia* in question were certainly of wood.

A cousin of the *subsellium*, smaller and used as both a bench and a footstool, was the *scamnum*. Varro (*Ling.* 5.168) describes the *scamnum* as a device to step on when getting into or out of a high bed; the passage indicates that the *scamnum* was higher than another type of stool known as the *scabellum*. By late Roman times the *subsellium* and

Fig. 11.5. A solium *of wicker with a seated female musician. She is seated next to a funerary bed, one turned leg of which is situated behind her right foot. From a sarcophagus in the Vatican Museums, inv. 9538-9. Ca.* A.D. *270. (Author, after Richter 1966, fig. 508)*

the *scamnum* were easily confused, if not interchangeable: "*subsellia*, or, as the common man says, *scamna*" (*Cod. Theod.* 3.1.2).

Mols suggests that the three simple wooden benches documented from Herculaneum are perhaps examples of *scamna*, if not *subsellia* (Mols 1999, 54) (fig. 11.6). The rectangular seats range in length from ca. 1.05 m to ca. 1.40 m; they stand between 36 and 39 cm in height. The wooden seats from the benches at Herculaneum are slightly concave on the surface for comfort; the forward edges tend to be rounded. Legs take the form of either two solid slabs attached to the underside of the seat or four individual supports. In either arrangement the forward profile of the front leg curved in an S form and was carved to represent a stylized animal leg; we have seen that this same support is used for a class of Greek-inspired *solia* (note, too, the same motif, even more naturally rendered, observed on three-legged tables, or *mensae*, discussed below). Restoration and subsequent deterioration of the recovered artifacts have made it difficult to analyze methods of attachment between the legs and the seat. The slab-style legs may have been attached with a tongue-and-groove joint, the individual legs with a mortise and tenon (Mols 1999, 183, 185). A small stool from Herculaneum, perhaps also in

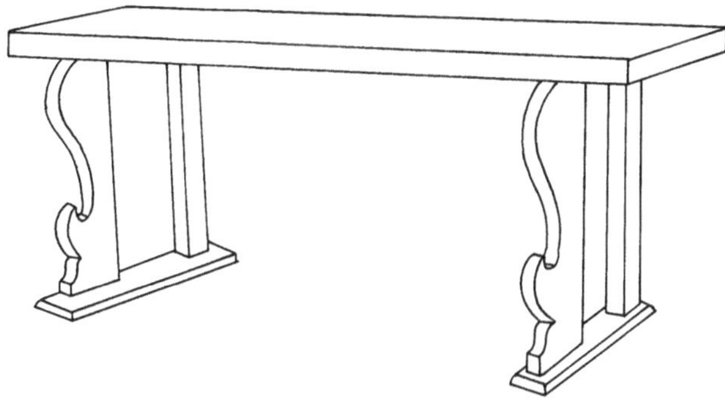

Fig. 11.6. A wooden bench from Herculaneum. Reconstruction drawing based upon carbonized fragments. Other examples from this site are made with one-piece slab legs. L: 1.05 m; H: 0.38 m. Herculaneum, inv. E-3153. First century. (Author)

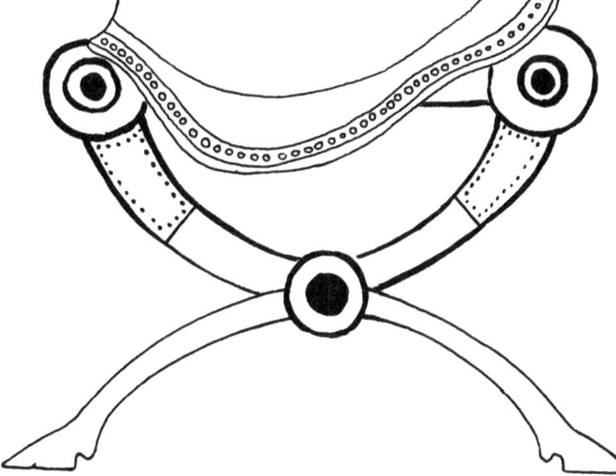

*Fig. 11.7. A folding chair (*sella*) depicted on a painted plaque from the Etruscan town of Cerveteri, now in the Louvre. The top surface of the rendition is obscured by the drapery of a seated aristocratic male. Ca. 550–540 B.C. (Author)*

the category of the *scamnum* or *scabellum* (see below), was found with a geometric inlay of veneers on its seat (see below for discussion of this example).

Another important group of backless chairs and stools was capable of folding, scissor fashion, for ease of transport. The form was of great antiquity and widely dispersed throughout the Mediterranean world, with examples known from dynastic Egypt, Assyria, and classical Greece (Liversidge 1955, 29) as well as Etruria (fig. 11.7). The earliest Etruscan depictions date from the late seventh to sixth centuries B.C. (MacIntosh 1974, 18). Romans believed it was Etruscan influence that led them to adopt this form of folding stool as an emblem of public office (Liv. 1.8.3).

In profile the Roman folding stools resemble the letter X, hinged by a pin that passes through the point of juncture of the legs. The legs tend to assume a slight serpentine profile. The head of the hinge-pin on both sides is often rendered as a prominent disk, or boss, a commonplace item of hardware that was used extensively by woodworkers for door and table construction as well. The basic chair requires only one pair of X-shaped supports, connected at the top (and sometimes at the bottom) with trans-

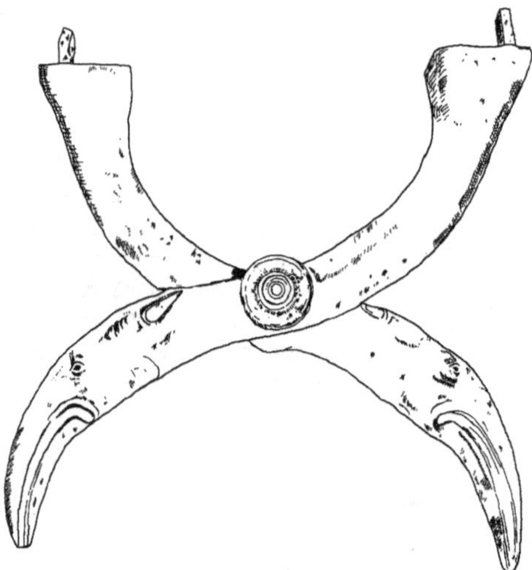

Fig. 11.8. A pair of bronze legs from a folding chair (sella curulis) found at Herculaneum. The circular boss that conceals the hinge is evident; the feet are rendered in the form of a fantastic creature. Note the prominent tenons used for the attachment of the horizontal stretchers. Naples Museum, inv. 73152. First century. (Author)

verse stretchers. Having no arms or backrests, the chair resembles a traditional camp stool (a term sometimes used in modern archaeological reports published in English to indicate this type of artifact).

Despite the simplicity of design and the promise of only minimal comfort, this type of folding chair, particularly that known by the name of *sella* (with a descriptive modifier, such as *curulis*), became emblematic of high public status and the attendant political office. One explanation may be the military associations this portable and utilitarian form of furniture represented (MacIntosh 1974, 20). It was on such a chair that Julius Caesar sat in the senate house during his dictatorship (Dio 43.14; Schäfer 1989, 114); lesser Roman officials paid for representations of the sella to be carved on their tombs; and republican moneyers stamped the *sella curulis* on their silver coinage.

Discoveries of actual folding stools are rare, as might be expected. The Naples Museum houses a unique example in bronze from Herculaneum, with characteristic bowed legs (still capable of folding) and feet carved with the heads of a fantastic beast, its beaklike nose supporting the weight of the chair (fig. 11.8). The practice of carving hooves, claws, or animal heads on the wooden feet and "knees" of chairs and tables was common to Mediterranean cultures; unusual and whimsical examples include ducks' heads on a wooden Egyptian folding stool and hooves and fetlocks on the stools depicted on the terra-cotta reliefs from Etruscan Murlo as well as on the painted example from Cerveteri illustrated in figure 11.7 (MacIntosh 1974, 19). The seats were presumably fashioned from leather, upon which a cushion could be placed. Some models may have had detachable rigid side rails that were connected to the front and back of the chair (Schäfer 1989, 47).

As a high-status item, a *sella curulis* of plain wood would have been embellished with appliqués or constructed entirely of more expensive materials: bronze (probably gilded) or even ivory (for example, Liv. 41.20.1).[1] Multiple, parallel legs configured like intertwined fingers could also provide a visually impressive variant. Such elaborate folding chairs had evolved from simple stools, many of which presumably had four straight, fixed legs. The word *sella*, when used alone, can also refer to such quotidian examples. Poles could be attached to the sides of the *sella* (along the rails) so that it could be carried by porters; wealthy Romans could thus be conveyed through the crowded streets of the city.

The *bisellium* was a broader version of the *sella* that could seat two individuals; it was a mark of honor for magistrates in provincial towns. A prominent example can be seen on a large cenotaph in the necropolis outside of the Herculaneum Gate at Pompeii. The monument was set up by a freedwoman named Naevoleia Tyche for herself and Gaius Munatius Faustus; the latter was awarded a *bisellium* for his good virtue (*ob merita eius*) as a citizen. The *sigma* and *stibadium* refer to a curved bench that could seat several persons. Such benches were popular both as garden furniture and as seating for convivial dining when a dining couch (the *lectus tricliniaris*) was not employed.

Tables

Tables include the *abacus* and the *mensa*; they are explicitly distinguished from one another in Latin texts (for example, Cato *Rust.* 11.3). Both types were also rendered in marble and, presumably, bronze. G. Richter prefers to use the general term *mensa* for all manner of tables, reserving *abacus* specifically for the equivalent of "sideboard" (1966, 116). Since the Latin *abacus* can also refer to a simple gaming table, it seems reasonable to suggest that the term could indicate a table of sober geometric linearity: a rectangular slab top and four or more straight legs. Because the definition is so general, we can imagine that *abacus* encompassed both utilitarian and high-status objects. A small fresco from Herculaneum, for example, depicts two *erotes* ("cupids") working at such a table making shoes (fig. 11.9). The figure on the left is seated at a stool; a large cupboard (*armarium*) (to which I shall turn shortly) dominates the right side of the composition. The table itself is supported by four legs, appearing square in section, that are stiffened by stretchers (horizontal pieces of wood usually located directly under the tabletop) placed at both the top and middle sections. A similar table is depicted in a tavern scene rendered in relief that comes from the Isola Sacra cemetery near Ostia (Kampen 1981, 44) (fig. 11.10). Two figures are seated behind the table, which is supported by six legs; the sculptor rather improbably shows the legs as being attached to the top of the table with a miter joint—unless what we see is a broad molding that runs up the legs and along the edges of the tabletop. From such examples we can identify a clear tendency to employ a stretcher directly under the tabletop to strengthen the joint with the legs. This will not be the case, as we shall see, with the smaller, circular tables used in dining rooms. The slab-top table served many other utilitarian purposes. For industrial applications a form of the *abacus* may have been used as a kneading trough; simple

Fig. 11.9. A frescoed scene of erotes *(cupids) working as shoemakers at a four-legged table (*abacus *or* mensa*). Also depicted are a simple stool, perhaps a* scamnum*, a shelf (*pegma*), and an* armarium *with three interior shelves and double folding doors. Originally from Herculaneum. Naples Museum, inv. 9179. (Author, reproduced by permission of the Soprintendenza per i Beni Archeologici delle province di Napoli e Caserta)*

tables are depicted for just such a purpose on the Tomb of Eurysaces (the "Tomb of the Baker") in Rome (late first century B.C.) (Zimmer 1982, 106). The *abacus* apparently also served a function similar to that of a modern sideboard, used particularly in spacious rooms like dining rooms (*triclinia*), namely, to display the fine silverware of the household (Cic. *Verr.* 4.35; Varro *Ling.* 9.46). Thus Richter identifies the piece of furniture holding various containers carved in the Simpelveld sarcophagus as an *abacus*, but perhaps what we see is the *urnarium* described by Varro (Richter 1966, 116; Varro *Ling.* 5.126) (see fig. 11.3). According to Livy (39.6.7) the *abacus* as a luxury item was introduced from Asia for the first time in the early second century B.C.

A variant of the slab-top table, low in height and placed next to a banquet couch, was built with three legs instead of four. The type is well documented in Greek and Etruscan settings. Jean MacIntosh suggests that the tables depicted in a banquet scene from Etruscan Murlo (ca. 575 B.C.) were fashioned from thin trapezoidal tops strengthened by horizontal rails and vertical struts (fig. 11.11). A single leg with a flared foot supported the narrow end of the table; a pair of legs with curved horizontal projections (in profile something like an inverted letter *y*) were placed at the broader end. The distinctive form of this kind of table may be a function of its desired portability. The overhang of the tabletop on only one side (that with two legs) and the gap between the longitudinal rail and the tabletop minimized weight and provided a form that was easy to carry to and from the banquet hall. The joinery of such a table would have depended primarily on the mortise and tenon. Bronze furniture fittings (for the *solium*, or "throne")

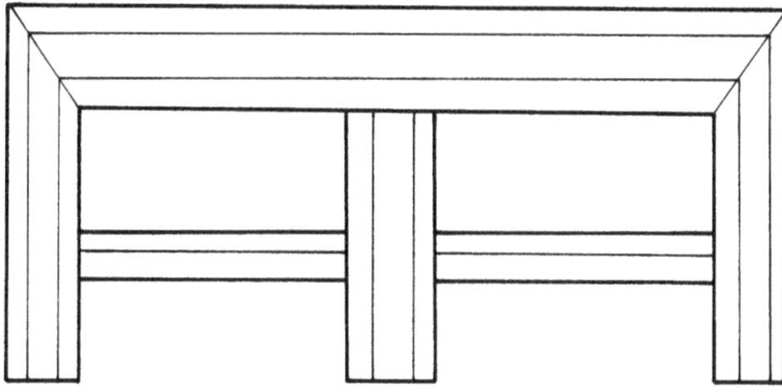

Fig. 11.10. A six-legged table, perhaps an abacus, as depicted on a funerary relief showing a tavern scene from the necropolis of Isola Sacra (Ostia). Second century. (Author)

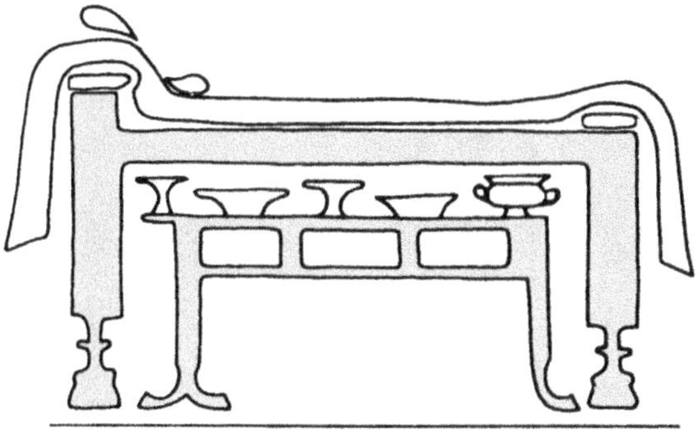

Fig. 11.11. Three-legged banquet table standing in front of a dining couch as depicted on a terra-cotta relief from Poggio Civitate (Murlo). Ca. 575 B.C. (Author, after MacIntosh 1974, 27)

of somewhat earlier date indicate that Etruscan furniture makers employed wooden pegs, square in section, to create strong joins (Pereti 1947, 240). Romans may well have imitated the Etruscan three-legged rectangular table. The form does appear in Roman frescoes, but the renditions may simply be copies of Greek paintings (Richter 1966, 110). A good candidate for such a table is prominently featured on the relief of the woodworker's shop from Rome (see fig. 3.36). While the representation is poorly preserved, the object depicted appears to be a low table supported at one end by two cabriole-style lion-headed legs linked by stretchers and a third, solitary leg on the right side. Here the prominent overhang is on the side with a single leg. The identification of this object with the Etruscan-style three-legged table seems better than that of a round table with a tripod configuration (but shown by the artist with a conventional skewed perspective).

The most recognizable form of the three-legged table was the *mensa delphica*, distinguishable by its round tabletop (*orbis*) supported by legs configured like those of a tripod. The type is apparently of Greek invention, appearing first in the fourth century B.C. and supplanting a rectangular predecessor (Richter 1966, 70). The round (or occasionally crescent) table form is also associated with eating and drinking, particularly as a serving table (Varro *Ling.* 5.121); it is commonly depicted as a central feature of funerary banquets (see fig. 3.28). Examples in marble are well known, but wooden fragments and fully preserved carbonized *mensae* have also been recovered from Herculaneum (fig. 11.12). Indeed, the *mensae delphicae* used for dining were, as far as we can

Fig. 11.12. Wooden mensa delphica *from Herculaneum. Carbonized wood on a modern metal frame. The top center element is carved with a griffin's head, the feet are lions' claws. Diameter of tabletop: 57 cm; total H: 64 cm. First century. (Mols 1999, fig. 94, reproduced by permission)*

tell from the evidence, always of wood; marble versions were too heavy to move about in the dining room (Mols 1999, 128; Moss 1988, 275). The legs (sing. *pes*, pl. *pedes*; Plin. HN 12.5) of the *mensa* were generally bowed outward and carved to represent the limbs of animals, often with clawed feet; a naturalistic head is carved at the knee near the top of the support. At Herculaneum both lion and dog feet are preserved, while the carved heads represent lions, dogs, griffins, and even humans or gods (Mols 1999, 44). These motifs were widely dispersed over the Roman world; note the fine example rendered in sandstone in the Simpelveld sarcophagus (see fig. 11.3). Not all woods are good for carving. Mols has noted that at Herculaneum the wood used for the tables was expensive hardwood; it is clear in these cases that solid wooden tables, embellished with carvings, were a luxury item (Mols 1999, 129).

Damage to the fragile carbonized fragments at Herculaneum makes comments about the specific joinery used difficult. It is interesting to note that the legs were usually joined directly to the tabletop. A mortise and tenon would be the only stable solution. If the mortise used was open, that is, bored right through the tabletop, the join could be hidden by a sheet of veneer. Without conventional stretchers the legs would have tended to wobble; this explains the tripartite "junction stretchers" that were used to strengthen the legs of some *mensae delphicae* (see fig. 3.28).[2] The ends of the three spokelike stretchers were connected to one another by being sandwiched between two wooden disks or metal bosses (Mols 1999, 49).

By far the most coveted tables of this type were the *mensae citreae*, those fashioned from the wood of the citrus (I discuss the Roman passion for this species in chapter 12). The Latin references do not tell us much about the precise forms of the citrus tables. Fine smaller tables were often made with round tops (*orbes*), but whether all were crafted with a tripod configuration is more difficult to say. The beautiful tables of bronze and marble supported by a single pedestal (*cartibula* or *monopodia*) that were recovered from Pompeii and Herculaneum may represent expensive variants of wooden models, but neither archaeology nor literary sources can verify for certain that this is so. Other woods with interesting grain patterns were also employed for *mensae*. Horace (*Sat.* 2.8.10) depicts a slave wiping down a maple-wood *mensa* during an elaborate feast. It is likely the poet has the highly valued bird's-eye maple in mind (Plin. *HN* 16.66).

Wooden Shelves

A basic wooden shelf, used for a variety of purposes, was called a *pegma*. Board shelving was frequently attached directly to a plastered masonry wall. This required some form of underlying reinforcing bracket. A good painted example can be seen in the workshop of the *erotes* from Herculaneum that has already been cited (see fig. 11.9). The brackets, depicted with a few quick brush strokes, are curvilinear in profile, a decorative flourish that can also make for easy attachment. Evidence for wooden shelving in Roman houses usually consists only of scars left in the plaster from the attachment points of the shelves and their supporting brackets. Undoubtedly the most famous example is to be found in the House of the Bicentenary at Herculaneum, where the excavator, A. Maiuri, interpreted a cross-shaped imprint standing immediately above a carbonized wooden cupboard as the earliest tangible evidence of Christianity in southern Italy (1939, 203). It is now accepted by most scholars, however, that the feature observed is characteristic evidence of wooden shelving; in this case a narrow shelf may have been supported by a single, centrally located bracket.

Chests, Boxes, and Cupboards

Chests, cupboards, and small boxes were particularly important in a world without closets or built-in cabinetry. These included the *arca*, *armarium*, *cista*, and *loculus*. All are based upon the basic form of an enclosed or semienclosed box (although the *cista* may be cylindrical in form). Chests were in such demand and built for such a variety of functions that a specialist, the *arcularius*, devoted his craft to their manufacture.

An upright chest placed on legs and provided with a hinged door becomes an *armarium*, the ancient equivalent to the wardrobe, which is still the preferred way of storing clothes in Italy today. *Arma* most commonly refer to weapons, and one is tempted to speculate that the name for the *armarium* is derived from a cabinet used to store the military equipment that distinguished the noble Roman family. *Arma* can also refer simply to "implements," thus the *armarium* is, quite simply, a place to store the stuff of the household. The *armarium* may be of Roman invention, since the Greeks were content to store their clothes in wooden chests (Mols 1993, 492). A variant form of the *armarium* also served to house the household gods (*Lares*) of the family; in this case the furnishing was called a *lararium* or, more generally, a *sacellum*. The *armarium* was a common furnishing in the Roman house; as such, the particular function of any excavated example is best understood by the kind of room it is found in.

Examples of *armaria* are found in Roman wall paintings, relief sculpture, and fragments found during excavations. Our exemplary painted depiction from Herculaneum also represents an *armarium* with paneled folding doors (see fig. 11.9). The carcase, containing three deep shelves and crowned by a molding, rests upon a base that projects beyond the plane of the doors; all is supported by a pair of slab legs which share the S-profile seen on the legs of benches. A fine example of an *armarium* has been reconstructed from traces found in the atrium of the Casa del Fabbro of Pompeii. The appearance of the standing cabinet has been deduced from the metal fittings that survived as well as from the imprint of the doors and moldings on the ash that engulfed it. A replica now stands in the Museo della Civiltá Romana in Rome (fig. 11.13). As reconstructed, the *armarium* is 1.40 m wide (1.50 with crowning molding) and stands 2.25 m tall. The two main doors were opened by bronze rings. Each door was divided into two panels decorated with a diamond-patterned grating (*clatri*) built from wooden strips connected by saddle joints and bronze pins. A lower compartment was closed by a row of four side-hinged doors with sunken panels framed by a simple mitered molding. The *armarium* carved within the previously cited Simpelveld sarcophagus again shows a pair of doors with double panels, and the center rail of each door appears to be raised so as to serve also as a handle mechanism (see fig. 11.3).

Among the pieces of carbonized furniture discovered by Maiuri at Herculaneum in 1958 was a combination cupboard-*aedicula* (shrine) that has been studied by Mols (1999, cat. 29) (fig. 11.14). Most of the bottom of the cabinet has been lost; Mols estimated that in its original form it stood ca. 1.65 m tall by 0.73 m wide. Statuettes of Hercules and Venus were found in the upper compartment; these, in conjunction with the presence of miniature columns, point to the partial use of the cabinet as a household shrine. The lower section of the cabinet, the cupboard, was closed by double-leaved folding doors, hinged with bone cylinders so that the inside leaves swung backward against the outer leaves, which could themselves be opened outward. This is precisely how the multileaved doors from the *armarium* in figure 11.9 are depicted in their opened configuration. The paired upper doors of the shrine folded back on themselves in the same way. The composition of rails, stiles, and sunken *tympana* is identical to the ar-

Fig. 11.13. Armarium from the Casa del Fabbro, Pompeii. First century. (After NSc *1934, 37)*

rangement that has been observed for other Roman doors. The folding panels call to mind the references to full-size doors described as *valvatae.* The base moldings of the cabinet on both levels were formed by layering individual boards whose edges had been shaped with a molding plane; the corners were mitered. This is the very practice employed to create the compound base and crowning moldings of the Black Sea sarcophagi (see fig. 4.1). A veneer of parquetry was applied to both the upper base molding and the architrave supported in part by the wooden columns. This took the form of thin strips and triangles of wood that describe a simple pattern of Xs separated by vertical bands. The design is similar to the form of a balustrade or fence that one might find protecting a balcony or the side of a bridge (the art of parquetry will be considered in more detail at the end of this chapter). While the top of this particular cabinet is crowned only by the uncomplicated horizontal molding that is a feature of the other examples depicted in these pages, a popular variant for cabinets acting as shrines—at least at Herculaneum—was the addition of a miniature triangular pediment, which evoked the facade elevation of a temple's gabled roof.[3]

The heavy chest, or *arca,* of the Roman house was well known for its prominent placement in the main hall (*atrium*), where valuables were safely stored (Plaut. *Aul.* 823) while at the same time in evidence to impress visitors in a quasi-public setting. Re-

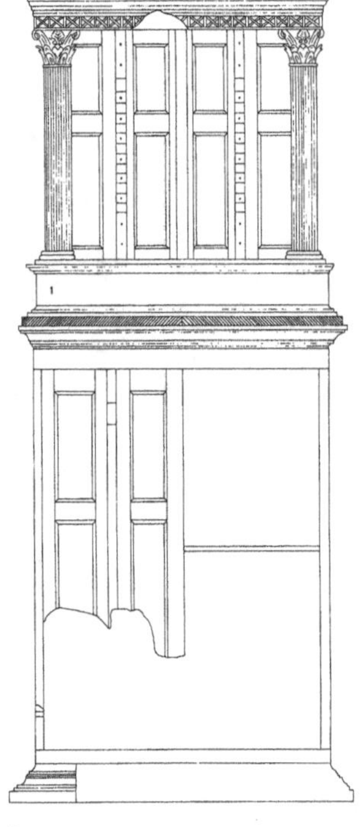

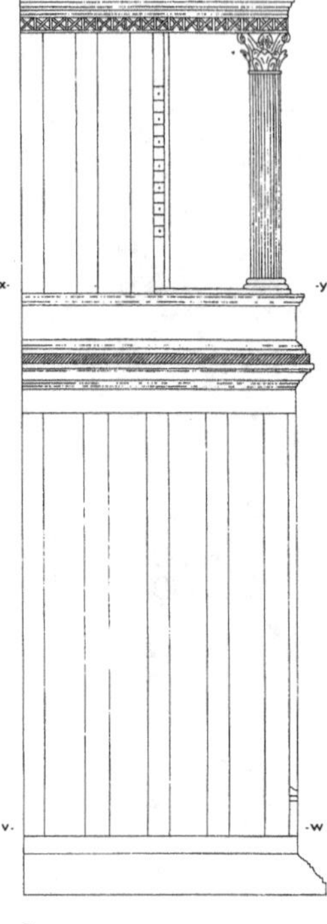

Fig. 11.14. A combination cupboard-aedicula (shrine) from Herculaneum. Casa del Sacello di Legno, Herculaneum. H: ca. 1.65 m. Ca. first century. (Mols 1999, fig. 145, reproduced by permission)

mains of *arcae* can be seen in several houses excavated at Pompeii and Herculaneum. These strongboxes, secured with locks, were built from wooden frames and planks, then sheathed in bronze or iron (*arca aerata* or *arca ferrata*, respectively) or decorated with bronze fittings. The entire unit could be bolted to the floor with an iron rod or even buried in the ground so that its lid was at the level of the floor (Joyce 1866, 411). It is this function that may associate the term *arca* with the Latin for citadel, or *arx*. That the basic construction of the *arca* relied upon the use of boards is clear from a comment by Varro: "Since the wagon was made from boards [*ex tabulis*], like an *arca*, it was called *arcera*" (Varro *Ling.* 5.140).

A smaller, lighter portable version of the ironclad domestic version, perhaps signified by the Latin *arcula*, was probably widely employed. The woodworker himself owned one to carry his valuable tools. The shipwright Publius Longidienus, for example, prominently displays his toolbox, its keyhole clearly depicted, in a funerary relief showing him hard at work (see fig. 3.9). The box serves double duty as a form of footstool, a portrayal that also alludes to its stout construction.

A substantial wooden chest with a hinged lid has survived in its carbonized form from Herculaneum (Mols 1999, cat. 41).[4] The rectangular container measured just under 50 cm high and was about 1 m in length and 63 cm deep. The box rested on two unadorned board runners that were placed at each side from front to back. These supported a baseboard (or boards glued edge to edge). The sides of the chest were also formed from flat boards 2.2 cm thick; it is unclear whether single boards or multiple glued-up planks were used, nor is it evident how these were attached to the base. The sides were joined at the corners by a finger or box joint similar to that illustrated in figure 4.6. The flat lid was slightly broader than the box itself. Wooden battens glued at each side and along the front formed a flange for a tight fit. The rear edge was rabbeted along its length to seat the hinging mechanism. The hinge itself was a Roman version of what is known today as a piano hinge. A series of wooden cylinders (ca. seven survive, none longer than 6 cm) were alternately pinned to the box and the lid with cylindrical pegs. The cylinders fit into one another by virtue of their alternating mating parts: one end was drilled with a small hole, from the other a small pin projected.

The small (10.4 cm long) wooden box (perhaps an example of a *loculus*) recovered from the Comacchio wreck on Italy's Adriatic coast is representative of a broad category of portable containers used to carry personal effects, including writing instruments and cash (Desantis 1990, 266). In this example the wood has been joined with an elegant dovetail joint, and the lid can be opened, closed, or entirely removed simply by sliding it into place from one of the short ends (see fig. 4.7). Grooves were cut on the inside faces of the long sides and into one of the short sides to firmly seat the lid, the edges of which were slightly beveled to create a tight fit. Boxes like this could easily fit into a pocket or the sleeve of a garment. Another well-preserved example has been found at Herculaneum.[5] The small case, measuring 12.6 cm long by 6.5 cm wide and 4 cm high, was carved from a single piece of wood. Like the example from Comacchio, the top of the box slid on grooves. In this case there was a locking mechanism as well. The outside of the box was carved with a floral motif and inlaid with silver.

Another type of small wooden container was based upon the cylindrical Greek pyxis (which was itself developed from earlier Bronze Age ancestors). Roman examples were formed and hollowed on a lathe from a single piece of wood, then provided with a wooden disk for a cover; here again the lathe was useful for the turning of an inside flange (Pugsley 2003, 67). Boxwood was favored; in this regard it is not mere coincidence that the Greek word for boxwood is *pyxos*. Roman forms include both simple cylinders and others with a convex or globular profile. Paola Pugsley provides an excellent description of turning techniques and summarizes extant forms in her study published in 2003.

A lightweight container, cylindrical in form, was favored by Romans for the transportation and storage of fragile papyrus scrolls, which when rolled could be stored in a vertical position. At least some of these vessels, known by the terms *capsae* (sing. *capsa*) and *scrinia* (sing. *scrinium*), were fashioned from thin sheets of wood. Usually one sheet, rolled into a cylinder, would form the main body of the container. The cylindrical (or perhaps oval) form was in part determined by the shape of the wooden base to which the lower end of the sheet was attached. The method of construction is strikingly similar to the famous Shaker-style box still manufactured by New England craftsmen. Traditional Shaker boxes were made from thin sheets (ca. 3 mm) of cherry or maple that were softened in trays of hot water, bent around wooden forms, and clinched with small nails. Eighteen hundred years earlier Pliny wrote that a thin sheet of beech was the only wood suitable for such work (HN 16.229). In fact, just about any species can be used as long as a radially cut, knot-free, and suitably thin piece of wood is properly prepared (with hot water or steam) for bending. There is no ancient description of such a process, but it is evident that Roman craftsmen were adept at bending wood and held opinions about the species best suited for the job (Plin. HN 16.227).

Considering the fragile nature of such boxes, it is surprising that any physical traces have survived. Pugsley cites fragmentary examples from German and French sites; these were found without lids and have been considered examples of dry measures, or *modii* (sing. *modius*) (Puglsey 2003, 97). The German containers were made from sheets of oak and ash. The nearly complete *modius* from Rezé lès Nantes exhibited a vertical seam covered by a wooden slat through which nails had been driven and clinched on the interior. None of the observable Roman-period seams exhibits the fingers characteristic of the New England Shaker tradition.

Beds and Couches

Roman beds and couches were closely related in terms of assembly and appearance, although, as we shall see, they were probably not interchangeable furnishings. The Latin *lectus* can refer to either type and so must be modified with an adjective to indicate specific use if application is not clear from the broader context of the ancient passage. The *lectus cubicularis*, for example, was for sleeping, the *lectus tricliniaris* for dining.

The fact that the excavations of Herculaneum have recovered nearly intact examples of *lecti cubiculares* but not any movable *lecti tricliniares* (only component and decorative parts, or fixed examples built from masonry) suggests that the latter, when made of wood, were of more fragile build, complex in design, and therefore more costly to make (Mols 1993, 490). This is not surprising considering the importance a Roman host placed upon impressing dinner guests with fine furnishings and delicacies.

The making of the Roman bed required a certain discrete set of skills and materials that resulted in the craft becoming a subspecialty within the field of woodworking. The *faber lectarius* presumably could build the framework and undertake the embellishment of both *lecti cubiculares* and *lecti tricliniares.* He would have owned and understood how to operate a lathe and been proficient at carving, inlay, and veneer work. Many of these same skills and joinery techniques were employed in the making of tables and benches; it is of course possible that the *faber lectarius* was fully capable of producing all such furnishings.

Nevertheless, despite its similarities in basic form to the long bench and the rectangular table, the bed frame and its supports differed in several fundamental ways. The appearance of the most common Roman bed used for sleeping can be described as a three-sided, open, rectangular box. One long side and two short sides of this box were defined by a combination of vertical or outwardly sloped planks. The fourth (long) side was open; this was the front of the bed and the only practical means of access. The base (*sponda*) of this open box held a mattress and was itself supported by at least four legs.

The simplest way to make the base of the bed would have been to edge-join several long planks to form a smooth, solid, rectangular surface. But even though details of the joinery vary, the discovery of actual Roman beds indicates that their *spondae* were always built as open frameworks of wooden (or, rarely, bronze) slats. Presumably this allowed the mattresses to "breathe" and also afforded a slight degree of flexibility for the comfort of the occupant. Fewer board feet of lumber were required to make a framed base than a solid one, but the practice also called for greater skill.

The base frame had to be strong enough not to sag (let alone break) under a person's weight or otherwise deform over time. Three sides would normally support the side walls of the bed; this join had to be accomplished without interior braces that would intrude upon the sleeper's comfort. No single solution to these challenges was universally adopted. The simplest method was to build an open, rectangular frame from two long rails and two stretchers to which a latticework of leather strips or woven hemp could be attached to support the mattress. Another solution involved inserting a latticework of thinner wooden slats joined with half-laps within the outer rectangular frame. A third solution focused attention on the points of greatest stress: the middle two-thirds of the bed. Here two additional stretchers were placed underneath the rails, and these were linked by a ladderlike grid of lighter battens (fig. 11.15). This treatment created a firm, strong platform, especially if reinforced with a pair of subsidiary legs placed at the center of the frame.

The legs themselves were an important decorative feature of most portable beds. Once again observable practice defies the simplest solution: since the Roman bed employed side walls, it would have been an easy matter to construct slablike legs and side walls from a single unit of joined planks. Moreover, this kind of joinery was practiced by builders of chairs and benches; there was nothing to stop the *lectarius* from adopting a similar tactic. The legs of Roman beds and dining couches, however, were distinct elements, invariably freestanding, vertical, and, unlike table legs, turned instead of carved freehand. This can be asserted not only from the limited evidence of finds from sites like Pompeii and Herculaneum, but also from the scores of examples of beds and couches represented on Roman funerary monuments: decorative inscribed markers, sarcophagi, and tomb covers (see figs. 3.28, 10.1, 11.3).

In some instances the turned legs of Roman beds exhibit such curvaceous profiles that it seems impossible they were capable of supporting their own weight. Here, clearly the woodworker has employed a kind of sleight of hand, for the secret of expecting such tenuous spindles to bear a heavy load was the insertion of an iron rod through the vertical axis of the turned leg. The difficulty of drilling a long and true hole through the fragile leg was avoided by turning and boring each element of the leg separately and

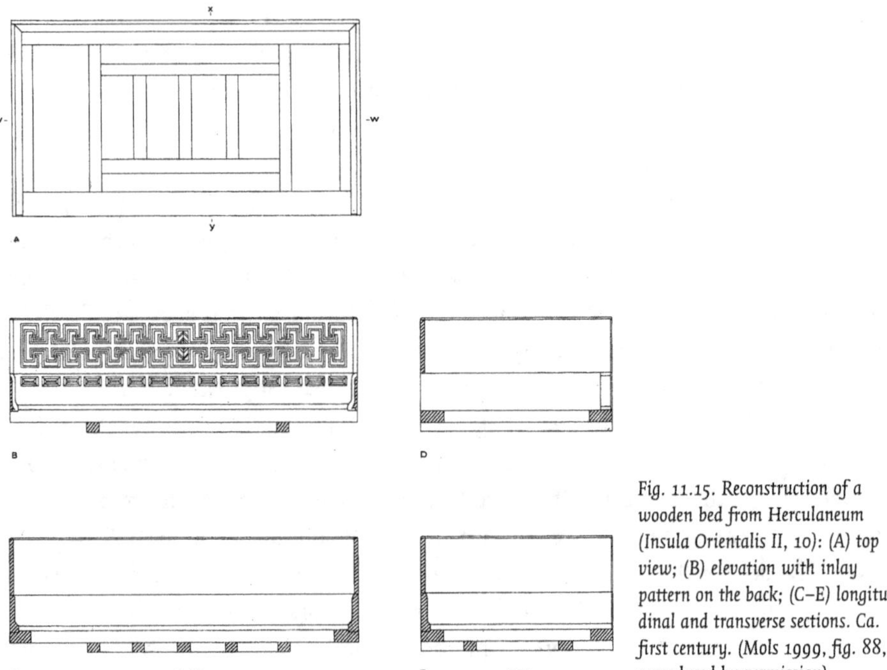

Fig. 11.15. Reconstruction of a wooden bed from Herculaneum (Insula Orientalis II, 10): (A) top view; (B) elevation with inlay pattern on the back; (C–E) longitudinal and transverse sections. Ca. first century. (Mols 1999, fig. 88, reproduced by permission)

then assembling the components like so many beads on a string. Physical evidence for this practice has existed since the nineteenth century (although not always recognized as such). In the summer of 1866, for example, two iron rods, the longer measuring just under 80 cm, were found in a tomb in Colchester, England, known from early publications as the Child's Grave. The rods, square in section and only a centimeter thick, have been interpreted as serving as the structural supports for the legs of a funerary bed (*lectus funebris*) (Eckardt 1999, 78). If this interpretation is correct, then it is likely the rod continued through the *sponda* into the side board of the bed, providing strength at a most vulnerable point.

Indeed, the most challenging join for the *lectarius* was the attachment of the side walls, or *plutei*, of the bed to the base. Greek-style beds employed a support (*fulcrum*) only on the ends of the bed; each could be treated as an extension of the piece of wood from which the leg was fashioned and thus retain strength. In most depictions of Roman three-sided beds the legs are placed at the extreme ends of the *spondae*, directly under the side walls. Iron rods like those found at Colchester could have added significant strength to the side pieces. Leaving aesthetic and practical considerations aside (the latter having to do with protecting the occupant from chilly drafts), the use of three *plutei* offered a good deal of strength to the assemblage, for the vertical joins at the corners served as braces for each of the constituent sides; these had to be strong enough to support the weight of the semireclined occupant.

The short side walls of the bed presented one exposed vertical edge. This could be rendered, most simply, as a straight vertical, as a slightly inclined plane (at Herculaneum up to 135 degrees), or as a serpentine profile (something like that of a traditional sleigh bed; the *pluteus* represented on the tomb cover from Isola Sacra is a good example (see fig. 3.28). A telling feature is the pronounced flare that is observable at the base of the shorter side walls. On the relief from Isola Sacra the broad base of each lateral *pluteus* is wide enough to accommodate at least two anchoring pins; there appears to be an intermediate plank between the *sponda* and the curved side piece to further strengthen the join. On the Simpelveld sarcophagus the side *plutei* of the bed depicted are reinforced by L-shaped side brackets that have been carved to resemble a fish or dolphin. On the Turin relief (see fig. 10.1) and the actual bed from Herculaneum depicted in figure 11.15, the sides of the bed are nearly vertical, and there is no flaring bottom element. Mols's reconstruction indicates that the bottom edge of the *pluteus* was fitted into a rabbet cut into the outer edge of the end rails; this join could be further reinforced with a molding to increase the gluing surface.

The construction of the side boards, or *plutei* themselves, may have also involved quite intricate joinery. The simplest solution was the edge-joining of squared and planed planks. From ancient depictions and actual examples, however, it is evident that the beds of highest quality included *plutei* with inserted panels (here the techniques would have been familiar to those who fashioned cabinet sides or doors, for example, fig. 11.3) or *plutei* with inlays of ivory, bronze, or wood veneers. An especially good example is depicted in the combination cabinet-*lararium* from Herculaneum discussed above, which employs a geometric pattern of wood veneers (see fig. 11.14).

Veneer and Parquetry

Pliny wrote, "The middle part of trees has a more wavy grain, and the nearer the root the patterns are smaller and more curled. This is the origin of trees as luxury goods, overlaying one with another and making 'bark' for a cheaper wood out of a more expensive one. In order that one tree might be sold multiple times, even thin layers of wood have been conceived" (*HN* 16.232).[6] By Pliny's day woods valued for their interesting grain patterns were commonly cut into thin sheets (*laminae* or *bratteae*) of veneer that were then glued to a base of supposedly inferior stock. Pliny mentions citrus, turpentine-tree, some maples, box, palm, holly, ilex (holm oak), the root of the elder, and poplar as the principal (*praecipua*) species used for veneering (*HN* 16.231). The most interesting patterns were acquired not so much from the main trunk of the tree as from the roots and the large burls that grew on the trunks of certain species, specifically, as reported by Pliny (*HN* 16.232), alder, citrus, and maple. These woods, of which citrus was the most highly regarded, came into such demand that other materials such as horn and tortoiseshell were dyed and cut into wafer-thin slices as faux wooden veneers, a practice looked upon with a jaundiced eye by Pliny (*HN* 16.233).

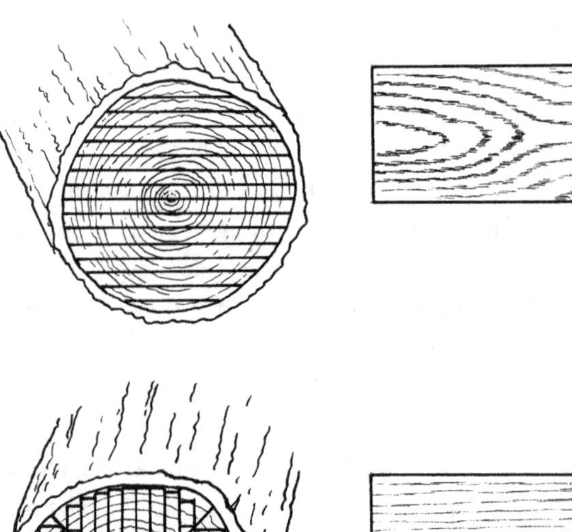

Fig. 11.16. Methods of sawing a trunk into boards: (A) *flat cut* or *plainsawn*; (B) *quartersawn*, so-named since the trunk must first be quartered before the boards can be cut. (right) Grain patterns produced by each sawing method. (Author)

Veneering is a highly economical way of using valuable and exotic species of wood. Great skill was needed to cut thin sheets of uniform dimensions from thicker planks. A thin blade held under tension in a frame saw would have been most suitable. The saw marks were removed from the exposed side of each sheet by gentle scraping of the surface with an iron blade held at a high angle (some modern cabinetmakers still use such scrapers). Sheets of veneer could be cut down to the required dimensions with a small, fine-toothed handsaw or a razor-sharp knife. Of prime importance was the pattern of grain, which is determined not only by the species of wood but also by the contrasting colors of the heartwood against the sapwood and by the way in which the board was initially cut from the round trunk of the tree. Flat cut or plainsawn boards yield a surface that is characterized by a wavy and elliptical grain pattern, perhaps the fennel pattern that Pliny cites (HN 16.226). Veneer from a quartersawn board will produce a pattern of parallel lines on the surface that correspond to the tree's annual growth rings (fig. 11.16). Minimal shrinkage during drying is a main advantage of the latter.

The art of veneering as it is understood today falls into three general categories. Basic veneering involves the art of covering broad surfaces of a piece of furniture (for example, a tabletop or drawer front) with a thin sheet of facing material. Marquetry is a form of veneer in which a variety of wood species cut into smaller pieces are used to create a picture on a piece of furniture. Parquetry is similar to marquetry, except that the image created from the veneering is an abstract, geometric pattern. Although ancient sources do not differentiate between these related practices, it is probable all three were

Fig. 11.17. Veneer work on a stool from Herculaneum. (top) *The stool;* (bottom) *the pattern of parquetry on the seat of the stool. The hatched areas indicate the direction of the grain. Ca. first century. (Author)*

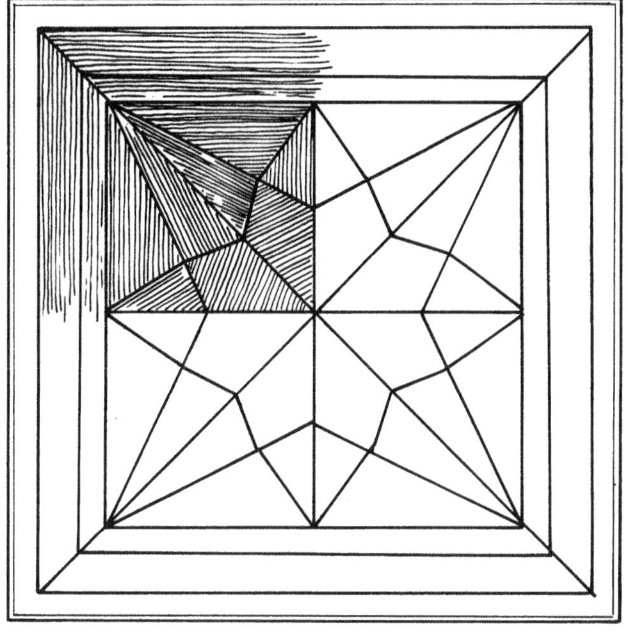

20

widely used. Figural images created from thin pieces of marble, for example, are well known from Roman contexts; this kind of *opus sectile* work may well have been inspired by the art of wooden marquetry. For clear examples of parquetry, on the other hand, we have direct evidence from the carbonized wooden furniture from Herculaneum.

The simple horizontal band of Xs and vertical strips applied to the surface of the cabinet-*aedicula* from Herculaneum has already been cited as an example of parquetry. The *lectus* illustrated in figure 11.15 is another outstanding example. Here both the rear and side backing boards (*plutei*) of the bed were decorated by a series of interlocking swastikas forming what is commonly known as a meander or Greek key pattern. Beneath this meander extended a row of rectangles, each of which was formed from nine independently shaped pieces. In all, the craftsman ended up covering over ten thousand square centimeters of the bed with parquetry. A somewhat less ambitious example, but nonetheless elegant in its execution, is the parquetry work on the seat of a small stool (perhaps a *scabellum*) discovered at Herculaneum in 1936 by Maiuri (fig. 11.17).

The nearly square seat of the stool (46.0×45.5 cm, ca. 1.5 Roman feet square) is supported by four straight legs placed at the corners and connected to one another by upper and lower wooden rails. The decorated surface, realized with thin, perfectly fitted sheets of wood cut into geometric shapes, covers the top of the seat. Maiuri dated the stool to the first century A.D.

While pieces of the parquetry have been dislodged, the original layout of the seat's decoration is clearly visible, as is the grain pattern of the individual elements still in place. The craftsman began by scoring the underlying area from corner to corner and from midpoint to midpoint; the pattern at this stage was composed of eight triangles. These were further subdivided to create two superimposed eight-pointed stars. Veneering sheets of a close-grained quartersawn wood were cut to fit the pattern (the process and the result resemble that of a patchwork quilt). Of particular interest is the careful attention given to the direction of the grain of the wood of each panel; it is ultimately the fine parallel lines of the grain that create the starlike pattern. Additional strips of wood were applied to the square borders; in one place an arrow-shaped splice is visible.

The interest in the decorative properties of the veneer's grain reminds us of the high premium Romans paid for furniture woods distinguished by a beautiful grain pattern. By gluing geometric shapes with contrasting grains, as is done in the examples from Herculaneum, the Roman craftsman could create artificial patterns which evoked the abstract designs supplied by nature.

XII Classification of Trees and Species of Timber

Cato, Vitruvius, and Pliny devote lengthy comment to the suitability of individual species for certain applications and the best times for harvesting wood. They also offer their opinions on how best to protect wood intended for construction from moisture and decay. Much of this knowledge was certainly passed from woodworker to woodworker, from father to son, or through apprenticeship, just as it is today. Indeed, it is surprising that so much practical information was committed to writing and subsequently preserved.

Roman craftsmen were attentive to selecting the most suitable timber for a given task, but when necessity prevailed they could compromise. Woodworkers were professionals. Few if any amateurs practiced the craft; the do-it-yourself mentality popular with today's middle class and the products that cater to this group did not exist in Roman times. Experience would have taught the Roman woodworker not to waste time and effort on unsuitable timber. Imported woods were usually destined for luxury applications. The trouble and expense of importing wood for common purposes were unnecessary since the uncultivated trees of Italian forests provided abundant and varied timber suitable for building projects large and small as well as for the specialized requirements of furniture making, carts, and all manner of common household objects.

Since the most important resource derived from most wild trees is timber, it should come as no surprise that Romans tended to categorize or classify trees by the relative properties of their wood. Our knowledge about this system or, more accurately, related systems of classification can be developed from the three Roman sources mentioned above and their Hellenistic inspiration, Theophrastus. We have seen that Theophrastus's treatise had tremendous impact on Roman-period writers, to the point where Pliny at times offers little more than a Latin translation of the philosopher's work.

According to Theophrastus, a carpenter considering the suitability of a species for a particular job would look at its density, its hardness, and its weight (*Hist. Pl.* 5, passim). Each of these qualities—and their opposites—was carefully defined by the philosopher with a specific term: density (*pyknotês*) vs. porosity (*manotês*), heaviness

Table 6. Classification of woods by Theophrastus, book 5 (T), followed closely by Pliny, book 16 (P), from dense and heavy (top, left) to porous and light (bottom, right)

ebony, boxwood (T, P)	heartwood of oak (T,P)	holm oak (P)	hornbeam (P)
heartwood of the cork tree (P)	laburnum (T)	olive, wild and cultivated (T, P)	poplar (P)
larch (P)	terebinth (T, P)	maple (T)	fir, elder, fig, apple (T)
lotus (or "nettle tree," "hackberry") (T, P)	dogwood (cornel) (P)	chestnut (P)	linden (lime) (T, P)

Note: Latin, English, and standard botanical names for most common species can be found in the alphabetical listings included later in this chapter.

(*barytês*) vs. lightness (*kouphotês*), hardness (*sklêrotês*) vs. softness (*malakotês*). Woods with greater density, that is, with a close grain, tend to be heavier and often harder than more porous types. Both Theophrastus and Pliny place ebony and boxwood at the top of their list of the densest and heaviest of woods and the linden (lime) tree (*tilia*) at the very bottom. It is difficult to arrange precisely either writer's list of woods in order because both tend to wander in their discussions, digressing on related topics such as color or tendency to polish well. Despite these drawbacks, it is possible to create a list of woods based upon their properties of grain density and weight (table 6).

Neither Theophrastus nor Pliny offers us anything close to a complete list of woods ranging from dense and heavy to porous and light. With the exception of the linden, included to provide stark contrast, all the woods listed are close-grained and relatively heavy. Relative strength is not an issue: fir, for example, is very strong, even though it is relatively light in weight. Similarly, woods that were considered the hardest (oak and dogwood) are not featured at the top of the list since their density, or closeness of grain, is not as pronounced as that of ebony or box. Finally, the woods listed in table 6 are arranged in order of mention, but in their brief discussion both Theophrastus and Pliny list several species in a single sentence; it is unlikely that their specific place within the discussion was intended to present a strict linear assessment of relative density.

There were other ways trees and their wood were classified. Theophrastus divided trees broadly between the wild and the cultivated (*Hist. Pl.* 5.3.1). The former, he theorized, generally produced wood that was more dense, harder, and heavier. Trees were often gendered (5.4.1). The capacity to bear fruit was yet another distinction (5.4.1). These kinds of classifications do not concern us here, for they were of less import to the woodworker choosing a particular species for a particular application.

A word should be said, however, about the elemental properties attributed to different species. More specifically, this concerns the relative amounts of the four elemental and constituent properties—earth, air, fire, and water—that were believed by some to determine the very qualities of hardness, density, or porosity just described. As a source, Vitruvius is most informative (2.9). His understanding of the elemental composition of wood allows him to explain with confidence the reasons that oak, for example, is dense and sustains prolonged exposure to moisture.

Vitruvius works backward, first noting the recognized qualities of a given wood and then hypothesizing about its elemental makeup. Consider the larch, a tree admired by Vitruvius for its strength and resistance to fire (2.9.14). Vitruvius supposed that the wood of this tree had an abundance of the elements water and earth and only little of fire and air. Lack of air was manifested by fewer pores (*foramina*) in the wood, reducing the ability of fire to penetrate. A direct opposite was the wood of the fir, which had abundant fire and air but scarce moisture and earth. In this case the dominant elements explained the fir's lightness (note the relative placement of larch and fir on the table above) but also explained its porosity and thus inflammability.

So goes the reasoning of Vitruvius. Woods that have an overabundance of the element water (for example, elm and ash) are pliant and therefore good for bending. Oak (*robur*) is characterized by earth; its lack of air results in few open pores, which, in this case, do not let in moisture and therefore explain the usefulness of oak in wet conditions—such as for sleeper beams in foundations. From such examples we might suppose that woods known to be without pores and therefore resistant to water penetration were resistant to rot from contact with the ground.

The logic of Vitruvius, however, may confuse the modern student. The alder, for example, has an abundance of air and fire, but only a little of water and earth. This is in fact a compositional state close to what is accorded the fir tree. We would expect that, having so much air, the alder would be too porous for wet applications. According to Vitruvius, however, it is because alderwood admits fluids that it remains "imperishable" when used in wet conditions, such as foundations. Remember, it is the supposed lack of pores in the oak that produces the same resistance to decay.

Species of Trees Used by Roman Woodworkers

Hunting and forestry are in my blood,
But I've been wondering what kind of wood
That shaft you hold is made of. If of ash,
It would be brown, or if of cornel-wood,
It surely would be knotted. Whence it comes
I cannot tell, but never have my eyes
Beheld a javelin more beautiful.

—Spoken by Phocus, in observing the javelin of Cephalus in Ovid's *Metamorphoses*, book 7, 675–80, translated by A. D. Melville (© Oxford 1986).

Our knowledge of the kinds of trees harvested by ancient Romans for woodworking is based upon three complementary sources of evidence that have informed so much of this book: literary, archaeological, and parallels observable in modern Mediterranean silviculture. In his work *About Trees* (*De Arboribus* 1.1), Columella reflected what must have been general wisdom among those who worked with wood when he said that wild trees were best suited for woodworking: trees which do not "come up by the help of man [are] better suited [*aptum*] for timber."[1] Cultivated trees, on the other hand, were generally used for food. This distinction between the wild and the cultivated was not without exceptions. Olive wood, for example, was valued for its hardness and therefore was employed for many small household items, as it is still today in rural Italy. Such wood presumably came from nonproductive trees in cultivated groves. Because the Romans preferred long, straight, single pieces of wood for structural applications such as cross-beams, the height and girth of certain species, such as the larch, frequently elicited special comment (see, for example, Plin. HN 16.200).

Following are descriptions of the most important species of trees used by Roman woodworkers (arranged alphabetically by their Latin names).

ABIES, FIR

The common European fir (*Abies alba*, or silver fir) was highly prized by Roman (and Greek) carpenters and shipwrights. Mature firs reach up to 45 m (147 ft.); these are the tallest and straightest of the native Italian species. Foliage is concentrated at the top of the tree, so that the trunk is relatively knot-free. European firs are found in the mountains, thriving at altitudes ranging between 800 and 1,800 m (2,600–5,900 ft.) (fig. 12.1). Different varieties of the firs have been identified as characteristic of geographic location. In central Greece the dominant variety is *Abies cephalonica*; in the mountains of Lebanon *Abies cilicica*, in the region of the Black Sea, *Abies nordmanniana* (Meiggs 1982, 43).

In Roman Italy firs would have been found along the length of the peninsula growing in the Apennine range; good stands existed as far south as Sicily, growing on the slopes of Mount Etna. *Abies alba* often grows alongside the European spruce (*Picea abies*),[2] although occasionally pure stands of silver fir are found: in the Apennines extant forests of mature trees can still be seen at Camáldoli and Vallombrosa; Latin literature does not differentiate between the two types. During the period of the Roman republic, fir logs, along with mountain (Corsican) pines (*Pinus nigra 'larico'*), could have been easily transported from the mountains of Etruria and Umbria by the Tiber River and its tributaries. Vitruvius (2.9.10) believed that the best firs came from the western slopes of the Apennines; here the climate was drier, and therefore the wood was harder.

Theophrastus held fir in high regard, considering its timbers most versatile, of highest quality, and of greatest size (*Hist. Pl.* 5.1.5). He described in some detail the careful scrutiny carpenters gave to each fir log; fissures around the heartwood prompted Theophrastus to categorize the trunks of fir trees as "four-cleft," "two-cleft," and "one-cleft"; the first of these was considered the best for carpentry, the second an inferior grade that tended to warp (5.1.9).

Fir wood was an important structural element in shipbuilding and architecture and was also used for the manufacture of many household objects. Livy's description of the preparations made for war against Carthage in 205 B.C. reveals the essential role

Fig. 12.1. The mountains of Italy and central Europe can be seen in terms of the distribution of the silver fir, a high-altitude species that was prized by Roman carpenters. A related species, Abies cephalonica, *grew in the mountains of mainland Greece (indicated in gray).* (Author)

fir played in the construction of the Roman fleet (28.45.18). Theophrastus considered fir excellent for the frames of roofs, believing it would resist warping when placed at a slant (5.6.1); the use of this wood for rafters immediately comes to mind. Great logs of fir would have been shaped by the Romans to bridge the largest spans required by the architects of the basilicas, temples, and imperial audience halls in Rome. Only larch (*Larix decidua*) from the alps produced longer beams.

On the rare occasions when fragments of wood are recovered by archaeologists, fir is often represented in samples. Silver fir was found in 60 percent of nearly two hundred samples of wood analyzed from the sites of Pompeii and Herculaneum (Fioravanti and Caramiello 1999, 85). While no Latin sources mention fir as a furniture wood, S. Mols (1999, 82) found that it was the dominant material used in the wooden furniture found at Herculaneum and suggests it was locally available from the higher slopes of Mount Vesuvius. Fir wood is not good for turning on a lathe or for carving; this may explain the reticence of the Latin sources, since the most distinctive features of Roman furniture, other than grain pattern, were produced by the work of the lathe or the carving knife. Wine barrels and buckets excavated at Silchester in England have been found to be made of fir staves (Boon 1974, 86, 263), and a spool used for weaving recovered from a disused drain under the Colosseum was also of fir (Follieri 1988, 346).

In the modern European timber industry, fir is used for paper products, general carpentry, and various other purposes, including shingles for roofing. Such multiple applications are reflected in the ancient literary record as well.

ACER, MAPLE

According to Pliny (HN 16.66), the maple was valued as a furniture wood. For this application the most valuable was "peacock maple" (*acer pavonum*), now described in North America as bird's-eye maple, a variation—not a different species—exhibiting an intricate wavy pattern in the grain. *Acer pavonum* enjoyed a status similar to that of precious *citrus*. Large burls on the trunks of some maple trees could also be fashioned into smaller implements, including cups and bowls. Theophrastus mentions the use of (ordinary) maple for yokes and bedsteads.

The most common form of Italian and European maple (including Britain) is *Acer campestre* (hedge maple, field maple). It can grow to a good size (15–20 m), although its trunk tends to be short and therefore not useful for longer timbers. The hard wood of this tree was probably favored by Romans for tool handles. Axe shafts made by Vikings in the tenth century were of *Acer campestre* (Malmros 1988, 375); the same species is still used today in Italy to fashion tool handles.

Acer pseudoplatanus (sycamore maple) is widely distributed in Europe (yet is not found in Britain), particularly in high, cool locations. It is often found in conjunction with silver fir and would certainly have been exploited by Roman foresters working in the Apennines. The tallest of the European maples, the sycamore maple in exceptional examples can reach heights of 40 m. Modern applications reveal the versatility of the wood produced by this tree: furniture, flooring, tool handles, and even musical instruments.

ALNUS, ALDER

The common, or black, alder (*Alnus glutinosa*) grows throughout peninsular Italy on both level and hilly terrain at altitudes up to 1,200 m, or 3,900 ft.; it is often found along riverbanks or swampy land. Mature alders reach a height of about 20 m (65 ft.) and have a straight but often divided trunk.

The alder was considered excellent for piles and for fence posts. The durability of its wood in wet or boggy conditions was described in enthusiastic terms by Vitruvius (2.9.10) and Pliny (HN 16.219); the timber is still valued for this characteristic. Alder was thus recommended by ancient writers for such diverse uses as water pipes and foundations. It is interesting that alder planking was installed at ground level for flooring at the site of Vindolanda, a notoriously soggy site. At this same Roman outpost excavations revealed the still-functioning alder pipes described in chapter 5.

Alder is relatively easy to work and does not readily split. In postclassical times alder was favored for wooden shoes in England. At the site of Roman Carlisle (Luguvalium), four soles of shoes made from alder have been excavated (Pugsley 2003, 49). The *tuber* of the alder, when cut into thin slices, exhibited an interesting grain that was valued for veneers (Plin. HN 16.231); the Latin *tuber* refers to the protuberance, or burl, that forms on the trunks of some species of trees. Roman woodworkers also used the *tuberes* of the citrus and the maple for veneer work.[3]

BETULLA, BIRCH

Birch trees are common throughout mainland Europe (except Spain and Portugal); trees generally reach about 20 m in height, although in exceptional cases may grow as high as 30 m. In Italy stands of *Betula pendula* (European white birch), the most common variety, can be found as far south as Sicily. The toughness and elasticity of the wood are still prized for specific applications. Today birch is used for rudders, wooden wheels, parts of barrels, wooden toys, plates, baseboards, and fine laminates. Well-seasoned birch, along with beech, does not taint food with odor or taste; in modern Sweden birch is still the favorite for wooden tableware. One can imagine a similar variety of applications for ancient objects. In fact, Pliny (*HN* 16.74–75) recognized the usefulness of supple birch branches for the manufacture of various hoops and basket frames. Curiously, we also hear from Pliny that birch was chosen for the rods carried by Roman magistrates (16.74–75). Excavations at Vindolanda have revealed that birch branches were an important component of wattling panels in the northern provinces (Birley 1977, 112).

BUXUS, BOX

In present-day Europe the evergreen box tree (*Buxus sempervirens*) is widely used for ornamental garden borders. Stands of wild box can still be found in parts of Spain, France, and central Italy. These small trees (in their natural state growing 5 to 8 m tall) were harvested for their close-grained, yellowish, hard, and dense wood, so dense that, according to Pliny (*HN* 16.204), it will not float. Box grows very slowly. Twenty years of growth may add only 5 cm to the diameter of the tree. A cargo of boxwood logs on the wrecked Roman ship discovered at Comacchio were on average 276 years old and only 17 cm in diameter (Berti 1990, 53; Pugsley 2003, 68). Boxwood was used for a variety of small implements and furniture inlays. In postclassical times the hard wood was favored by printers for high-quality woodcut engravings.

Box is mentioned frequently in Latin sources, recognized for its resistance to splitting and decay (Plin. *HN* 16.212), its value as a veneer (*HN* 16.231), its suitability for combs, furniture, writing tablets, auger handles, mallets, and even for less visible uses such as providing the ties to support a false vault of lathe and plaster (Vitr. 7.3.1). Boxwood can be turned on a lathe with excellent results. Most of the turned elements from the group of wooden Kertch (Black Sea) sarcophagi have been identified as boxwood (Vaulina and Wasowicz 1974, 23). Pyxides are discussed above (p. 231).

Ovid (*Met.* 4.30) happens to praise box for making flutes; the wood is still used in the manufacture of musical instruments as well as for buttons, handles, chess figures, and similar small items. Flutes of boxwood, along with a writing tablet, were recovered in 1982–83 from the Giglio shipwreck (ca. 600 B.C.) off the coast of Tuscany (Bound 1991, 28). Indeed, the oldest writing tablets found in the Mediterranean, from the Bronze Age shipwreck at Uluburan off the coast of Turkey (fourteenth century B.C.), were of boxwood (Pulak 1993, 7).[4]

The grain of box is so tough that a fine saw can cut the teeth of a comb from a single blank of wood without any breakage. The resistance of the wood to splintering has the added benefit of not catching individual hairs when in use. Dating from the Neolithic period, the oldest combs discovered so far in Europe (Charavines, France) are of box. The teeth were cut by repeated scoring of the wood with a sharp piece of flint (Pugsley 2003, 14). By the historical Roman period, *buxus* alone could mean "comb." Boxwood combs have also been found at the British sites of Chew Valley Lake, St. Magnus House, London, Portchester, and Vindolanda (Cunliffe, Henig, MacGregor 1996, 98). Indeed, so many such combs have been found in Britain alone that Paola Pugsley (2003) has proposed a typology based upon variants of the basic H-shape of the artifacts. Another fine comb with an inscription recovered from the site of Kertch is now in the collection of the Hermitage Museum (Vaulina and Wasowicz 1974, no. 81, pl. 130). The standard appearance of the boxwood comb is prominently depicted on the funerary relief of P. Ferrarius Hermes in Florence (see fig. 3.4). Boxwood was used for the legs of a banquet table found in the royal tomb of Gordion (eighth century B.C.) as well as for the faces and side pieces of two small serving tables (Simpson 1992, 14).

CASTANEA, CHESTNUT

The European chestnut (*Castanea sativa*), of the same botanical family as the beech (the *Fagaceae*), was a common hardwood in the forested hills of ancient Italy, the rest of southern Europe, including all of Greece, and vast zones in western and northern Asia Minor. Chestnut trees can thrive at relatively high altitudes: up to 900 m in most of southern Europe and even higher (1,400 m) in the mountains of Sicily. Today many of the wild trees have been lost, as is the case of the species native to North America, because of a fungal blight. While rarely as tall as the beech, a mature chestnut can nevertheless reach a height in excess of 20 m. The timber, included in Pliny's list of durable woods (although Pliny confuses chestnut and walnut), is of medium quality, elastic, and used for everything from structural beams to barrel staves, limited only by its tendency to split and fracture. Trees were commonly coppiced for the production of posts. In one of the rare stories preserved for us concerning a building's structural collapse (the tragedy of Fidenae has already been mentioned—see chapter 6), we hear of chestnut beams cracking and failing in the roofing of a bathhouse built at the Greek site of Antandros. Pliny (HN 16.223) relates the story as originally recorded by Theophrastus. In this case the groaning of the timbers before they collapsed gave the bathers enough time to escape safely.

CEDRUS, CEDAR

The cedar is still distinguished by its two great properties: the pleasant fragrance of the wood and its resistance to decay. Pliny also praised cedar for its resistance to splitting and cracking (HN 16.212). Given these properties, it is little wonder that cedarwood was used extensively for Egyptian coffins, even though the cedar was not native to Egypt (Meiggs 1982, 292). Five thousand years ago extensive forests of cedar (*Cedrus libani*) covered the hills of Lebanon. Logs culled from these stands were shipped along the eastern Mediterranean coast as early as the Bronze Age (Pulak 1994, 15). Flat panels of cedar were still the medium of choice for the painters of Roman-period Egyptian funerary masks. Planks of Lebanese cedar were shaped and sewn together to create the magnificent funerary boat discovered in 1954 in a forty-five-hundred-year-old pit next to the Great Pyramid of pharaoh Cheops. Bronze Age merchant ships also employed hulls of cedar.

Since the cedar is not native to Italy—despite its popularity in Italian parks and gardens today—the lack of literary mention of the cedar as an important structural wood in Italian structural contexts is not surprising. From Pliny, via Theophrastus (*Hist. Pl.* 5.42), it is clear that beams of cedar were integral to some of the most celebrated temples of the eastern Mediterranean. To illustrate the properties of durability, for example, Pliny relates that the beams (*trabes*) of the Temple of Apollo at Utica, constructed of Numidian cedar, lasted for 1,178 years (*HN* 16.216). Both Pliny (*HN* 16.213) and Vitruvius (2.9.13) cite the use of cedar for the ceiling beams and coffers of the Temple of Artemis at Ephesus, one of the most admired structures of the ancient Mediterranean. Josephus praised the ceilings of cedar used in Solomon's temple in Jerusalem (*BJ* 5.191). Pliny also mentions the oil derived from the tree (*HN* 15.28).

CITRUS, CITRUS, SANDERAC, THUJA

Citrus was the Latin name for a tree native to Mauretania in northwest Africa; the region now includes northern Morocco and the western parts of Algeria. Modern identification of the prized wood has proven elusive. Confusion about the modern identity of the tree stems from the number of names under which it is known and how many such variants appear in modern translations. The Latin *citrus* is in all likelihood identical to the Greek *thuon* or "thyine" wood. The Greek name suggests the fragrance of the wood, small pieces of which may have been burned during sacrifice. It is mentioned in the New Testament (Revelation 18.12) as *thuinos*. Citrus belonged to the botanical family of the *Cupressaceae*. The variant botanical names *Thuja articulata*, *Callitris quadrivalvis*, and *Tetraclinis articulata* all seem to refer to the same tree, which is also known in English as the sandarac, the thuya tree, the red cedar of Morocco, and gum juniper.

The intricate grain of citrus so highly prized was presumably obtained from burls or similar deformation of the trunk. Pliny (*HN* 13.94) says that the most valuable burls were taken from large growths (*tuberes*) on the roots of the tree. With its fine grain and fragrance, citrus was in high demand for the circular tabletops of the popular *mensae delphicae* discussed in chapter 11. Before the first century B.C. the citrus forests of coastal north Africa spread from Libya westward, but aggressive cutting decimated the forests (see Lucan 9.428). By the latter half of the first century B.C. the last trees were found in the area of Morocco (Meiggs 1980, 186). Pliny writes that in his day the most celebrated forests, on Mount Ancorarius in Mauretania Citerior, were exhausted (*HN* 13.95). This would mean that mature trees suitable for furniture making had all been cut. In fact, the tree still grows in Morocco and southern Spain.

From Pliny's descriptions of citrus wood it appears the qualities of the grain were similar to those of good bird's-eye maple. Pliny compares the patterns to the eyes of a peacock (presumably the pattern on the bird's tail feathers), curving waves, leaves of parsley, and clusters of grain (HN 13.96–97). One can still purchase burls of citrus (often marketed as thuja) imported from Morocco. These are fashioned by modern craftsmen into bowls, pens, trays, and similar small objects. Citrus was the most precious wood used on the wooden Kertch sarcophagi of the Roman period (Vaulina and Wasowicz 1974, 23).

Working with citrus required the special talents of the *citrarii*. Citrus wood for tables produced "table mania" (*mensarum insania*), a particularly male affliction which, according to Pliny (HN 13.91), Roman women used as a retort to the men against the charge of extravagance in pearls. Pliny, our single most important source for the use of citrus in Roman times, reports that Cicero, consul and orator of the mid–first century B.C., paid half a million sesterces for a citrus table (HN 13.91). The largest table of citrus wood was made for Ptolemy, king of Mauretania, out of two semicircular planks of wood joined together: it was 4.5 Roman feet in diameter (1.33 m) and a quarter foot (*quadrantalis*) in thickness (7.5 cm). The emperor Tiberius (A.D. 14–37) owned a round table nearly as large, but Pliny points out that it was only covered with a veneer of the precious wood (HN 13.94). The wood was so prized by emperors that by the late first century it was included in the accounts of the imperial treasury (Stat. *Silv.* 3.3.94).

Citrus wood was considered ideal for serving wine, according to Pliny, for spillage would not harm the surface (this was probably because of the high natural oil content in the wood; the oil of citrus was itself prized: HN 15.28). As might be imagined, woodworkers soon realized that furniture of solid citrus wood wasted valuable stock; by the first century veneering was apparently common (Plin. HN 16.231). The surface of citrus wood was prone to checking; proper seasoning was essential, and exotic remedies were reported (see below).

CORNUS, CORNEL-TREE, CORNELIAN CHERRY, OR DOGWOOD

Although the cornel-tree (*Cornus mas*) does not grow to any great size and its wood does not form a good bond when glued to other woods (Plin. HN 16.226), Roman woodworkers valued cornel for the fashioning of wheel spokes, pins and dowels, wooden wedges, and presumably similar small implements that required a tough, dense wood (Plin. HN 16.206). Virgil makes reference to a spear of cornel thrown with deadly accuracy by the Rutulian augur Tolumnius in the *Aeneid* (12.267).

CORYLUS, HAZEL

The tree that bears the hazelnut grows only a few meters in height and has the appearance of a large shrub. Thus its wood was suitable only for small implements; Pliny specifically mentions the use of hazel for torches (HN 16.75). A barrel excavated in 1897 at the site of Roman Silchester in Britain was reported to have bands of hazelwood (Boon 1974, 263).

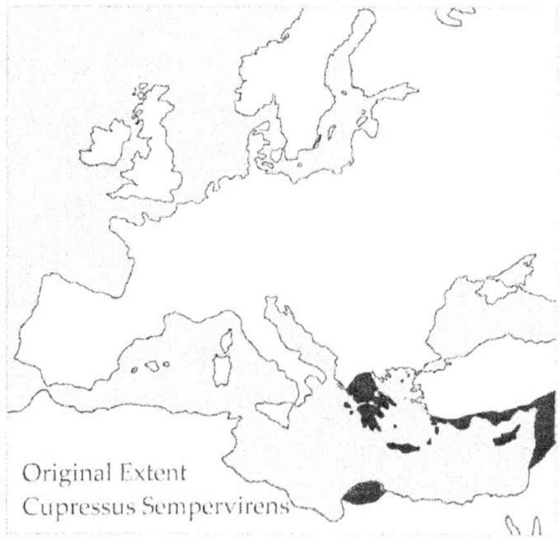

Fig. 12.2. The original range of cypress in the Mediterranean. (Author)

CUPRESSUS, CYPRESS

Originally a native of Asia Minor and Greece (fig. 12.2), the cypress is now a familiar element of the Italian landscape; the trees were introduced in Roman times and are especially suited to the Mediterranean climate, requiring little rainfall and thriving in heat (Meiggs 1982, 46). *Cupressus sempervirens pyramidalis* (Italian cypress) grows straight and tall, its branches rising vertically close to its trunk. *Cupressus sempervirens horizontalis* grows with its branches extending outward from the trunk.

Prized for their durability and resistance to decay (Theophr. *Hist. Pl.* 5.42; Plin. *HN* 16.212) as well as their color, suitability for carving, and high polish, cypress planks were fashioned into many of the greatest monumental wooden portals found in ancient Greece and Italy. Among the most famous of these were the doors of the enormous Temple of Artemis at Ephesus, which were over four centuries old in Pliny's day (Plin. *HN* 16.215). The doors of Old St. Peters in Rome (fourth century) have long since vanished, but one can still observe the carved cypress doors (fifth century) of Santa Sabina on the Aventine Hill, now fifteen centuries old and protected only by the covered narthex of the church. These two Roman examples are quite late; Russell Meiggs can cite no example of cypress being used for structural purposes in pre-Christian Rome (Meiggs 1982, 242). Samples have been identified from Pompeii and Herculaneum, but the context of these has not been published (Fioravanti and Caramiello 1999, 85). Cypress was the most important wood used for the Roman-period sarcophagi excavated in and around Kertch on the Black Sea (Vaulina and Wasowicz 1974, 23).

Since cypress wood was known to have a fine luster, it was apparently used for some of the earliest cult statues revered in the shrines of Greece and Rome. Pliny (*HN* 16.216) specifically mentioned that the statue of the god Veiovis, the "little Jupiter," which stood on the Capitoline Hill of Rome and had been put in place in 193 B.C., was still on view in his day over two hundred years later.[5] The cypress tree was also valued as a source of oil employed as a preservative (Plin. *HN* 15.28).

FAGUS, BEECH

The European beech (*Fagus sylvatica*), which was and is still common in Italy, is valued for its strength, durability, and versatility. This same availability, however, precluded its membership in the group of high-status furniture woods. Beech trees grow throughout the Italian peninsula, central Europe, southern England, and northern Greece. Mature trees reach heights of 30 m (100 ft.). Stands of beechwood are still easily found in the vicinity of Rome. A typical beech forest (*faggeto*) covers the slopes of Monte Artemisio to the east of Lake Nemi, just thirty kilometers to the east of the capital. Place names like Monte Faggeto in southern Latium attest to the pervasiveness of the beech in central Italy. In Rome, Jupiter was worshiped, in addition to other places, at the Beech-tree Shrine, the *Fagutal* (Varro *Ling.* 5.152). Pure stands of mature beech trees can be visited today in the Camáldoli forest, north of Arezzo.

Theophrastus (*Hist. Pl.* 5.8.3) recognized the beech as one of the most notable and useful trees of Italy's Latium; he adds that timbers of beech would be most suitable for the keels of ships but also for household furniture, such as bedsteads (5.6.4). The identification of a beechwood turned bed leg from Pompeii confirms the ancient account (Mols 1999, 81). Beech must have been used for all sorts of building timbers and household utensils; it is one of the most common of woods identified at Pompeii and Herculaneum (Fioravanti and Caramiello 1999, 85). A rare Roman plane with its stock of beechwood nearly intact was discovered at the bottom of a well in 1907 at Saalburg, and the same site produced several examples of wooden shoes made of beechwood (Gaitzsch and Matthäus 1981, 214; Pugsley 2003, 42).

As Ovid suggests in his poetic description of the primitive hut of Baucis and Philemon (*Met.* 8.669–70), beechwood was likely the source of everyday plates and cups used by rural Italians. When seasoned, the wood does not impart an unpleasant taste or odor to food. Beechwood burns well and must have been an important source of fuel for cooking, heating, the baths, and pottery kilns. Beechnuts were used as fodder for swine. Since Pliny confuses the Latin *fagus* with the Greek *phegos* (used by Theophrastus to designate the oak), Pliny's list of uses of beechwood (for example, HN 16.229, where he mentions the use of *fagus* for boxes, desks, and veneer) must be considered accordingly.

FICUS, FIG

Although the wood of the fig tree is not generally considered suitable for woodworking because of its weakness, the wide availability of fig wood in central Italy resulted in its being used anyway. Samples from the hulls of two ships excavated at Pisa revealed, much to the surprise of researchers at the Istituto per la Ricerca sul Legno (the institute for the study of wood in Florence), eight examples of fig (Giachi, Lazzeri, Paci 2000, 82).

FRAXINUS, ASH

The tough, close grain of the ash makes it suitable for a variety of household implements. The wood of the ash was immortalized by Homer, who described it as having been used for the spear of Achilles (*Il.* 20.277); Pliny himself also mentions this story as he extols the ash as the most productive of all timber-bearing trees (*HN* 16.62; 16.228).

Ancient Greeks and Romans knew of several species. The common European ash (*Fraxinus excelsior*) is a tall tree (45 m, 150 ft.) found throughout Europe and Britain. A smaller species (25 m, 80 ft.), found along the Mediterranean shore and North Africa, is characterized by narrower leaves (*Fraxinus angustifolia*). Another Mediterranean species, the Manna ash (*Fraxinus ornus*) has been used traditionally in Sicily and Italy as a mild laxative (Allaby 1999, 130). The ash of Macedonia in northern Greece was celebrated for its suppleness; the combination of light weight and suppleness that characterized the wood of the Gallic ash (*Fraxinus gallica*, as described by Pliny in *HN* 16.228) was well suited to the requirements of the war chariot. Pliny presumably had the wheels of the chariot in mind. Until the advent of the metal wheel in the nineteenth century, ash was favored by wheelwrights for the curved rim sections, or felloes, of the wheel. The springy nature of the wood was ideal against hard cobbles, paving stones, and potholes.

Vitruvius (2.9.11) similarly praises the wood of the ash, noting its flexibility while green but its stiffness upon being air-dried and good gluing characteristics. Today ash is popular for baseball bats.

HEBENUS, EBONY

Ebony (*Diospyros ebenum*) was and remains one of the world's most exotic woods. In Roman times it was imported to Italy from Africa and southern India; Virgil (*Georg.* 2.117) credited India alone as a source for ebony. Lucan (10.117) associated ebony with "mareotica," a district in Egypt. Pliny (*HN* 12.20) tells us that Pompey exhibited ebony at Rome following his victory over King Mithridates in 63 B.C. Herodotus cited Ethiopia as a major source of the wood, relating that the Ethiopians paid a tribute of two hundred logs of ebony, gold, and ivory every three years to the kings of Persia (Herod. 3.97; the story is also told in Plin. *HN* 12.17). It is possible that the word *hebenus* came to be associated with more than one species of tropical hardwoods that were imported from exotic places (Rackham 1945, 12).

One might well wonder if a Roman ever saw a living ebony tree. Pliny tells us that the emperor Nero sent an exploratory force of praetorian troops into the region of Ethiopia; the tribune in charge reported they had found only desert and palm trees (*HN* 12.18). It is reasonable to suspect that the purpose of the trip was to assay the possibility of establishing imperial control over the sources of ivory and ebony. The nearly black wood was highly prized for fine furniture and implements. Nevertheless, in Pliny's day, ebony is mentioned more for its medicinal properties than for its use in woodworking

Fig. 12.3. The distribution of holm oak in the Mediterranean. (Author)

(Meiggs 1982, 286); it is safe to assume that its use by Roman woodworkers in Italy was quite rare.

ILEX, HOLM OAK

The ilex is still one of the most distinctive trees growing in the parks and green spaces in and around the modern city of Rome. The dark green leaves of this evergreen oak resemble those of the holly in color and texture; in fact the two trees belong to the same botanical family. Pliny (HN 16.19) divided the ilex into two categories: the Italian tree has a leaf that resembles the olive leaf and was called *milax* by ancient Greeks; the second, provincial type of ilex has pointed leaves (perhaps the *aria* of Theophrastus). Mature trees reach heights of 20 m and thrive in all areas of the coastal Mediterranean, surviving at altitudes up to 700–1,000 m (fig. 12.3).

Ilexes do not produce long straight timbers, but the squat, thick trunks can produce beams two or three meters long. It is easy to imagine thick architraves of ilex over the colonnades of Rome's oldest temples. The ilex was a favorite of the Roman poets: Virgil depicts the doomed Carthaginian queen Dido constructing a funeral pyre of ilex and pine (*Aen.* 4.505). Pliny (HN 16.229) mentions other more creative uses: the dense, hard wood of the ilex, characterized by a complex grain, can be cut into thin sheets for veneers; he considered the wood to be resistant to friction and thereby suitable for wheel axles and for the handle of a forcefully twisted auger. These same properties led Theophrastus to declare ilex one of the most difficult of woods for the craftsman to work with (*Hist. Pl.* 5.5.1). Even so, it appears to have been used prominently in boat construction; both fishing and cargo vessels excavated in 1959 at Rome's port (built by the emperor Claudius) were found to have employed ilex (Scrinari 1979). Branches unsuitable for woodworking could be converted to the charcoal that warmed Roman homes during chilly winters. The ilex can grow in dry, hot lowlands. The dense shade it casts provides relief in pastureland.

IUGLANS, WALNUT

The Latin *iuglans* is derived from *Iovis* ("Jupiter") and *glans*, a seed or nut. The walnut tree was not indigenous to Europe but was gradually introduced from a period of great antiquity, probably from eastern Asia.

The walnut is not a particularly large tree: only rarely do specimens grow much higher than 15 m; it does not, therefore, figure prominently in ancient texts as an important structural timber. Pliny (*HN* 16.223) does mention, as noted, that beams of walnut creaked before breaking in a notorious accident at Antandros. The story is one that Pliny retold from the commentary of Theophrastus (*Hist. Pl.* 5.6.1), in which the wood in question is thought to have in fact been chestnut. Today the wood of the walnut is especially valued as a furniture wood. In the ancient Mediterranean world walnut appears to be especially associated with tabletops. Mols reports that one of the round tables recovered from Herculaneum was of walnut (1999, 81). In addition, the top of a banquet table found in the royal tumulus at Gordion was of walnut, as were the carved legs of two smaller serving tables (Simpson 1992, 14). Walnut oil was known as *caryinum* (Plin. *HN* 15.28).

IUNIPERUS, JUNIPER

The common juniper (*Juniperus communis*), an evergreen member of the cypress family, continues to thrive in all but the northernmost regions of Europe, Britain, and Ireland. The most common forms in the Mediterranean grow little more than 6 m tall, but the occasional juniper growing in higher altitudes may reach a height of 20 meters or more. Nevertheless, juniper wood is rarely associated with structural timbers. Pliny (*HN* 16.216) made the unusual claim that beams of juniper were used to roof the Temple of Diana at Saguntum in Spain, and he marveled at their ability to survive in this application for many years. The use of the wood for roofing beams and the claims of resistance to decay are usually reserved for massive timbers of cedar; Pliny may be confused here. As has been pointed out, Greek and Latin texts confuse cedar and juniper (Meiggs 1982, 410).

Unlike softer cedarwood, however, the wood of the juniper is extremely hard and tough, with a fine grain. Vitruvius (7.3.1), for example, prescribes juniper for the wooden ties used to suspend a false vault of lathe and plaster over the ceiling of a domestic room. The wood is good for turning on a lathe, meaning the use of juniper for the legs of tables and chairs is highly probable, although not documented in literary sources.

Here again the discoveries from the eastern Mediterranean at Gordion and Kertch can be considered to provide analogies as to how juniper may have been used in Italian contexts. Two serving stands found in the royal tumulus at Gordion were decorated with inlays of juniper that had been set into a face of boxwood (Simpson 1992). The moldings on a Roman-period wooden sarcophagus discovered in 1860 on the shores of the Black Sea were carved from fine-grained juniper (Vaulina and Wasowicz 1974, 25).

LARIX, EUROPEAN LARCH

Pliny writes, "What is considered the largest tree ever seen at Rome down to the present time was that marvel Tiberius Caesar exhibited on the bridge where mock naval battles are held. It had been imported to Rome with the rest of the timber, and it lasted until [it was employed for] the amphitheater of the emperor Nero. It was a beam of larch, 120 feet long and of a constant thickness of two feet, from which could be surmised the almost incredible height of the rest [of the tree] by estimating its length to the top."[6] If Pliny's story is true, the beam of larchwood so described must have come from a truly gigantic tree. Few modern larches are found that exceed a total height of 40 m (135 Roman feet); the length of a beam two feet square in section from such a tree would be considerably less. We hear that the larches for the project Pliny speaks of were harvested in the alpine province of Raetia. Simply transporting such a beam as Pliny describes from the high elevations of the Alps to Rome must have been an arduous undertaking. The long, straight timbers obtainable from the trunks of larch were a raw material valued for the roofing of large structures and for the masts of ships, although for the latter fir was preferred because of its lighter weight (Plin. HN 16.195).[7]

Vitruvius, a great admirer of the larch, mistakenly believed that the tall, straight trees grew not only in the Alps but also along the banks of the Po River valley and the shores of the Adriatic (2.9.14), yet larch grows only at high altitudes (800–1,200 m, or 2,600–4,000 ft.) and its range is fairly limited (fig. 12.4). The best stands were in the European Alps, while other extensive forests grew in the remoter Balkans. If the larch had grown closer to Rome, along the slopes of the Apennines, doubtless it would have been used in the capital at least as much as the silver fir. Vitruvius wished that the roofing of all the buildings of Rome had been framed in larch, for the wood was considered both tough and fire-resistant. He credited Julius Caesar with discovering larch and its fire-resistant qualities when Caesar's legionaries besieged the fort of Larignum (the name was taken from the larchwood of which its walls were built) and unsuccessfully attempted to burn their way in (Vitr. 2.9.15–16). Larch could be transported south via the river Adige from along the steep slopes of the Dolomites and eastward along the river Po. Beams of larch were available in the Adriatic cities of Ravenna and Fanum (Vitr. 2.9.16). Fanum is the site of the basilica designed by Vitruvius; one is tempted to speculate that larch may have been used for the roof beams described with such pride in his architectural treatise (see above, pp. 144ff.).

Pliny's accounts do make it clear that for applications requiring unusually long, straight timbers, beams of larch could even be shipped to Rome, presumably by sea (samples have been identified from excavations of Pompeii and Herculaneum as well) (Fioravanti and Caramiello 1999, 85). Pliny marvels at another beam of amazing proportions: it was 100 feet (30 m) long and 1.5 feet (45 cm) thick and was left over from

Fig. 12.4. The distribution of larch in Europe. (Author)

the roofing of a building known as the Diribitorium, next to the Saepta Iulia, or polling place, of the capital (*HN* 16.201). Although we do not know what kind of wood was used to span the broadest rooms ever constructed—the public rooms in the palace of Domitian on the Palatine Hill—larch would be a good candidate.

Pliny recognized larch as a wood well suited to damp conditions; it was found to be nearly as resistant to groundwater as the alder (*HN* 16.218). This reckons well with the trade in larch that Vitruvius reports was active in Ravenna, a city where many of the buildings were supported on wooden pilings constantly subjected to groundwater (2.9.16). Larchwood is still prized in Europe today for this quality.

Excavations of both prehistoric and Roman sites in the mountainous regions of northern Italy have verified that larch was used extensively in a variety of structural applications, confirming the ancient accounts. Larch was used for pile-dwellings at Bressanone-Stufles, dating from the sixth–fifth centuries B.C., and for floorboards at ground level in both early (for example, Sanzeno-fondo Gremes, sixth century B.C.) and imperial Roman (Mezzocorona) contexts (Cavada 1994, 209). These finds indicate that the water-resistant qualities of larch were understood from prehistoric times.

The bark of the larch can be employed for tanning leather, and its resin is valued as a treatment for respiratory ailments.

OLEA, OLIVE

The olive tree (*Olea europaea*), sacred to Minerva, was cultivated primarily for its fruit and the oil derived thereof. The wood of the olive, however, was appreciated for its density, strength, and resistance to decay. While the tree does not grow either tall enough (maximum 10 m) or straight enough to produce long timbers, and so figured little in Roman construction, Vitruvius (1.5.3) suggests that charred olive wood beams be employed in earthwork fortifications to tie the walls together. Apparently his specification is based upon the longevity of the wood in such an application. Indeed, most ancient recommendations for the use of olive wood emphasized strength and durability: tool handles and the wooden hinges used on doors (Plin. *HN* 16.230), suspended ceilings (Vitr. 7.3.1), and the like. The hardest wood came from wild trees, the leaves of which were used as crowns for victors in the Olympic games.

PINUS, PINE

Next to firs, pines were capable of supplying the longest beams available from Italian trees; the tallest mountain varieties (*Pinus nigra 'larico,'* popularly known as Corsican pine because of the well-established forests on the eastern shores of the island) could furnish beams up to 18 m in length (Meiggs 1980, 189). On the mainland, *Pinus nigra* (European black pine), growing alongside silver fir in the central Apennines, was more easily accessible and also of good quality. Other varieties of pine produce inferior timber; the best known of these are undoubtedly the characteristic umbrella pines (*Pinus pinea*) or similar types (for example, Aleppo pine, or *Pinus halepensis*, and *Pinus pinaster*) that characterize the rural landscapes of the Italian coastal plains. These inferior grades were valued for their resin, but even their wood was used by carpenters, as proved by samples found in architectural contexts (Fioravanti and Caramiello 1999, 85). Vitruvius (2.9.12) exhibits little interest in the pine, criticizing the tree for its tendency to warp even though acknowledging its resistance to decay. Other writers, such as Seneca (*Ep.* 90.9) and Juvenal (3.254–56), make passing reference to the use of both pine and fir for structural roofing beams. Virgil imagines the pines of primeval Italy as "rising to the stars" (*Aen.* 11.136).

Since pine has a high resin content and was considered to be resistant to rot, it was used in applications requiring good moisture resistance but little strength. At least two of the ships discovered at Pisa use pine (*Pinus pinaster* and *Pinus pinea*) planking (Giachi, Lazzeri, Paci 2000, 81). Pliny (*HN* 16.224) stated that pine, because of its softness, was easy to hollow and thus was useful for the construction of wooden pipes to carry water. A third-century well in Roman Cologne was lined with pine boards that were still intact when excavated in the mid–twentieth century (Schoppa 1951). Split pine served well for shingles (*HN* 16.36). Cato (*Rust.* 18.8) regarded the pine as something of an all-purpose wood, suggesting that the anchor posts for an olive press made from the softwood served as well as oak.

PLATANUS, PLANE

The plane tree was admired by the Romans, as it is by modern Europeans, for its spreading branches and comforting shade. Pliny states that the tree was not native to Italy but imported first into the eastern Mediterranean and from there to Sicily and the peninsula. It was one of the first foreign species to reach Italy (Plin. *HN* 12.6). Traditionally the tough wood of the *Platanus orientalis* has been favored in Italy for the chopping blocks used by butchers, but it is not mentioned in the ancient sources as a wood favored by Roman woodworkers and is not considered in depth here.

POPULUS, POPLAR

The poplar was sacred to the hero Hercules (Plin. *HN* 12.3), an ironic association considering the rather weak qualities of the wood. The poplar and the elm are the only two species Cato recommends for planting around farmsteads (*Rust.* 6.3). Poplars are commonly found growing along roadways and streambeds. Poplar wood is poor for building, yet the inferior quality of the wood is offset by its fast growth and impressive size (adult trees reach 30 m, 100 ft.). Pliny mentions three varieties of poplar: the white (*alba*), the black (*nigra*), and the Libyan (*libyca*); this last has not been identified.

Populus nigra is easy to recognize from a distance, as the branches grow like those of the cypress, vertically and close to the trunk of the tree (a profile known as columnar). The similarity is reflected in the popular Italian name for the tree: *pioppo cipressino.* This distinctive profile may explain why black poplars were planted around the site of the funeral pyre (*ustrinum*) of Augustus in Rome (Strabo *Geo.* 5.3.8) (the cypress was associated with cemeteries, as it is still today in Italy). Surprisingly, Pliny includes poplar wood in the class of materials most suitable for veneer work (*HN* 16.231). Presumably he means the root of the poplar, for in the passage in question, he has just mentioned the suitability of the roots of the elder (*sabucus*). Italian woodworkers today still prize the roots of the black poplar for veneering. Vitruvius praises the light color of poplar wood and cites its convenience for carving (2.9.9).

QUERCUS, OAK

When forests covered most of Italy the slow-growing oak was one of the most common of native species; oaks can still be found throughout most of peninsular Italy (except for Puglia) and Europe. The common, or Valonia, oak (*Quercus robur*) flourished from the toe of Italy to European regions far northward, including all of the British Isles (fig. 12.5). Because of the oak's slow growth (trees may grow for five hundred years or more), oak trees were not cultivated but harvested from forests. Along with the beech they were plentiful in Latium and were doubtless one of the most important sources of hardwood for the indigenous population.

Remains of oak have been found in some of the earliest archaeological contexts connected with the history of Rome. We have already seen that oaken logs were used whole as coffins in the Iron Age. The ashes found in the archaic foundation pit (ca. 600 B.C.) of the shrine of Vesta, arguably the most sacred spot in the Roman forum, have been identified as the remains of oak (Quilici 1979, 78). The presence of vast oak forests in Italy had corollary effects upon the environment. In Strabo's words (5.1.12), "The forests [of northern Italy] have acorns in such quantities that Rome is fed mainly on the herds of swine that come from there."

Vitruvius (2.9.8) discusses three species of deciduous oak in his commentary: common, or Valonia, oak (*Quercus robur:* "strong"), *quercus aesculus*, which does not correspond to a single modern botanical equivalent, and turkey oak (*Quercus cerris*).[8] Of

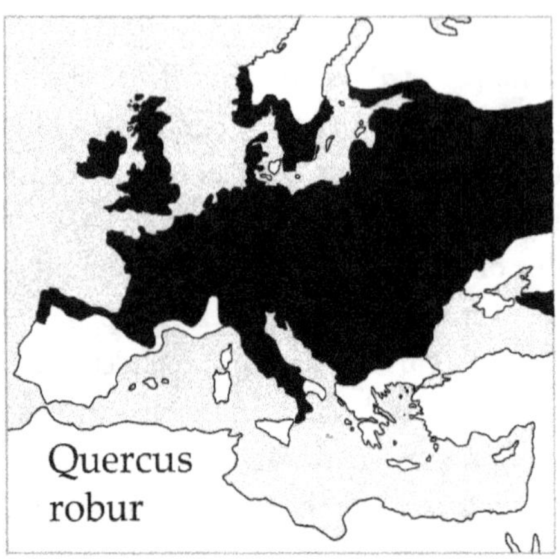

Fig. 12.5. The distribution of Valonia oak in Europe. (Author)

these the turkey oak is the least desirable for woodworking purposes of any kind, for although the tree grows rapidly and high (up to 35 m), it produces inferior timber. Valonia oak produces a tough, dense wood but does not grow very tall (20 m) and has a rather short trunk. *Aesculus* was the best for building (Vitr. 2.9.9) because the quality of the wood was good and the trunks of the tree grow quite long and straight. It is Aesculus that Pliny says was sacred to Jupiter (HN 12.3). Certainly it must have seemed eternal, some specimens living for as long as one thousand years. Because of the general toughness of all varieties, oak is considered one of the most difficult woods to fashion with hand-tools.

Both Pliny (HN 16.218) and Vitruvius (2.9.8) considered oak excellent for foundations; we have seen that massive piles of oak were discovered over a century ago more or less intact along the banks of the Tiber. Oaken planking, both split and sawed, was favored for lining wells at Camulodunum (Colchester) (Hawkes and Hull 1947, 126). Oak's tendency to warp was its main fault (Plin. HN 36.187); nevertheless, oaken timbers must have formed the beams and columns of many old Roman buildings and when used for the hulls of ships, for example, in the vessels of the Gauls described by Julius Caesar, provided strength against "any violence and buffeting" (BGall. 3.13.3). Oaken keels provided the resistance to splintering required when ships were hauled up on land (Theophr. Hist. Pl. 5.7.2). Oak has been identified as the planking material in at least two ships excavated at Pisa (Giachi, Lazzeri, Paci 2000, 81) and was found in another boat discovered still sitting on Herculaneum's ancient beach during the excavations of 1982 (Ferroni and Meucci 1989).

Oak was favored for the spokes of wheels and was used for centuries for this purpose in England, as it had the reputation of rarely breaking. Despite its inherent strength, oak can be split easily; Pliny considered shingles from Valonia oak superior (HN 16.36). While oak does not show up in the ancient sources as an important furniture wood, Mols discovered that it had been used in the construction of a cradle, now heavily restored, from Herculaneum (Mols 1999, 81).

SALIX, WILLOW

While the familiar weeping willow was introduced to Europe from central Asia only in the eighteenth century, other varieties, such as the white willow (*Salix alba*) are native to the continent. Willows do not grow much higher than 15 m and produce a lightweight wood with little strength. Wooden shoes are still fashioned from the easily worked wood; this tradition may date back to ancient times.

There is some evidence for the use of the willow by Romans; craftsmen found many uses for the thin, tough, flexible branches cut from the tree. Cato (*Rust.* 1.7) believed the willow could be useful to the rural farmstead in a number of ways, and later Pliny (*HN* 16.174) would echo this praise. A well-provisioned farm would include a grove of willows (*salictum*). Wands of willow were turned into vine trellises, baskets, and beehives (Columella *Rust.* 2.2.90; Verg. *Georg.* 4.34). The medium-sized vineyard would employ a willow specialist (*salictarius*) (Cato *Rust.* 11.1). Willow, along with birch, could be woven into panels for wattle work, of which medieval-period examples have been discovered in Britain (Hall 1982, 236). Finally, Pliny (*HN* 16.174) praises the willow for its suitability for wicker chairs, popular and comfortable furnishings in Roman houses if sculpted representations are any gauge (see chapter 11).

TEREBINTHUS, TEREBINTH OR TURPENTINE-TREE

More of a large shrub than a tree, the terebinth rarely grows beyond 5 m in height and produces wood with a fine grain, valued only for veneers and small utensils, such as the goblets of terebinth described by Pliny (*HN* 16.205). Theophrastus claimed that in Syria the wood was prized for dagger handles and that cups turned from the wood on a lathe looked like pottery (presumably for the high burnish) (*Hist. Pl.* 5.3.2). As a furniture veneer, we can imagine terebinth being used much like boxwood (Verg. *Aen.* 10.136).

TILIA, LINDEN, LIME

The linden tree, sacred to Germanic tribes because of its association with fertility, is discussed primarily by Pliny, who is indebted to the comments of Theophrastus. Known as the softest of woods used by carpenters, linden was ideally suited to lathe work; it is still employed for the making of inexpensive furniture in Europe. Considered one of the hottest woods, it was blamed for the rapid blunting of tools used to cut and shape it (Theophr. *Hist. Pl.* 5.3.3; Plin. *HN* 16.207). Vitruvius cites the *tilia*, along with poplar and willow, as a source of good carving wood (2.9.9). In North America basswood, a close relative, is equally prized. Otherwise the medium-quality linden was used for low-status applications such as crates. The stringy bark could be woven into baskets or employed as cheap shingles (Plin. *HN* 16.35).

ULMUS, ELM

Elm, along with the poplar, was considered by Cato (*Rust.* 6.3) so useful to the farmer that he recommended both types of trees be planted around farmsteads. The wood of

Table 7. Nomenclature for trees used by Roman woodworkers

Latin Name	*Botanical Family*	*English*	*Italian*
abies	Pinaceae	fir	abete
acer	Aceraceae	maple	acero
alnus	Betulaceae	alder	ontano
betula	Betulaceae	birch	betulla
buxus	Buxaceae	box	bosso
carpinus	Corylaceae	hornbeam	carpino
castanea	Fagaceae	chestnut	castagno
cedrus	Pinaceae	cedar	cedro
citrus	Cupressaceae	thuja/sanderac	tuia
cornus	Cornaceae	cornel/dogwood	corniolo
corylus	Corylaceae	hazel	nocciolo
cupressus	Cupressaceae	cypress	cipresso
fagus	Fagaceae	beech	faggio
ficus	Moraceae	fig	fico
fraxinus	Oleaceae	ash	frassino
hebenus	Ebenaceae	ebony	ebano
ilex	Fagaceae	holm oak	leccio
iuglans	Juglandaceae	walnut	noce
iuniperus	Cupressaceae	juniper	ginepro
larix	Pinaceae	larch	larice
olea	Oleaceae	olive	olivo
palma	Palmae	palm	palma
pinus	Pinaceae	pine	pino
populus	Salicaceae	poplar	pioppo
quercus	Fagaceae	oak	quercia
salix	Salicaceae	willow	salice
taxus	Taxaceae	yew	tasso
terebinthus	Anacardiaceae	terebinth	terebinto
tilia	Tiliaceae	linden, lime	tiglio
ulmus	Ulmaceae	elm	olmo

Note: With the exception of citrus, the Latin name corresponds to the modern *genus* classification.

the elm is strong and durable, although notoriously difficult to split and very difficult to work with hand-tools. Elm has been recovered from archaeological sites of great antiquity; a prehistoric roadway of elm branches, for example, has been discovered in Britain (Orme 1982). Elm has some of the same characteristics of ash in terms of flexibility and toughness, although elm is heavier (Plin. *HN* 16.229). It may have been used by wheelwrights; traditionally wheelwrights in England, for example, always used elm for the wheel hubs, or naves, of wooden wheels because it was so difficult to split. Elm destined for this use was generally seasoned for four or five years.

Drying, Preservatives, and Conditioners

The tendency of wood to decay over time is due primarily to invasive fungi which require moisture to live. Wood in contact with the ground will absorb water by capillary action no matter how well it has been dried. Ancient builders suggested some specific remedies for preserving wood from rot, but these would have had only limited success in retarding the process of decomposition. The natural resistance to decay varies between species of trees which produce timber suitable for building. An oaken post ca. 10 centimeters in diameter would last about thirty years before losing all of its structural value; pine of the same dimension might last only ten years (Hanson 1978, 296).

While air-drying was and still is the preferred method of seasoning wood, the practical craftsman knew that stacking wood in a warm place would accelerate the process. Ancient references to the practice are rare. Columella (*Rust.* 1.6.19), in his thorough discussion of rural industries, advised that green wood could be stacked in the smoke room (*fumarium*) of a farm complex for curing. The passage indicates that it is the heat, not the smoke, that is of value here.

Farmers in the countryside of Latium still char the lower ends of fence posts used to border their fields. If done properly, a thick layer of inert carbon protects the inner core of the wood. The method is an ancient one; Vitruvius (1.5.3) recommends tying the outer and inner walls of a defensive circuit with "timbers of charred olive wood . . . in order that both facades of the wall . . . may have everlasting strength. For such timber cannot be injured by decay." The timbers Vitruvius has in mind were logs of tough olive, interred completely in the stones and earth used to fill the gap between the walls and therefore exposed constantly to dampness. Eighteenth-century coopers, it should be noted, burned bungholes and tapholes with a special cylindrical iron (a bung burner) to prevent rot at these vulnerable points on a cask (Salaman 1989, 35).

A penetrating oil brushed or sponged onto bare wood could impede water and perhaps repel some species of insects. That Romans were aware of such remedies is clear from Pliny (*HN* 16.198), who stated that when timber was coated with cedar oil (*cedri oleo peruncta*), it would not decay. The oil itself was obtained by the distillation of pitch or by similarly processing waste wood recovered from areas of harvesting (Baker 1978, 138). The wood of cedar trees, as noted, was prized for its durability.

While cedar oil was somewhat exotic and not always easy to obtain, *amurca*, the liquid by-product of olive oil production, was widely available. Cato (*Rust.* 98.2) suggested that rubbing furniture with *amurca* not only prevented decay but added a fine luster to the wooden surface. We can imagine that the same relatively plentiful substance was used on rough timber.

Some timber was apparently seasoned by being smeared with dung or buried in dung heaps. This practice may have been intended to allow green wood to dry evenly and slowly, preventing the surface of the beam from drying faster than the interior and thereby promoting cracking or checking (Hanson 1978, 296). Pliny (*HN* 16.222) claims that architects (*architecti*) advised that certain woods be smeared with dung in order to make them resistant to the atmosphere (*adflatus*)—presumably he means rapid drying by sun and wind. Unfortunately Pliny does not specify which species are to be so treated. Cato's (*Rust.* 31.1) advice that the *fibulae*, or wooden pins, of an olive press should be treated by immersing them in a pile of dung indicates that this was a rustic practice employed in a broad array of applications.[9]

According to Pliny (*HN* 13.99), ancient beachcombers observed that the wreckage of ships, having been soaked with seawater and then dried, solidified "with a hardness that resists decay." The "pickling" of wood by immersing in saltwater was also known to Theophrastus (5.4.3). Palladius, a late source, relates that the Sardinians buried their timber in sand by the sea for a year to season it (12.5.3). Cato (*Rust.* 31.1) implies this treatment in his discussion of the preparation of wood that will be used for the pins of an olive press, but he does not specify that the water is saline.[10] Hanson reasonably suggests that this kind of treatment of wood should be associated with shipbuilding activities (Hanson 1978, 296).

Pliny (*HN* 13.99) is our single source for the practice of drying wood by burying it in piles of grain or earth. He says that "carpenters lay [citrus wood] in piles of grain for periods of a week with intervals of the same period, and it is surprising how much its weight is drawn off this way."[11] The dried grain apparently can wick the moisture from the wood at a controlled and steady rate. In the same passage Pliny reports that some *barbari* treat citrus wood by coating it in wax and burying it in the ground. While wax on the end-grain of the cut planks can retard uneven drying (and thus checking), it would make little sense to coat the entire plank with wax and then bury it in moist ground.

XIII The Forests of Italy

The Romans are afforded a wonderful supply of materials by the large number of mines, by the timber, and by the rivers which bring these down.—Strabo, *Geography* 5.3.7

Nor do we venerate images gleaming with gold and ivory more than the groves and the silences they hold. —Pliny *HN* 12.3

It is beyond the scope of this book to offer a full discussion of the forests of Italy or of the immense tracts of timber found in other parts of the Roman world that were harvested to the benefit of Roman woodworkers. Nevertheless, it would be remiss not to offer some comments about the nature of the forests, especially in the regions around Rome, which provided the raw materials for Roman carpenters and craftsmen.

The dominating presence of the Mediterranean Sea around Italy and its maritime neighbors has created a climate that produces fairly stable temperatures and predictable, if intermittent, patterns of rainfall. The forests that once covered most of Italy exhibited remarkable variety not because of their relative location in terms of latitude, but more because of relative altitude, exposure, and soil composition. Furthermore, as Meiggs has observed, flat, level land was, from an early date, appropriated for intensive agricultural use (1982, 40).

Logging was associated with the mountains of Italy, a country in which there are few regions far removed from steep terrain (see fig. 12.1). The high ridge of the Apennines dominates the landscape from the Dolomites to the toe of the peninsula and emerges again from the sea across the landscape of Sicily. The scattered towns situated far from good forests, such as the prosperous villages of the vast Po watershed, were served by navigable rivers that could transport fine timbers for hundreds of kilometers (Verg. *Georg.* 2.451). Even if freshwater transport was not feasible, most major Italian towns were near the sea, and timber could be floated along the coastline to any number of ports. Ovid makes the association between high terrain, timber, and its transformation into seafaring vessels vivid in his depiction of humankind's first "golden" age: "Not yet had the pine, felled on its mountains, descended to the waterways to visit foreign lands" (Ov. *Met.* 1.94–95).[1]

The hot, drought-prone lowlands of Italy were suited to seasonal crops and hardy trees and shrubs: evergreens, including the introduced species of cypress and cedar, coastal pines with little value other than for their nuts and resins, and the ilex, or holm oak, which, as we have seen, must have figured prominently in the buildings of early Rome (Meiggs 1982, 42). Deciduous hardwoods, dominated by forests of oak, but also stands of beech, maple, and elm as well as pine preferred higher ground, generally thriving in areas at least two or three hundred meters above sea level, but not growing on terrain much above one thousand meters. Higher still, on the slopes of the Apennines, Roman loggers sought the tallest conifers: mountain pines and firs, but also good stands of beech, although the beech grew at lower elevations as well and was of such quality that they are specifically mentioned by Theophrastus in his discussion of Latium.

Within these bands of elevation, it is clear that certain species thrive on soils that discourage other growth. Thus within a fairly small area with diverse geological characteristics the Roman woodsman would have found concentrated stands of individual species. Caroline Malone's and Simon Stoddart's recently published survey of the Gubbio basin, for example, shows that the white oak (*Quercus pubescens*) thrives in places with a substrate of limestone and turkey oak (*Quercus cerris*) is found growing over substrates of sand and mudstones, while sessile oak (*Quercus petraea*) prefers soils with clay and sand (1994, 42). For the ranges of specific trees used by the Romans the reader is referred to comments on individual species made in chapter 12.

Within an area of fifty to sixty kilometers of Rome there were four areas of high ground that could provide good hardwoods and even a few stands of taller pine and fir (fig. 13.1). These areas would have been logged from an early date but were never entirely depleted. The land is so rugged in places that it would have been easier, as Rome's territory expanded, to look for more distant sources which imposed fewer obstacles to transport.

Closest to Rome, only twenty-five kilometers to the southeast, lies the steep terrain of the Alban Hills, a ring of small mountains surrounding the dormant crater of an extinct volcano that once measured nearly twenty kilometers in circumference. At the heart of this formation was the town of Alba Longa, in Roman legend the seat of the descendants of Rome's first king, Romulus. Deep forests covered the hilltops, including the two highest peaks, Monte Cavo (the revered Mons Albanus, 949 m), and Monte Iano (938 m), as they still do today. The predominant species are oak and beech. Timbers from these woods would have been hauled down the mountainsides to the flat plains and brought along the Via Appia (built in 312 B.C.) to the city.

While the Alban Hills are closest to Rome, literary sources suggest that the cut-

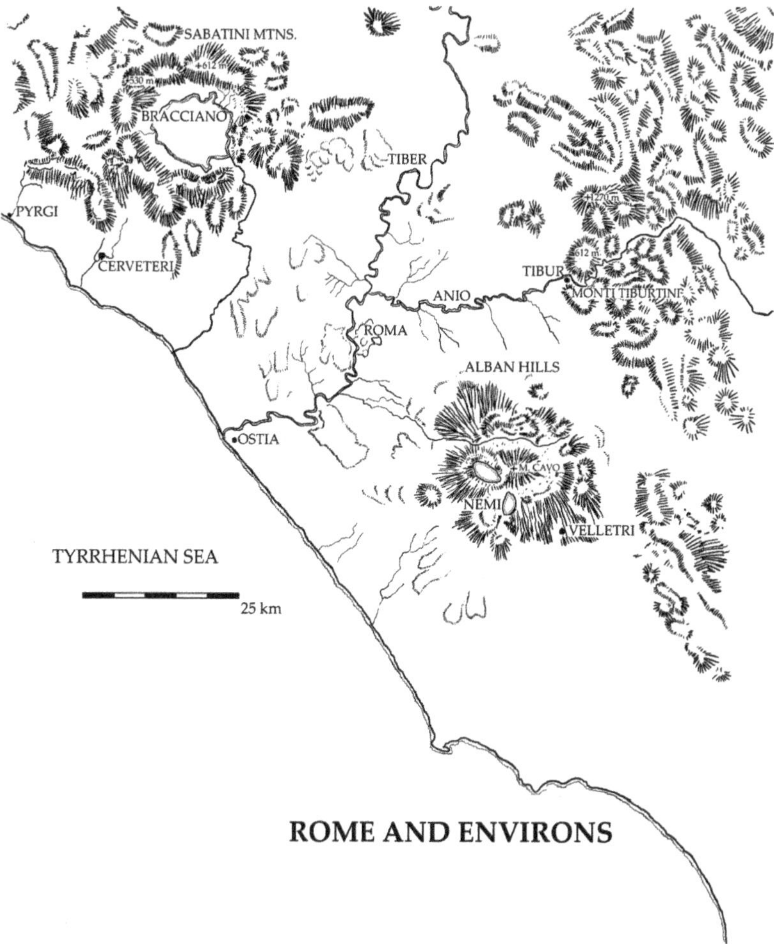

Fig. 13.1. Rome and its neighboring topography in the Tiber and Anio watersheds. (Author)

ting of timber there would have been highly restricted. Some of what must have been the choicest stands were completely off-limits to woodcutters because the forests were considered to be under divine protection. Pliny, for example, mentions a giant holm oak that stood in a grove sacred to the goddess Diana on Monte Corne, the peak that now overshadows the town of Frascati, near the ancient site of Tusculum (*HN* 16.44/242). The tall deciduous oaks that covered Monte Cavo were protected as a forest sacred to Jupiter Latiaris (Edlund 1987, 44). The densely growing timber around the shores and slopes of the volcanic Lake Nemi, named for the ancient Nemus ("sacred grove") Aricinum, was forever protected for the cult of Diana. There were other ancient stands, one sacred to Ferentina, and, at Alba Longa itself, a forest under the protection of Mars, for it was here that Romans believed Rhea Silvia (her name means "Rhea of the Forest") was impregnated by the War God and as a result gave birth to Romulus (Quilici 1979, 80).

Thirty kilometers to the north and west of the city were the wooded hillsides that skirted another immense volcanic lake, known today as Lake Bracciano, to Romans the Lacus Sabatinus. The terrain here is not quite as elevated as that of the Alban Hills; the highest summit is Monte Rocca Romana, a modest 612 m. Here, too, were extensive stands of hardwoods, with an abundance of tall oaks. Transport of timber from this area would have been cumbersome for there were no waterways that led to the city. Besides, this was an area firmly under control of powerful Etruscans and Faliscans, bitter enemies of Rome until a relatively late period. It was not until the first decades of the third century B.C. that the forest of Bracciano was accessible to the Roman axe.

Covering the foothills to the east and north of Lake Bracciano was the fabled Ciminian forest, or Silva Cimina, remembered by later Roman historians as an impenetrable and forbidding place. Little of this terrain rises above 200 m, and the timber, while plentiful, would not have been of the quality sought out at higher elevations. An exception is the terrain of the steep slopes around the volcanic Lago di Vico (Ciminius Lacus); one can get a sense even today of the once vast forest on Monte Fogliano ("leafy"; 965 m), where beech and chestnut are still logged for woodworking and carpentry. With the restoration of the old Via Cassia in 117–107 B.C., the best wood felled from here would be destined for Rome.

Slightly more distant but also more promising than either the Alban Hills or the forests of Lake Bracciano were the forested mountains of the Anio River valley, due east of the city. The first truly high peaks, ranging up to 1,200 m, are those of the Monti Prenestini and the Monti Lucretili, about forty kilometers from Rome. These elevations are high enough to support stands of mature beech, fir, and mountain pines. Better yet, the mountain slopes lead more or less directly to the banks of the Anio, a river broad enough to accommodate lengthy logs to be floated downstream. It is probably no coincidence that Strabo, in his comments about the quantity of timber and building stone that arrived in Rome during the early years of the empire, places the Anio at the top of his list of rivers important for such transport (5.3.7). The mountains in this region are steep and in some places inaccessible; we can imagine the slopes closest to the river being logged first while others were simply passed over. Good stands of timber can be found even today in some of the remoter valleys, which still preserve the aura of the primeval Latial forest.

This leaves us with the forests of the Monti Sabini, which lie on the first high terrain easily accessed by the Tiber, about fifty kilometers to the northeast of Rome. For the most part these high peaks (up to nearly 1,300 m) rise on the east side of the great river and thus have the southwestern exposure that Vitruvius felt produced the soundest timber. Many small tributaries run into the Tiber from these hills, some too small to be much help in the transport of large logs, others crucial access-ways to remoter stands.

With the final subjugation of the Etruscans over the course of the third century B.C., the forested hillsides of Etruria, many still accessible from the sinuous loops of the Tiber, produced the timbers necessary for the largest of public building projects. Within sixty kilometers of the capital–and well beyond–there were valuable forests on

*Fig. 13.2. The straight trunks of mature silver firs (*Abies alba*) rise in the forest of Vallombrosa, in the Apennines of central Italy (ca. 950 meters above sea level). (Author)*

either side of the river. In winter months, when the level of the river was high, the Tiber was navigable for one hundred and fifty miles (Plin. *Epist.* 5.6.12). Strabo tells us specifically that "the wooden material for the buildings [of Rome], in beams that are very straight and very long, is for the most part supplied by Tyrrhenia, since by means of the [Tiber] river it can be brought down directly from the mountains" (5.2.5). It was here, in the mountains of Tuscany, that Pliny the Younger owned a cherished country villa, its neighborhood characterized by a "spreading plain ringed round by mountains, their summits crowned by ancient woods of tall trees" (*Epist.* 5.6.7). The area was not only a good source of wood but a region celebrated for its tradition of craftsmanship. I have already turned to the evidence of rock-carved Etruscan tombs to document the importance of wood for framing and ceilings. The district of Nuceria, including the territory around present-day Orvieto, was celebrated for the production of fine wooden vessels (Strabo 5.2.10). Further to the north the Arno cut westward across a region rich in timber. Even today one can still see stands of mature silver firs in the Camàldoli national forest near Arezzo (part of the Parco delle Foreste Casentinesi, Monte Falterona, Campigna) and in the Vallombrosa forest near Florence (fig. 13.2).

The expansion of Rome's control over central Italy and eventually the entire peninsula granted access to new sources of raw material that would never be fully exploited. The construction of the Via Appia opened the forested ranges south of the Alban Hills: the Monti Lepini, Ausoni, and Aurunci as well as the inland mountains accessible by the Liris and Volturnus rivers. In the first century B.C., we hear, Marcus Agrippa ordered the forest around Lake Avernus felled (Strabo 5.4.5). This may have been one of the last pristine areas around Naples to be logged, since the forest had probably been protected by the oracular cult long associated with the site, much as the trees of Lake Nemi were sacrosanct. Further south was the immense Sila forest of Calabria, known to Romans for its stands of tall pines and firs, but because of its inaccessibility not exploited until the Middle Ages and, indeed, modern times (Dion. *Antiq.* 20.15; Meiggs 1982, 462).

The last extensive wooded areas on the west side of the Apennine range in peninsular Italy were the forests around and to the north of Pisa. The discovery of a source of high-quality marble in the mountains near Luni in the first century B.C., and the proximity of these mountains to the sea, guaranteed that this area would be heavily exploited by Rome, beginning with Rome's first emperor, Augustus. Strabo tells us the forests around Pisa had once provided Rome with warships second to none (5.2.6). The fortuitous discovery of eight well-preserved Roman ships near Pisa in 1999 during the construction of a railway station attests to this local tradition. As the area around Pisa fell under Roman sway, the supplies of timber were shipped down the coast to Rome. Timber and marble were stockpiled at the coastal port of Ostia and eventually towed up the Tiber on barges to Rome.

As the extent of Roman territories grew, new species were imported for specialized use. The larch, for example, found only in the Alps, was, as noted, prized for its strength and resistance to fire. Likewise the mighty cedars of Lebanon, as mentioned earlier, were valued for their large scale, straightness, and resistance to rot. Rome's ultimate expansion throughout the ancient Mediterranean world guaranteed broad exposure to a variety of raw materials, tools, and techniques that spanned the hot, dry climates of north Africa and the Middle East to the temperate forests of northern Europe and England.

Glossary of Roman Woodworking Terms

Contents

Introduction

Individual terms in the glossary generally correspond in order of presentation to the discussion in the main text. Thus chapter 8, "Roofing and Ceilings," in the main text is complemented in the glossary under section VIII. Whenever possible the definitions of terms are located under the appropriate Latin word. If the Latin term is not known, the definition can be found under its English name. Each term is defined with a selection of Latin passages. The abbreviations of ancient authors follow those used in the *Oxford Classical Dictionary*. The English translations of the Latin passages are based upon those used in the Loeb Classical Library unless otherwise noted.

Individual entries in the glossary generally adhere to the following format:

(1) Plin. *Epist.* 5.6.38: (2) (Pliny's country villa): (3) there is a bed here [in the alcove] and windows on all sides" (4) (lectus hic et undique fenestrae).

1. Plin. **Epist.** 5.6.38: the location of the ancient passage;
2. **(Pliny's country villa):** a brief context for the passage;
3. **[in the alcove]:** words in brackets within the English translation have been supplied so that the passage makes better sense. These are usually based upon interpretation of specific words or phrases from neighboring lines in the ancient source;
4. **lectus hic et undique fenestrae:** the Latin text.

I. General Woodworking Terms

ars fabrica: the art or skill of making something: craft, art, or structure. The exact meaning depends upon context, as indicated by the following example:

> Plin. *HN* 16.225: "Fir wood is very suitable . . . for the panels of folding doors and any kind of interior work whether in the Greek or the Campanian or the Sicilian manner of the craft" (abies . . . valvarum paginis et ad quaecumque libeat intestina opera aptissima, sive Graeco sive Campano sive Siculo fabricae artis genere).

There is not enough evidence to suggest how one type or style of carpentry differs from another. This passage may indicate either a difference in decorative style, proportion, or a variation of technique. Cato (*Rust.* 18.9) also refers to the Punic method (here the style may refer specifically to wheel construction using dovetailed planks instead of spokes and felloes). For this passage, see *coagmentum* under "IV. Joints."

ars fabrilis: carpentry, woodwork. See also, in this section, *materiarius*, *materiatio*, *opus fabrile*, and *secamentum*.

carpenter: see *faber carpentarius*, *faber tignarius*, *lignarius*, and *tignarius* under "II. Areas of Specialization."

carpentry: see, in this section, *ars fabrilis*, *materiarius*, *materiatio*, *opus fabrile*, and *secamentum*.

fabrica materiaria: see, in this section, *materiarius*.

firewood: see, in this section, *lignum*.

***lignator -oris* (m.):** a collector of firewood. Thus the verb *lignor -ari*: to collect firewood.

> Caes. *BGall.* 5.26.2: "after the [detachments of] wood gatherers had been unexpectedly overwhelmed they [the enemy] came to attack the camp with a large force" (subitoque oppressis lignatoribus magna manu ad castra oppugnatum venerunt).

***ligneolus -a -um* (adj.):** something made of wood.

> Apul. *Mun.* 27: "wooden figures of men" (in ligneolis hominum figuris).
> Lucil. 224: "a wooden tray" (scutam ligneolam).

See also, in this section, *ligneus*.

***ligneus -a -um* (adj.):** something made of wood.

> Cato *Rust.* 10.5 (concerning the furnishings necessary for a farmhouse): "a wooden mortar" (pilam ligneam).

Cat. *Rust.* 11.5: "wooden scoops" (palas ligneas).
Cat. Rust. 13.3: "wooden ladles" (trullas ligneas).
Cic. *Inv.* 2.170: "it is a given fact that wooden construction material is able to be burned" (uri posse flamma ligneam materiam necesse est).
Mart. 14.44: for text and translation see *candelabrum* under "XI.6. Miscellaneous Household Furnishings."
Vitr. 5.5.7: "all public wooden theaters have many wooden floors" (omnia publica lignea theatra tabulationes habent complures).

See also, in this section, *ligneolus*.

***lignum* -i (n.):** a general term for wood, including that used for fuel. Wood as it exists in nature. For wood intended for carpentry, see, in this section, *materia*.
Cato *Rust.* 55: for text and translation, see *tabulatum* under "VII. Flooring."
Plin. *Epist.* 2.17.26: "The neighboring woods provide plenty of firewood" (suggerunt adfatim ligna proximae silvae).

***materia* -ae = *materies* (f.):** the term usually employed for wood destined for the workshops of carpenters, furniture makers, and builders (Meiggs 1982, 359).
Caes. *BCiv.* 1.58.3: "for the boats were made hurriedly from green [lit. 'wet'] lumber" (factae [naves] enim subito ex humida materia).
Cato *Rust.* 14.3 (regarding the contractor hired to build a new farm): "he will fell, hew, square, and finish the lumber" (materiem . . . succidet, dolabit, secabit facietque conductor).
Vitr. 2.9.4: "when these [trees] are felled for timber" (cum eae [arbores] ad materiam deiciuntur).

Compare the meaning of *lignum* and see also *perdolo* under "XIII.2. Felling and Rough Cutting."

***materiarius* -a -um:** an adjective used to associate something with wood or woodworking. Thus, *fabrica materiaria* means, literally, "woodcraft" or "carpentry."
Plin. *HN* 7.198: "the craft of working with wood [was invented by] Daedalus" (fabricam materiariam Daedalus).

***materiatio* -onis (f.):** wood intended for building or other woodwork.
Vitr. 4.2.1: "In all buildings wood, called by various terms, is used in the upper parts" (in aedificiis omnibus insuper conlocatur materiatio variis vocabulis nominata).

***opus fabrile*:** in context of woodworking, "carpentry."
Plin. *HN* 16.47 (of resinous trees): "the wood of the male is hard and when used in carpentry splits unevenly" (lignum maribus durum et in fabrili opere contortum).

***secamentum* -i (n.):** a general term for a work of carpentry or of joinery.
Pliny *HN* 16.42: "The wood . . . of the pitch pine is useful for . . . a few other implements of joinery" (materies . . . piceae ad . . . pauca alia secamenta).

See also *picea* under "XII.2. Species of Trees."

silviculture: the exploitation of forests for wood products.

wood: see, in this section, *lignum*, *materia*.

wooden: see, in this section, *ligneolus*, *ligneus*.

woodwork, -ing: see, in this section, *ars fabrilis*, *materiatio*, *opus fabrile*, *secamentum*,

II. Areas of Specialization

***abietaria negotia*:** the timber trade. The term implies that fir (*abies*) was synonymous with timber sales.

Paul. *Fest.* 25L: "the timber trade used to be called *abietaria* which we now call *materiaria* (*negotia*)" (abietaria negotia dicebantur, quam materiariam nunc dicimus).

***abietarius*:** a dealer in wood; specializing in fir sales.

CIL 6.9104 (from Rome, funerary inscription): G. ROIUS ABIETARIUS

***arcularius*:** a maker of chests or boxes.

Plaut. *Aul.* 519: "the offering collectors, the weavers, the lacemakers, and the *arcularii* stand in [your] halls" (stant thylacistae in atriis textores limbularii, arcularii).

bed maker: see, in this section, *faber lectarius.*

box maker: see, in this section, *arcularius.*

cart maker: see, in this section, *faber carpentarius.*

contractor: see, in this section, *conductor, redemptor.*

couch maker: see, in this section, *faber lectarius.*

chest maker: see, in this section, *arcularius.*

***citrarius* = *citriarius* -i (m.):** a dealer in citrus wood; an artisan who works with citrus wood.

CIL 6.9258 (from Rome, Via Appia, funerary inscription): FELICI IUN[IORI], FELICI SATURO SEN[IORI] GENNA AEREA NEAPOLITANORUM CITRARIORUM

CIL 6.33885 (from Rome; regulations connected with a guild): NEGOTIATOR EBORARIUS AUT CITRIARIUS

***conductor*:** the term for a general building contractor.

Cat. *Rust.* 14.3: see *materia* under "I. General Woodworking Terms."

Cic. *Q.Fr.* 3.1.5: "then he replied to me that he himself had been the contractor of the work for 16 sestertia" (tum is mihi respondit, se ipsum eius operis HS xvi conductorem fuisse).

dealer in wood: see, in this section, *abietarius, lignarius*

faber carpentarius or ***carpentarius*:** a builder of the two-wheeled cart (*carpentum*).

Plin. *HN* 16.34: "[cork trees] are used especially by the cart makers [in southern Greece]" (in carpentariis praecipue fabricis).

See, in this section, *faber plaustrarius.*

***faber intestinarius*:** a specialist in finish carpentry, interior woodwork.

CIL 6.8173 (from Rome, funerary inscription from the tomb of the freedmen of Quintus Sallustius): Q. SALLUSTIUS SERMO Q. SALLUSTI ABINNAEI FILIUS FABRI INTESTINAR V. A. V.

CIL 10.1922 (from Puteoli, funerary inscription of Gaius Atilius Fortunatus): G. ATILIVS FORTUNATUS FABER INTESTINARIUS

See, in this section, *faber subaedanus.*

***faber lectarius*:** a craftsman who makes beds and couches.

CIL 6.7882 = *ILS* 7719 (from Rome, funerary inscription of Lucius Hostilius): L. HOSTILIUS L. L. AMPHIO FABER LECTARIUS AB CLOCA [*sic*] MAXIMA SIBI ET . . .

***faber limarius*:** a craftsman who specializes in making files.

CIL 12.4475 = *ILS* 7720 (from Narbonne, funerary inscription of Quintus Baebius): (Q.) BAEBIUS Q. L. TERTIUS FABER LIMARIUS IN SUO HIC REQUIESCIT

faber naupegus* -i** (or simply ***naupegus): a shipwright.

***faber navalis*:** a shipwright.

> *CIL* 11.139 = *ILS* 7725 (from Ravenna, funerary inscription of P. Longidienus): P. LONGIDIENUS P. F. CAM. FABER NAVALIS

(fig. 3.9)

> *CIL* 14.168 (from Ostia, from a guild [here *corpus*] to their patron): C. IULIO PHILIPPO EQUITI ROMANO CORPUS FABRUM NAVALIUM OSTIENS QUIBUS EX S. C. COIRE LICET S. P. P.

***faber pectinarius*:** a maker of combs, usually of boxwood. Possibly a maker of combs used for carding wool.

> *CIL* 5.2543 (from Alessium, funerary inscription): L. OCTAVI SERVANDI PECTINARI

faber plaustrarius (or ***plostrarius***)**:** a builder of wagons (*plaustra*, and wagon wheels).

See, in this section, *plaustrarius.*

***faber subaedanus*:** a craftsman concerned with the interior work of houses. Depending upon specialty probably similar to the *faber intestinarius. CIL* 6.7814 indicates a man who specialized in interior marble work (i.e., for floors and cornices).

> *CIL* 12.4393 (from Narbo, dedicatory inscription to a patron): FABRI SUB AEDIANI NARBONENES PATRONO OB MERITA EIUS
>
> *CIL* 6.7814 (from Rome, funerary inscription): L. VALERIUS L. L. PHARNACES MARMORARIUS SUBAEDANUS

***faber tignarius* (= *faber tignuarius*):** a carpenter, one who works with beams (*tigna*) used in framing buildings. At Ostia the so-called Caseggiato dei Triclini (I, 12, 1) was probably the seat of the guild (*schola*) of carpenters (*fabri tignuarii*) by the third century (Blake 1973, 175).

> Cic. *Brut.* 257: "I would rather be a Phidias than the best of carpenters" (ego me Phidiam esse mallem quam vel optimum fabrum tignarium).

Phidias was a celebrated sculptor from Athens.

> Cic. *Rep.* 2.22.39 (concerning reforms undertaken by the Roman king Servius (sixth century B.C.) and his organization of the citizenry into groups, or centuries, within a centuriate assembly): "and there was the first class with the addition of one century which was composed of carpenters because of their very great usefulness to the city" (et prima classis addita centuria, quae ad summum usum urbis fabris tignariis est data).
>
> *CIL* 6.6364 (from Rome, funerary inscription from the Tomb of the Statili): ANTEROS FABER TIG.
>
> *CIL* 6.9405 (from Rome): "the carpenters' guild" (COLLEGIUM FABRUM TIGNUARIORUM)
>
> *CIL* 6.9409 (from Rome, funerary inscription of Marcus Allius Apollonius): M. ALLIUS APOLLONIUS FABER TIGNUARIUS
>
> Gaius *Dig.* 50.16.235: "we use the term *fabri tignarii* not only for those who cut wood, but for all who are builders" (fabros tignarios dicimus non eos dumtaxat, qui tigna dolant, sed omnes, qui aedificant).

file maker: see, in this section, *faber liminarius.*

finish carpentry: see, in this section, *faber intestinarius.*

***intestinarius*:** see, in this section, *faber intestinarius.*

***lectarius*:** see, in this section, *faber lectarius.*

***lignarius*:** a dealer in wood. *Lignarius* may also serve as the most general term for one who works with wood (at least by Isidore's day; see also Richter 1926, 157; Aldred 1956, 233; Meiggs 1982, 359). In Rome, as in other cities, dealers may have congregated at one place (see Livy 35.41.10).

> *CIL* 4.960 (from Pompeii, electoral endorsement from the wood dealers): CUSPIVM PANSAM AED[ILEM] LIGNARI[I] UNIVERSI ROG

Isid. *Orig.* 19.19.1: "an artisan of wood is generally called *lignarius*" (lignarius generaliter ligni opifex appellantur).

Liv. 35.41.10 (relating events in Rome in 192 B.C.): "the same men built a portico outside the Porta Trigemina in [the quarter of] the *lignarii*" (iidem porticum extra portam Trigeminam inter lignarios fecerunt).

For additional comment on this passage, see Blümner 1875, 240, n. 3.

See also, in this section, *abietarius.*

lignarius plostrarius: a man who transported wood in wagons (perhaps wood for fuel or log-length wood). (Cf. Meiggs 1982, 359).

CIL 4.485 = *CAP* 252 = *ILS* 6417b (from Pompeii, electoral graffito): "the wood-haulers ask you to vote [for . . .]" (LIGNARI PLOSTRARI ROG[ANT]).

lignarius universus: a general term for wood-handlers? The term is known from a graffito at Pompeii (Meiggs 1982, 359).

CIL 4.960 (electoral endorsement): see, in this section, *lignarius.*

***naupegus* -i (m.):** a shipwright.

***officina* -ae:** the general term for a workshop.

Plin. *HN* 36.90: for text and translation, see *tornus* under "III. Tools."

pectinarius: see, in this section, *faber pectinarius.*

***plaustrarius* -a -um:** in the masculine, perhaps one who is a cart or wagon builder (Blümner 1879, 325).

CIL 4.485 (from Pompeii, electoral endorsement): for text and translation see, in this section, *lignarius plostrarius.*

CIL 13.11861 (from Mogontiacum [Mainz] Germany, funerary inscription): D. M. M. M. M. VETERANO LEG XXII P P F NAUPEG ET FIL IANUARIO P CURAVIT

scandularius: one who shingles roofs.

***tector* -oris (m.):** a general term for a builder (cf. the passage from Seneca cited here). The term also refers specifically to a plasterer.

Sen. *Ep.* 90.9: "believe me, that age was blessed before there were architects, before any builders" (mihi crede, felix illud saeculum ante architectos fuit, ante tectores).

tignarius: see, in this section, *faber tignarius.*

workshop: see, in this section, *officina.*

III. Tools

TERMS

A-level: see, in this section, *libella.*

adze: see, in this section, *ascia.*

***ascia* -ae (f.) (also *ascea* -ae):** an adze, a (small) axe or hatchet (n.b. Italian *ascia*), a mason's trowel.

Plin. *HN* 7.198: "carpentry [was invented] by Daedalus, and with it the saw, the *ascia*, the plumb line, the drill, glue and fish-glue" (fabricam materiariam Daedalus [invenit], et in ea serram, asciam, perpendiculum, terebram, glutinum, ichthyocollam).

Plin. *HN* 16.207 (blade blunted by the linden tree): for text and translation see *tilia* under "XII.2. Species of Trees."

Tert. *Apol.* 12.4: for text and translation see, in this section, *runcina.*

See also, in this section, *bipennis, securis.*

auger: a T-shaped boring tool that is turned by hand to create a hole in the workpiece. See, in this section, *terebra.*

axe: see, in this section, *ascia, bipennis, securis.*

bench-stops: see, in this section, "dogs."

binding: iron plates of various shapes that were used to connect pieces of wood or to reinforce points of stress (Manning 1972, 190). See also *fibula* under "IV. Joints."

***bipennis* -is (f.):** a double-bladed axe. The term is used especially in Latin poetry in connection with woodcraft. (fig. 3.12)

> Verg. *Aen.* 11.135–36: "the tall ash rings out under the double-bladed axe" (ferro sonat alta bipenni / fraxinus).
> Hor. *Carm.* 4.4.57: "the holm oak is shorn of its boughs by heavy double axes" (duris ut ilex tonsa bipennibus / . . . frondis).

blade: see, in this section, *lamina.*

bore: see, in this section, *foro.*

bow drill: a drill bit rotated by a thong attached to a piece of wood under tension (the bow). See, in this section, *terebra.*

bow saw: see, in this section, *serra.*

caliper: a hinged tool with two curved legs for taking both inside and outside measurements. Examples from Pompeii are depicted in Adam (1981, 102, fig. 26). (See figs. 3.36, 3.44). Closely related are compasses. See, in this section, *circinus.*

carpenter's bench: the work table used by a carpenter. Carpenter's benches may have become standard with the invention of the bench plane, which was introduced by the Romans (if not the Greeks). (See discussion in Goodman 1964, 183). Thus perhaps the origin of the German term for a carpenter's bench, the *Hobelbank*, or "plane-bench."
(fig. 3.17)

chalk line: a taut line of string, coated with pigment and snapped over the surface of a long piece of wood to create a straight line to use as a cutting guide. (fig. 3.45)

> Sen. *Ep.* 90.9: for text and translation, see, in this section, *serra.*

chisel: see, in this section, *scalprum.*

***circinus*:** compasses or "dividers." Compasses were used both for scribing circles and for transferring measurements from the *regula* (ruler) to the workpiece (Goodman 1964, 200). The Roman tool was usually distinguished by straight, pointed legs. The stationary leg of compasses was referred to as the *centrum*. In modestly appointed toolkits, a caliper could have also served the purpose of a *circinus.*

> Ovid *Met.* 8.247–49: "he [Perdix] was also the first / to bind with a joint two metal arms / so that, separated by a fixed distance, / while one stood fixed the other scribed a circular path" (primus et ex uno duo ferrea bracchia nodo / vinxit, ut aequali spatio distantibus illis / altera pars staret, pars altera duceret orbem).

compasses: see, in this section, *circinus.*

***cuneus* -i (m.):** a wedge of iron or hardwood used for cleaving wood, tightening clamps, or holding centering in place.

> Cato *Rust.* 10.3 (from a list of equipment needed to manage an olive farm): "three axes and three wedges" (securis III, cuneos III).

A similar list is found in Cato *Rust.* 11.4.

Verg. *Georg.* 1.144: "early men cleft easily split wood with wedges" (nam primi cuneis scindebant fissile lignum).

The passage is also quoted in Sen. *Ep.* 90.9.

See also *scindo* under "XIII.2. Felling and Rough Cutting."

decido -ere: see "XIII. Harvesting of Trees."

dividers: see, in this section, *circinus.*

dog, joiner's: a large metal staple used to join two pieces of wood (Manning 1972, 184).

dogs, or bench-stops: dogs keep the workpiece from sliding while it is being planed. Examples are known from Saalburg, Germany (Goodman 1964, 52, fig. 54).

dolabra: (pick)axe; see, in this section, *ascia, securis.*

dolo -are: see under "XIII. Harvesting of Trees."

drill (tool): see, in this section, "bow drill," "strap drill," *terebra.*

drill (verb): see, in this section, *foro.*

figo -ere: to hammer a nail.
Vitr. 7.1.2: for text and translation see *clavus* under "IV.2 Joints."

foro -are: to drill, to bore. Thus the adjective *forabilis -e* meaning "suitable for drilling."
Plin. HN 16.227: "[woods] which are slightly moist are suitable for boring and sawing, for dry woods break beyond the part which you bore or saw" (forabilia ac sectilia quae modice umida, arida enim latius quam terebras aut serras cedunt).

glue: see, in this section, *gluten.*

gluten -inis, also ***glutinum -i* (n.):** glue used for the purpose of joining pieces of wood together. *Glutinatio* refers to the process of gluing, e.g., for the attachment of veneers. References here include bull-hide (*gluten taurinum*) and fish-based glues (*ichthyocolla*).
Lucr. 6.1069: "wood is joined together with bull's glue, so that the grain of [joined] boards more often splits apart than the glued bonds fail their joints" (glutine materies taurino iungitur una ut vitio venae tabularum saepius hiscant quam laxare queant compages taurea vincla).
Plin. HN 7.198 (*glutinum* and *ichthyocolla* as inventions of Daedalus): for text and translation see, in this section, *ascia.*
Plin. HN 11.231: "glue is produced by boiling the hides of cattle; the bull's hide makes the best" (boum coriis glutinum excoquitur, taurorumque praecipuum).
Plin. HN 16.215 (Temple of Artemis at Ephesus): "it is worth noting that the doors were held for four years in a gluing-frame" (id quoque notandum, valvas in glutinis compage quadriennio fuisse).
In glutinis compage means literally "in a frame of glue."
Plin. HN 16.225 (on the suitability of using fir for glued joints): for text and translation see *abies* under "XII.2. Species of Trees."
Plin. HN 16.226: "gluing is also important for veneering things with cut sheets of wood" (magna autem et glutinatio propter ea quae sectilibus laminis . . . operiuntur); later in the same section: "some woods cannot be joined by gluing either to their own type or to other species, such as oak" (quaedam et inter se et cum aliis insociabilia glutino, sicut robur).
Plin. HN 28.236 (medicinal use of glue): "the best glue is derived from the ears and genitals of bulls, and there is no better remedy for burns" (glutinum praestantissimum fit ex auribus taurorum et genitalibus, nec quicquam efficacius prodest ambustis).

hammer: see, in this section, *malleus*.

hatchet: see, in this section, *ascia*.

iron (of a plane): see, in this section, *planum*.

jack plane: see, in this section, *runcina*.

***lamina -ae* (f.;** also ***lammina, lamna*):** a blade.

> Verg. *Georg.* 1.143 (on early technological development): "then came the stiffness of iron and the blade of the shrill saw" (tum ferri rigor atque argutae lammina serrae).
>
> V. Flaccus 1.123–24 (ripping logs into planks): "now [she sees] Thespian [Argus] slicing the pines with the slender blade" (iam pinus gracili dissolvere lamna Thespiaden).

The *lamna gracilis* makes reference to the thin blade of a frame saw.

See also *lamina* under "IX. Interior Woodwork."

lathe: see, in this section, *tornus*.

level: see, in this section, *libella*.

***libella -ae* (f.):** a term used to denote a plummet and line or an A-shaped level. Thus the expression *ad libellam*, which means "perfectly upright (or level)."

> Lucr. 4.513–15: for text and translation see, in this section, *norma*.
>
> Vitr. 1.6.6: "let there be placed perfectly level a marble *amusium*" (conlocetur ad libellam marmoreum amusium).

Amusium may refer to a "benchmark" (Rowland 1999, 30) or "dial" (Granger 1983, 59).

> Plin. *HN* 36.172 (on squaring and leveling walls): for text and translation see, in this section, *norma*.

Cf. also, in this section, *linea, perpendiculum*.

***libro -are*:** to make horizontal or level.

> Cat. Rust. 22.1 (on leveling a mill): "level it so that [the stones] are set equidistant from the rims" (librator uti statuatur pariter ab labris).

***lima -ae* (f.):** a carpenter's file.

> Varro *Ling.* 7.68: "a file belongs to the gear of the carpenter" (lima enim materia⟨e⟩ fabrilis est).

See also, in this section, *scobina*.

***linea -ae* (f.):** a plumb line. Thus, "in a straight line (*ad lineam*)."

> Cato Rust. 14.3 (concerning the construction of a new farm): "the owner will furnish the timber . . . and one plumb line" (materiem . . . dominus praebebit . . . lineam I).

See also, in this section, *libella*.

mallet: see, in this section, *malleus*.

***malleus -i* (m.):** mallet, hammer.

> Liv. 27.49.1 (Punic wars): "they used to carry a carpenter's chisel and a mallet [with which crazed battle elephants could be quickly killed by their handler-riders]" (fabrile scalprum cum malleo habebant).

See also, in this section, *scalprum*.

nail: see *clavus* under "IV. Joints."

***norma -ae* (f.):** a carpenter's (or framing) square, also used by masons, most commonly an L-shaped metal implement.

Isid. *Orig.* 19.18.1: "*norma* comes from a Greek word, without it nothing can be made straight. It is built from three slats (of wood), in such a way that two of the pieces are two feet long, the third measures two feet and ten *unciae;* one joins them, smoothed to an equal thickness, at their ends so that they form a triangle. This is the *norma*" (norma dicta graeco vocabulo, extra quam nihil rectum fieri potest. Conponitur autem tribus regulis, ita ut duae sint binum pedum, tertia habeat pedes duos, uncias decem, quas aequali crassitudine politas extremis cacuminibus sibi iungit ut schemam trigoni faciant. Id erit norma).

Lucr. 4.513–14 (comparing the careless measuring of a house under construction to poor reasoning): "finally, as in construction, if the original ruler [*regula*] is warped, if the square [*norma*] is defective and deviates from straight lines, and if the level [*libella*] is off to one side by the least bit in any part . . ." (denique ut in fabrica, si pravast regula prima / normaque si fallax rectis regionibus exit, et libella aliqua si ex parti claudicat hilum . . .).

Plin. HN 36.172: "[masonry walls] should be made square and level, and ought to conform to the plumb bob [i.e., be vertical]" (ad normam et libellam fieri, ad perpendiculum respondere oportet).

Vitr. 3.5.14: "[the flutes of a column] are carved so that if a *norma* is put in the hollow of the flute and rotated, the point of the *norma* will touch the right and left sides of the curve as it rotates" (ita excavatae, uti norma in cavo striae cum fuerit coniecta, circumacta anconibus striarum circum rotundationem tangendo pervagari possit).

Vitr. 9.pr.6: "Pythagoras showed how to make a *norma* without the specialized instruments of a craftsman [by using rods three, four, and five feet long]" (item Pythagoras normam sine artificis fabricationibus inventam ostendit).

***perpendiculum* -i (n.):** a plumb line and bob.

Plin. HN 2.87 (regarding the calculation of celestial distances): "as though the measure of the heavens were merely regulated by a plumb line" (tamquam plane a perpendiculo mensura caeli constet).

Plin. HN 7.198 (on the invention of the *perpendiculum*): for text and translation see, in this section, *ascia.*

Plin. HN 36.172 (testing walls): for text and translation see, in this section, *norma.*

Vitr. 8.5.1 (a leveling instrument, or *chorobates*): "[the chorobate has] plumb lines hanging from the plank on each side, one for each end" (pendentiaque ex regula perpendicula in singulis partibus singula).

The instrument described took the form of a long table, the top of which was a straight plank (*regula*); the plumb lines at the ends were used to level the device.

See also, in this section, *linea* and *libella.*

plane: see, in this section, *runcina.*

***planum* -i (n):** a flat element, an iron (or sole?) of a plane.

Arn. 6.14: for text and translation see, in this section, *runcina.*

plumb line: see, in this section, *linea*, *libella*, and *perpendiculum.*

***regula* -ae (f.):** a straight piece of wood; a ruler, or straightedge for drawing or sighting lines.

Lucr. 4.513–15: for text and translation see, in this section, *norma.*

Vitr. 8.5.1 (a description of a leveling device, or *chorobates*): "the *chorobates* is a *regula* [plank] about twenty feet in length" (chorobates autem est regula longa circiter pedum viginti).

rule (or ruler): see *regula.*

***runcina* -ae (f.):** a carpenter's plane. Thus "to plane": *runcinare.*

Arn. 6.14 (on the making of cult statues): "those images which intimidate you . . . smoothed with the irons of planes" (simulacra ista quae vos terrent . . . runcinarum levigata de planis).

Plin. HN 16.225: "under brisk planing [fir] makes pretty curly shavings, always twisting in a

spiral like the tendrils of a vine" (spectabilis ramentorum crinibus, pampinato semper orbe se volvens ad incitatos runcinae raptus).
Tert. *Apol.* 12.4 (on the making of a cult statue): "but upon your gods, over every limb . . . fall hatchets and planes and rasps" (at in deos vestros per omnia membra . . . incumbunt asciae et runcinae et scobinae).
Varro *Ling.* 6.96: "[the verb] to plane [is] from *runcina*, of which *rhykane* is the Greek source" (ut runcinare, a runcina, cuius rhykanê origo graeca).

saw: see, in this section, *serra*.

***scalprum* -*i* (n.):** a chisel (or chisel-like tool).
Liv. 27.49.1: for text and translation see, in this section, *malleus*.

***scobina* -*ae* (f.):** a rasp. The sharp teeth of the rasp were used for shaping wood, including grooves (Mols 1999, 86). The sawdust from filing is called *scobis*.
Isid. *Orig.* 19.19.15: "a rasp [*scobina*] is called such because it makes sawdust [*scobem*] by rubbing" (scobina dicta quod haerendo scobem faciat).
Plin. *HN* 11.180 "[the stomach of a sea turtle is] as rough as the rasp[s] of the craftsman" (asperitas ut scobina(e) fabrilis).
Tert. *Apol.* 12.4: for text and translation see, in this section, *runcina*.
Varro *Ling.* 7.68: "*scobina*, 'rasp,' from *scobis*, 'sawdust' " (scobinam a scobe).

See also, in this section, *lima*.

***securis* -*is* (f.):** axe.
Luc. 9.429: for text and translation see *citrus* under "XII.2. Species of Trees."
Pliny *HN* 16.47 (on the difficulty of using axes on "male" pitch-bearing wood): "they kick back and fall with a louder crash and are pulled out of the wood with greater difficulty" (hae [secures] . . . respuuntur et fragosius diut, aegrius revelluntur).

See also *ascia*. For the axe-shaped tenon of the dovetail joint, see *securicula* under "IV. Joints."

***serra* -*ae* (f.):** the saw.
Cato *Rust.* 14.3 (concerning materials to be supplied for a new farm): "one saw [for the contractor]" (serram I).
Plin. *HN* 16.198 (knots and twisted grain in wood are detrimental to saws): for the text and translation see *centrum* under "XII.1. Parts of Trees."
Sen. *Ep.* 90.9: "[all] these were born when luxury was being born: cutting timbers square and cleaving a beam with a steady hand as the saw ran over the marked-out [line]" (ista nata sunt iam nascente luxuria, in quadratum tigna decidere et serra per designata currente certa manu trabem scindere).
Vitr. 2.7.1: "[certain stones] are cut, like wood, with a toothed saw" (. . . quod etiam serra dentata uti lignum secatur).

See also the following entry, *serrula*.

***serrula* -*ae* (f.):** a small (hand?) saw, perhaps more useful for agricultural activities than for woodworking. A serrated cutting edge.
Cic. *Clu.* 180 (concerning the use of such a saw to cut the bottom of an *armarium* in order to steal the money within; perhaps a version of a modern hole-saw?): "a [small] saw with teeth all around and bent" (ex omni parte dentatam et tortuosam . . . serrulam).
Varro *Rust.* 1.50.2 (a tool like a hand scythe used for cutting grain): "they use a curved wooden handle, in the end of which is a small iron saw" (ligneum habent incurvum bacillum, in quo sit extremo serrula ferrea).

square: see, in this section, *norma*.

staple: see, in this section, "dog, joiner's."

strap drill: a powerful two-man drill. The bit is twirled by wrapping a thong, perhaps of leather, around the shank.
(fig. 3.24)

***terebra* -*ae* (f.):** a drill or auger; *terebro -are:* to drill a hole; *terebratio:* the act of drilling a hole in something, the state of being pierced with holes.
(fig. 3.10)

> Vitr. 10.16.5 (on the difficulty of drilling successively larger holes): "a hole is bored by a *terebra*, half a digit, one digit, one and one-half digits [in diameter]" (terebratur terebra foramen semidigitale, digitale, sesquidigitale).

See also, in this section, "bow drill."

toolbox: see *arcula* under "XI.3. Boxes, Chests, Cupboards."
(fig. 3.9)

***tornus* -*i* (m.):** a lathe for turning wood; *torno, -are:* to turn on a lathe.

> Lucr. 4.361 (in regard to the optical effect that makes sharp edges appear rounded at a distance): "as though the stone structures [squared towers] were rounded off with a lathe" (fit quasi ut ad tornum saxorum structa terantur).
> Plin. HN 16.205: "there is a celebrated [craftsman] named Thericles who used to make goblets of turpentine-tree wood on the lathe . . ." (celebratur et Thericles nomine calices ex terebintho solitus facere torno).
> Plin. HN 36.90 (the stone columns of the Lemnian labyrinth): "the drums of which hung so well balanced in the workshop that they could be turned on the lathe by a child" (quarum in officina turbines ita librati pependerunt ut puero circumagente tornarentur).

Presumably these drums are for columns of stone. That stone vessels were turned on the lathe is clear from the following:

> Plin. HN 36.159: "[on the island of Siphnos] there is a stone that is hollowed out and turned on the lathe to form vessels . . ." (lapis est qui cavatur tornaturque in vasa).
> Verg. *Georg.* 2.449–50: "so, too, the fine-grained lindens and boxwood, smoothed on the lathe, take form and are hollowed out by the sharp blade" (nec tiliae leves aut torno rasile buxum / non formam accipiunt ferroque cavantur acuto).
> Vitr. 9.1.2 (description of the entire universe as turning about on an axis which gyrates on wheels): "as if turned by a lathe" (uti in torno perfecit).
> Vitr. 10.15.4: "into this wood[en frame] were [inserted] two axles fashioned on the lathe" (in ea materia fuerunt ex torno facti axiculi duo).

wedge: see, in this section, *cuneus.*

IV. Joints

IV.1. GENERAL TERMS

***catenatio* -*onis* (f.):** a general term for any type of physical join, including those between two pieces of wood. The best example of the use of this term in the context of woodworking is found in Vitr. 2.9.11. For the text and translation of this passage see, in this section, *coagmentum*.

***coagmentum* -*i* (n):** a joint; verb: *coagmento -are:* to join. *Coagmentum* is commonly used for stonework; the term can also apply to a wooden joint. When used with other similar terms (cf. Vitr. 2.9.11 below), *coagmentum* may refer specifically to a vertical joint.

> Cato *Rust.* 18.9 (the construction of a wooden disk for an agricultural press): "make [it] with Punic joints" (Punicanis coagmentis facito).

Vitr. 2.9.11 (regarding elm and ash): "when they are milled for construction they are pliant and because of the weight of the moisture they have no rigidity and quickly sag [or warp] . . . they become harder [by drying] and on account of their suppleness they produce strong joints in both horizontal(?) [*commissuris*] and vertical connections" (sunt in operibus, cum fabricantur, lentae et ab pondere umoris non habent rigorem et celeriter pandant . . . fiunt duriores et in commissuris et coagmentationibus ab lentitudine firmas recipiunt catenationes).

See also, in this section, *catenatio* and *commissura*; and *alnus* under "XII.2. Species of Trees."

***commissura* -*ae* (f.):** a joint. In Vitr. 2.9.11, it is used with *coagmentum* (q.v.); perhaps the latter indicates a vertical joint, while the *commissura* indicates a horizontal (edge-to-edge) joint (in support of this, see the passage from Pliny cited here). Cato (*Rust.* 135.4) uses the term to indicate the "joining," or splice, in a rope. Such a splice in wood would indicate a scarf (q.v.).

Plin. *HN* 16.158 (caulking joints in planking with reeds): "[the reeds are] pounded and stuffed between the joints of ships" (contusa et interiecta navium commissuris).

Vitr. 2.9.11: for text and translation, see, in this section, *coagmentum*.

***compactura* -*ae* (f.), or *conpactura*:** something that has been joined together, a coupling. A composite of wooden elements.

Vitr. 4.7.4 (discussion of the "coupling" of heavy architrave beams): for text and translation see, in this section, *securicula*.

See also *trabs* under "VIII. Roofing and Ceilings."

***compages* -*is* (f.):** a joint. The term is not exclusive to woodworking. The verb *compingo* was used to denote, among other things, the joining of two or more pieces of wood. The term *in compage* (Plin. *HN* 16.215) may refer to the frame used for gluing together two pieces of wood.

Lucr. 6.1069: for text and translation see *gluten* under "III. Tools."

Plin. *HN* 16.215: for commentary on this passage, see *gluten* under "III. Tools" and *valvae* under "IX.3. Doors and Shutters."

Paul. *Fest.* 398–400 L (seams on a ship's planking): for text and translation see, in this section, *subscus*.

Suet. *Aug.* 43.5 (regarding an accident involving the emperor Augustus at the games): "it so happened that the joints of the curule chair loosened, and he fell flat on his back" (evenit ut laxatis sellae curulis compagibus caderet supinus).

Verg. *Aen.* 1.122–23 (a shipwreck): "with the joints in their sides loosened, all [the ships] admit the deadly water, and they gape at the seams" (. . . laxis laterum compagibus omnes / accipiunt inimicum imbrem rimisque fatiscunt).

The terms *commissura* (q.v.) and *iunctura* (q.v.) can also refer to the joints in the hull of a ship.

See also, in this section, "scarf."

***iunctura* -*ae* (f.):** any join of wood, such as that found in wooden statuary, in furniture, or in ship construction.

Caes. *BGall.* 4.17.6 (on the building of Caesar's bridge over the Rhine): the passage is translated in chapter 5 of the text (p. 81).

Plin. *HN* 16.214 (in reference to the cult statue of Artemis at Ephesus): ". . . nard [is poured through the apertures of the statue] so that it may keep the joints together" (nardo . . . ut . . . teneatque iuncturas).

Plin. *HN* 13.93 (a citrus table made out of two pieces of wood): "it is more a marvel of art because of its hidden joint than it could have been if a product of nature" (maiusque miraculum in ea est artis latente iunctura quam potuisset esse naturae).

Sen. *Ep.* 76.13: "a ship is said to be good . . . when it is with joints that keep out water" (navis bona dicitur . . . iuncturis aquam excludentibus).

Compare this particular usage of the term with *compages* and *commissura*, both in this section.

Verg. *Aen.* 2.464 (joints in floorboards): for text and translation, see *tabulatum* under "VI. Framing."

joint: the attachment point between two pieces of wood. Latin terms which describe the joining of two objects are usually not exclusive to woodworking but depend upon literary context. See, in this section, *catenatio, coagmentum, commissura, compages, iunctura.*

***nexus* -*us* (m.):** a bond or joint. Tacitus may use the term to refer to either the joints or the hardware used to secure a wooden framework.

Tac. *Ann.* 4.62 (on the collapse of the wooden amphitheater at Fidenae): "Atilius . . . neither laid the foundations on solid ground nor built the wooden superstructure with firm joints" (Atilius . . . neque fundamenta per solidum subdidit, neque firmis nexibus ligneam compagem superstruxit).

***occludo* -*dere*:** to fasten (two pieces of wood together).

Cato *Rust.* 18.9 (from the description of the construction of an olive press): "when you have fitted them [the boards] together, fasten [them] with dowels of cornel wood" (eas ubi confixeris, clavis corneis occludito).

See also *clavus* under "IV. Joints"; *cornus* under "XII.2. Species of Trees."

IV.2. JOINTS: TECHNICAL TERMS, METHODS, AND FASTENERS OF WOODEN JOINERY

butt joint: a joint formed when two timbers are placed end to end. The butt joint is practical for vertical loads.

***cardo* -*inis* (m., f.):** a point of juncture or rotation. In joinery the term can refer to a tenon or a mortise and tenon. A *cardo masculus* refers specifically to a tenon, a *cardo femina* indicates a mortise. The term is also used to mean "hinge" or "pivot": most Roman doors turned on vertical pins (see *cardo* under "IX.3. Doors and Shutters.").

Cato *Rust.* 18.1 (regarding the construction of an olive press): "[use] thick beams, two feet thick and nine high, including the tenons" (arbores crassas P. II, altas P. VIIII cum cardinibus).

"classical" joint: see, in this section, "mortise and tenon."

***clavus* -*i* (m.):** a bronze or iron nail, commonly square or rectangular in section, with a head, rounded or flat. The *clavus* can also be of wood, in which case it is best translated as "pin" or "dowel." A heavy spike used for connecting beams was called a *clavus trabalis.*

Caes. *BGall.* 3.13.3 (on the construction of Gallic ships): "the crosspieces . . . were attached by iron nails with the thickness of a thumb" (transtra . . . confixa clavis ferreis digiti pollicis crassitudine).

Cat. *Rust.* 18.9 (on the construction of a wooden disk for a pressing room; here a reference to nails (or dowels) fashioned both from cornel wood and iron): "when you have fastened them [the planks] together, join them with dowels of cornel wood . . . [attach the crossbars] with iron nails" (eas ubi confixeris, clavis corneis occludito . . . clavis ferreis).

Note the similarity of construction to that of wooden wheels fashioned for carts.

Isid. *Orig.* 19.19.7: "dowels [*epigri?*] and nails [*clavi*] are those things by which wood is fastened to wood" (epigri et clavi sunt, quibus lignum ligno adhaeret).

Plin. *HN* 16.198 (comparing knots in wood to nails, in terms of hardness): for text and translation see *centrum* under "XII.1. Parts of Trees."

Plin. *HN* 16.206 (dowels of cornel wood): for text and translation see *cornus* under "XII.2. Species of Trees."

Plin. *HN* 16.207: for text and translation see *cedrus* and *quercus* under "XII.2. Species of Trees."

Vitr. 7.1.2 (discussion of floor construction): "then a pair of nails should be driven into each joist through the ends of the plank" (deinde in singulis tignis extremis partibus axis bini clavi figantur).

See also, in this section, *figo* and "nail heading tool."

***complector*:** a verb possibly meaning "to dovetail." See, in this section, *securicula*.

***cultellus* -i (m.):** a wooden pin or dowel.

Vitr. 7.3.2 (concerning the preparation of reeds on the underside of a hanging or false vault that will be covered with plaster): "these [the reeds] are to be fixed with cord to the laths, as already described, and wooden pins are to be driven through them" (hae ad asseres, uti supra scriptum est, tomice religentur cultellique lignei in eas configantur).

Cultellus is used more rarely than *clavus* or *subscus*.

See also, in this section, *clavus*, *fibula*, mortise and tenon, and *subscus*. For a discussion of lathwork, see *asser* under "VIII. Roofing and Ceilings."

dovetail joint: see, in this section, *complector*, *securicula*.

dowel: see, in this section, *clavus*, *cultellus*, *epigrus*, *subscus*.

draw-tongue joint: use of a "false tenon," often pegged with a wooden dowel (*cultellus*, q.v.). This method allows the joint to have greater flexibility, an important issue for shipwrights and wheelwrights. (fig. 4.3)

***epigrus* -i (m.):** possibly a dowel. The meaning is obscure. In Seneca (*De Beneficiis* 2.12.2) the term appears to indicate the hobnails of a boot. Paired with *clavus* in the cited passage from Isidore, the term may refer to a dowel.

Isid. *Orig.* 19.19.7: for text and translation see, in this section, *clavus*.

false tenon: see, in this section, "draw-tongue joint."

***fibula* -ae (f.):** a kind of (diagonal?) brace or pin used to reinforce a wooden joint. Some *fibulae* may have been of metal, but from the examples offered below, particularly Cato *Rust.* 31.1, it is clear that such braces could also be made of wood.

Cato *Rust.* 3.5 "[an olive press should have] six double sets of *fibulae*" (fibulas duodenas).

Cato *Rust.* 31.1: "have dry oak, elm, nut, and fig sticks for making *fibulae* buried in the dunghill or in water, and make *fibulae* from them when needed" (fibulae unde fiant, aridae iligneae, ulmeae, nuceae, ficulneae, fac in stercus aut in aquam coniciantur; inde, ubi opus erit, fibulas facito).

Caes. *BGall.* 4.17.6: a brace used to stabilize the piles of Caesar's bridge over the Rhine. The passage is translated in chapter 5 of the text.

Vitr. 10.2.1 (the top juncture point of two heavy beams used for a lifting crane): "[the beams] are set up, joined at the [upper] ends with a *fibula*, and splayed at the base" (a capite a fibula coniuncta et in imo divaricata eriguntur).

halved joint: see, in this section, "saddle joint."

hinges: see *cardo* under "IX.3. Doors and Shutters."

mediterranean joint: see, in this section, "mortise and tenon."

miter: generally a 45-degree join between two elements to form a right angle. Possibly indicated by the phrase *in ungue* in Vitr. 4.6.2.

Vitr. 4.6.2 (the moldings of a door frame): "they [the upper corners] are joined to the cymation itself *in ungue*" (in ungue ipso cymatio coniungantur).

In general, *in ungue* (or, more commonly, *in unguem*) means simply "precisely" or "to an exact measurement." Since Vitruvius is describing the intersection of two moldings at a 95-degree angle, each connecting face would require a mitered cut of 45 degrees.

mortise and tenon (or classical joint, mortise joint, mediterranean joint): see, in this section, *securicula*, *subscus*, and *cardo*.

rabbet: a groove cut in one timber to receive another.

saddle joint: lumber is notched to fit snugly into the notch of a second piece. The saddle joint is also called a halved joint.

scarf: a (usually horizontal) joint that connects two planks end to end in order to create a longer piece.

See also, in this section, *commissura*.

***securicula -ae* (or *securicla*, f.):** a "little axe," thus the characteristic shape of a dovetail joint, probably referring specifically to the mortise. If only one side is cut with a dovetail, this is called a half-dovetail (or blind dovetail). Weakness lies at the shoulders, where the grain is cut at an angle.

> Vitr. 4.7.4 (the architrave of a temple): "above the columns [horizontal] beams are to be placed fastened together . . . and they are to be so connected with *subscudes* and *securiclae* that the coupling [*conpactura*] allows a space of two fingers between the beams" (supra columnas trabes compactiles inponantur . . . et ita sint conpactae subscudibus et securiclis, ut conpactura duorum digitorum habeant laxationem).

Here the *securiclae* probably refer to the mortises because the *subscudes* (q.v., *subscus*) generally refer to tenons.

> Vitr. 10.10.3 (on catapult construction): "the thickness of the *buccula* [side piece]—it is also called *camillum* or, as some would say, *loculamentum* [little box]—fixed with dovetailed *cardines*, is one hole" (crassitudo bucculae, quae adfigitur -vocitatur camillum seu, quemadmodum nonnuli, loculamentum -securiclatis cardinibus fixa, foraminis I).

Camillum is translated as "little bench" (Rowland 1999, 128) and as "chamber" (Granger 1983, 329). *Cardinibus* is probably best translated as "tenons." The term is discussed in this section and under "IX.3. Doors and Shutters."

> Vitr. 10.11.8: "the *replum*, that is, the 'cover' [of the trigger of a *balista*, or heavy catapult] . . . is set into the shaft of the ladder with a dovetail" (replum, quod est operimentum, securiculae includitur K scapo . . .).

Scapus and *replum* are further discussed under "IX.3. Doors and Shutters."

sewing: the joining of wooden planks with twine or animal gut.

***subscus -udis* (f.):** a tenon. In general usage perhaps simply a pin or a dowel. The wedge-shaped tenon of a dovetail joint.

> Cato *Rust.* 18.9 (the construction of a disk for an olive press with boards): "insert *subscudes* of holm oak" (subscudes iligneas adindito).

Hooper (1993, 35) translates *subscudes iligneas* as "dovetailed oak," but the term may refer to pins, dowels, or tongued tenons that were used for edge-to-edge joinery (such as for wheels).

> Paul. *Fest.* p. 398–400L: "the small wedge-shaped pieces of wood by which planks are joined together are called *subscudes* because the sections into which they are fitted correspond (*succuditur*). Pacuvius used the term in the *Niptra*: 'not a single *subscudus* holds a joint of the boat together,' Plautus in his *Astraba*: 'let it be bored well and add the *subscudes*'" ([subscudes ap]pellantur cune[a]tae tab[ellae, quibus] tabulae inter se con[figuntur, qui] a, quo eae immittuntur, [succuditur. Pa]cuvius in Niptris (250): "nec ulla subscus cohibet compagem alvei," Plautus in Astraba (13): "Terebratus multum sit, et subscudes addite)" (trans. Mols 1999, 95, n. 596).
>
> Vitr. 4.7.4: for text and translation see, in this section, *securicula*.

tenon: see, in this section, *cardo*, "mortise and tenon," and *subscus*.

V. Foundations

V.1. GENERAL TERMS

***fundamentum* -i (n.):** general term for the substructure of a building.

Cato *Rust.* 14.4: "[for a stone or concrete foundation extend] the foundation [one] foot above the ground" (fundamenta supra terram pede).

Vitr. 1.5.7 (concerning military defenses): "now when the foundations are laid out at such a distance from each other, then between these there should be placed other transverse [walls] joined to the outer and inner foundation, arranged comblike, as the teeth of a saw are disposed . . . then the greatness of the bulk of the earth being thus distributed into small parts, will not be able to press with its whole weight, so as to thrust out the foundations of the wall" (cum autem fundamenta ita distantia inter se fuerint constituta, tunc inter ea alia transversa, coniuncta exteriori et interiori fundamento, pectinatim disposita quemadmodum serrae dentes solent esse conlocentur . . . tunc ita oneris terreni magnitudo distributa in parvas partes; neques universa pondere premens poterit ulla ratione extrudere muri substructiones).

See also, in this section, *substructio.*

***substructio* -onis (f.):** general term for the substructure(s) of a building. In most cases these refer to stone foundations.

Caes. *BCiv.* 2.25.1 (description of the theater of Utica) "with the massive substructures of this work" (substructionibus eius operis maximis).

Caesar may refer to the retaining walls of the theater, not to its subterranean foundations.

Vitr. 1.5.7: "the foundations of a [fortification] wall" (muri substructiones).

For full text and translation see, in this section, *fundamentum.*

V.2. TECHNICAL TERMS

***arca* -ae (f.):** the box of a cofferdam. The *arca* was put in place for the construction of the piers of bridges. Vitruvius describes double-shelled construction that is made watertight by the packing of clay between the planked walls of the structure.

Vitr. 5.12.5: "double[-walled] *arcae* bound together with planks and reinforced with chains" (arcae duplices relatis tabulis et catenis conligatae).

While "chains" makes sense as a reinforcing element in this context, note that Vitruvius also uses this term to indicate wooden reinforcing ties (cf. Vitr. 7.3.1 under the term *asser* under "VIII. Roofing and Ceilings").

cutwater: a structure built upstream of the piers or piles of a bridge to protect the bridge from floating debris or rapid water. Caesar describes the construction of such "protectors" (*defensores*) in his description of the bridge over the Rhine.

Caes. *BGall.* 4. 17: for text and translation see chapter 5 of the text.

dwarf wall: a low wall of masonry, usually too narrow to have been built to a full wall-height, upon which a sill plate (q.v.) is set and timber framing used for the superstructure. Also called a sill wall or plinth wall. Used especially in damp areas to avoid damp rot.

***moles* -is (f.):** any massive structure, but in architectural terms, a foundation, dam, or pier. Commonly associated with masonry construction; wooden pilings may also support a masonry superstructure. See discussion above in chapter 5 of the text and also reports on sites in London (Milne 1985), Aldwincle (Jackson and Ambrose 1976), Laupen (Cüppers 1969), and Mainz (Fehr 1981).

Tac. *Ann.* 13.31: "the foundations of [Nero's] amphitheater" (molem amphitheatri).

In this passage Tacitus mentions the beams of wood employed for the project.

***palus* -i (m.):** a pile or a post. Wooden supports placed vertically into the earth, especially in boggy ground. See also, in this section, "post pit."
(fig. 6.3)

Vitr. 2.9.11 (on Ravenna, a low-lying coastal site): "all projects both public and private employ piles of this type [alderwood] under the foundations" (omnia opera et publica et privata sub fundamentis eius generis habeant palos).
Varro Rust. 1.14.2: "the second type [of fence], the 'rustic,' is made of wood but is not living. It is built either of posts placed closely and intertwined with brush or made of posts with holes bored through, and rails, usually two or three to the section" (secunda saep[e]s est agrestis e ligno, sed non vivit: fit aut palis statutis crebris et virgultis implicatis aut latis perforatis et per ea foramina traiectis longuris fere binis aut ternis).

See also, in this section, *sublica* and *tignum*; and *alnus* under "XII.2. Species of Trees."

pier: see, in this section, *moles.*

piles: see, in this section, *palus, sublica.*

plinth wall: see, in this section, "dwarf wall."

post: see, in this section, *palus.*

posthole: a hole dug to receive a wooden beam, or post, such as a *palus* (q.v.). The posthole itself may have been described by the common word for a hole: *foramen.*

post pit: a hole significantly greater in diameter than the post placed within. The excess space is usually packed with stones or gravel. Cf. Maxwell (1976).

post trench: a foundation dug as a trench, into which posts are set.

sill plate in "sleeper trench": a heavy wooden beam laid in a foundation trench, to which are joined vertical posts (or wall studs, perhaps *arrectaria*).

sill wall: see, in this section, "dwarf wall."

sleeper beam: see, in this section, "sill plate."

***sublica* -ae (f.):** a pile employed to support a heavy load, such as a bridge or building. Piles may also be used to buttress retaining walls.

Caes. BGall. 4.17 (Caesar's bridge over the Rhine): for text and translation see chapter 5 of the text.
Liv. 1.33.6 (Pons Sublicius): "[the river was to be spanned] by a pile bridge, the first to be built across the Tiber" (ponte sublicio, tum primum in Tiberi facto).
Plin. HN 34.22: "when he alone [M. Horatius Cocles] fended off the enemy from the pile bridge" (cum hostes a ponte sublicio solus arcuisset).

***tignum* -i (n.):** a general term for a beam of wood.

Caes. BGall. 4. 17 (Caesar's bridge over the Rhine): for text and translation see chapter 5 of the text.

The term *tignum* is also discussed under "VII. Flooring," "VIII. Roofing and Ceilings," and "XIV. Finished Lumber."

VI. Framing

***antepagmentum* -i (n.):** door and window frames. A board covering the exposed beam ends of a roof.

See this term under "VIII. Roofing and Ceilings" and "IX.3. Doors and Shutters."

architrave: see, in this section, *epistylium.*

***arrectarium* -i (n.):** general term for a timber used as a vertical support, such as a wall stud. In this sense, the term is similar to *postis* (q.v.).

Vitr. 10.15.2 (on the walls of a siege machine, or *testudo*, of Hagetor): "the uprights, which were placed upon the framework, were four in number, they were built from two beams" (arrectaria, quae supra compactionem erant quattuor conlocata, ex binis tignis fuerant compacta).

For additional comment on this passage see, in this section, *compactio*.

Vitr. 2.8.20: "for [wattle-and-daub construction] also makes cracks in the plasterer's work because of the arrangement of the uprights and the crosspieces" (etiamque in tectoriis operibus rimas in his faciunt arrectariorum et transversariorum dispositione).

See also, in this section, *stipes, pila, postis, cratis*.

cantilever: a projecting bracket or beam which supports a balcony or room.

See *maenianum* under "IX. Interior Woodwork."

cladding: the exterior skin (*extrema structura*) of a building. In the case of timber construction, wooden planks.

Caes. *BCiv.* 2.9.1 (concerning the construction of a defensive tower): "the ends of the floor joists are hidden by the outermost sheathing of the walls" (capita tignorum extrema parietum structura tegerentur).

colonnade: see, in this section, *porticus*.

column: see, in this section, *porticus*.

compactio* -*onis (f.), or *conpactio*: a general term for a framework of a structure. In the two passages from Vitruvius cited below, the *compactio* refers to the base, or deck, of the siege machine under construction. The term *compactio* is similar in meaning to *compactura* (see this term under "IV. Joints") and is derived from the verb *compingo*, to put (something) together.

Vitr. 10.14.2 (walls for a siege machine): "on top of this [base]frame let composite uprights be raised" (insuper hanc conpactionem exigantur postes compactiles).

Vitr. 10.15.2: see, in this section, *arrectarium*; and *laterarium* under "VIII. Roofing and Ceilings."

***cratis* -*is* (f.):** wattle work, woven panels of twigs, or staves of wood, usually covered with plaster. *Opus craticium* (from adj. *craticius* -*a* -*um*) refers to a gridded wooden frame that is interfilled with concrete.

Vitr. 2.1.3: "at first, with forked sticks placed upright and twigs interspersed, they [primitive men] plastered their walls with mud" (primumque furcis erectis et virgulis interpositis luto parietes texerunt).

Vitruvius does not use the term *cratis* per se but clearly describes earliest practice.

Vitr. 2.8.20 (on the firetraps wattle works create): "truly I wish that walls of wattle work had never been invented" (craticii vero velim quidem ne inventi essent).

However, in the same passage Vitruvius concedes that wattled walls do have the advantage of being speedy to erect.

For applications other than walls, see also, in this section, *furca*; and *cratis* under "VII. Flooring."

cripple stud: see, in this section, *postis*.

deck, wooden: see, in this section, *compactio, tabulatum*.

door frame: see *coassamentum* and *antepagmentum* under "IX. Interior Woodwork."

doorjamb: see, in this section, *postis*.

epistylium*, -*ii (n.), or *epistylum*: the horizontal beams (of wood or other material) above a colonnade (*porticus*). Thus, the term refers to the architrave. Roman builders made architraves of wood, stone, and fired brick. For the construction of wooden epistyles, see *trabes* (*conpactiles*) under "VIII. Roofing and Ceilings."

Vitr. 3.5.8 (on the Ionic order): "the dimensions of the architraves should be treated as follows . . ." (epistyliorum ratio sic est habenda . . .).

***furca* -ae (f.):** a prop. A stick with a V-shaped end. Thus, the *furca* is one of the most primitive forms of wooden supports.

> Ov. *Met.* 8.700: "columns took the place of the forked [wooden] props" (furcas subiere columnae).
>
> Varro *Ling.* 5.117: "the forked ends of sticks have the shape of the letter 'V'" (extrema bacilla furcillata habent figuram litterae V).
>
> Vitr. 2.1.3 (primitive application): for text and translation see, in this section, *cratis*, *pila*.

***iugumentum* -i (n.):** a lintel. Perhaps a more general term for a horizontal framing piece used in an exterior wall. In Cato *Rust.* 14.4, the term refers to a (horizontal?) framing device inserted into a mud-brick wall.

> Cato *Rust.* 14.1: for text and translation of this passage, see chapter 5 of the text.

In this passage Cato refers to both *limina* (q.v.) and *iugumenta*, which probably indicate, respectively, thresholds (or sills) and lintels.

> Cato *Rust.* 14.4: for text and translation of this passage see *antepagmentum* under "IX.3. Doors and Shutters."

See also, in this section, *limen*, *superlimen*.

***limen* -inis (n.):** lintel or threshold. The horizontal beam that spans the opening of a door or window. *Limen* can also refer in general to the entrance of a building. Note the adjective *liminaris* -is -e, "of or belonging to a lintel."

***liminaris*:** see, in this section, *limen*.

> Vitr. 6.3.4 (discussion of the lintels of the *alae* in an atrium of a house): "the lintel beams of these should be placed high" (trabes earum liminares ita altae ponantur).
>
> Vitr. 6.8.2: for the text and translation of this passage, see, in this section, *postis*.

See also, in this section, *iugumentum*, *superlimen*.

lintel: see, in this section, *iugumentum*, *limen*, *supercilium*, *superlimen*.

log-built structures: logs laid in a square pattern at right angles to one another and built up accordingly to form walls. According to Vitruvius, the roofs of such structures built by the Colchi (a race living near the Black Sea) were created by decreasing the log lengths, creating a pyramidal covering.

> Vitr. 2.1.4: "among the race of the Colchi in Pontus . . . once two entire trees [*arboribus perpetuis*] have been laid on the ground to the right and the left, two other trees are laid transversely on top, across the space between [the first pair], opened as far as the lengths of the trees allow, and placed right at the ends; these [four trees] enclose the central space for the house. Then, joining together the corners with alternate beams placed on the four sides, they build the walls with trees, rising perpendicularly from the bottom . . . the gaps which are left because of the thickness of the wood they plug with wood chips and clay" (apud nationem Colchorum in Ponto . . . arboribus perpetuis planis dextra ac sinistra in terra positis, spatio inter eas relicto quanto arborum longitudines patiuntur, conlocantur in extremis partibus earum supra alterae transversae, quae circumcludunt medium spatium habitationis. Tum insuper alternis trabibus ex quattuor partibus angulos iugumentantes et ita parietes arboribus stautentes ad perpendiculum imarum educunt . . . intervallaque quae relinquuntur propter crassitudinem materiae, schidiis et luto obstruunt).

The words *alternis trabibus* probably refer to other unfinished tree trunks, not squared timbers.

See also "half-lap" under "IV. Joints."

***longurius* -i (m.):** a long pole. The horizontal element of a simple structure like a fence.

> Varro *Rust.* 1.14.2: for text and translation see *palus* under "V. Foundations."

***palus*:** see this term under "V. Foundations."

pier: see, in this section, *pila*.

pila -ae (f.): a vertical support, or pier. A *pila* could be wooden, although the use of the term by itself does not signify the material used. Given the thick dimensions (4 Roman feet square) reported by Vitr. 5.1.9, the *pilae* described at Fanum were presumably of brick or stone. The simplest of *pilae* were upright forked sticks (*furcae*, q.v.). For wooden uprights, the terms *postis* (q.v.) or *arrectarium* (q.v.) were preferred.

> Vitr. 5.1.9 (description of Vitruvius's basilica at Fanum): "above these beams, over each capital and column shaft, piers three feet high and four feet square are placed" (supra trabes contra capitula ex fulmentis dispositae pilae sunt conlocatae altae pedes III, latae quoqueversus quaternos).

Here *trabes* may refer to the blocks (or wooden beams) of the architrave or even to the whole entablature. The *fulmentae* apparently refer to the shafts of the columns.
(fig. 8.19)

> Vitr. 6.8.1 (on the stability of buildings): "the walls [of structures], the piers, and the columns should be positioned vertically [lit: "to the plumb line"] in the middle of the substructures" (parietes, pilae, columnae ad perpendiculum inferiorum medio conlocentur).

Here, as in other passages, the use of *pila* refers to any vertical pier of unspecified material.

plate: the horizontal timber at the top or bottom of a timber frame. See also "sill plate." Walls of Roman buildings in Italy were usually not framed with timber (although doors and windows did employ timber framing).

porticus -us (f.): a colonnade of wood or stone.

> Vitr. 5.1.2 (porticoes and their upper floors of wood): for text and translation see *coaxatio* under "VII. Flooring."
> Vitr. 6.7.3 (fine woodwork in the porticoes of wealthy Greek mansions): for text and translation see *lacunar* under "VIII. Roofing and Ceilings."

post: see, in this section, *arrectarium*, *furca*, *postis*.

postis -is (m.): a vertical timber used in a wooden-framed building. A wall stud, but also a "cripple stud," a modern term used to describe a shorter piece of wall stud used to support a window frame in between two full-sized studs. The term can also indicate a doorjamb or one of the vertical stiles of a door.

> Vitr. 6.8.2 (on stabilizing openings in walls): "for when lintels and beams are weighed down by the walls, sagging in the middle they fracture the masonry by settling; when, however, the *postes* are propped underneath and shimmed, they do not permit the beams to settle nor to damage the masonry" (limina enim et trabes structuris cum sint oneratae, medio spatio pandantes frangunt sub lysi structuras; cum autem subiecti fuerint et subcuneati postes, non patiuntur insidere trabes neque eas laedere).

The word *lysis* apparently refers to the action of weakening or loosening, thus here, *sub lysi* (or *sublisi*) probably refers to "settling."

> Vitr. 10.14.2: "composite posts [of more than one timber]" (postes compactiles).

For full text and translation see, in this section, *compactio*.

See other meanings of this term below in "XIV. Finished Lumber." For the use of composite supports (*postes compactiles*), see *trabes conpactiles* under "VIII. Roofing and Ceilings."

stake, wooden: see, in this section, *stipes*, *sudis*.

stipes -itis (m.): a piece of wood, sometimes simply translated as "stick" or "stake," which can be used in construction. Since *stipes* can also refer to the trunk or stump of a tree, the term may refer to a heavier piece of wood in its natural state (such as a log).

> Cato *Rust.* 18.5 (concerning anchor posts for a pressing room): "make a foundation five feet [deep] . . . and set the timbers [*stipites*] there" (fundamenta P. V facito . . . ibi stipites statuito).
> Caes. *BCiv.* 1.27.3: "[when Pompey abandoned Brundisium he] dug trenches across the roads fixing stakes [*sudes*] and sharpened logs [*stipites*] in them" (fossas transversas viis praeducit atque ibi sudes stipitesque praeacutos defigit).

This action was done to slow down the advance of Caesar's troops. It is difficult to draw much of a distinction between *stipes* and *sudis* in this passage. *Sudis*, however, does not appear in the context of building; it is generally taken to indicate a sharpened stake.

See also *stipes* under "XII.1. Parts of Trees."

stud: see, in this section, *arrectarium*, *postis*.

***sudis*, -is (f.):** a wooden stake.

> Caes. BCiv 1.27.3: for text and translation see, in this section, *stipes*.

***supercilium* -i (n.):** the uppermost horizonal molding of a door frame, placed on the wall above the structural horizonal member (lintel). Or the lintel itself.

> Vitr. 4.6.2: "to the right and left of the *supercilium*, which rests upon the frame of the door, extensions should be made" (supercilii, quod supra antepagmenta inponitur, dextra atque sinistra proiecturae sic sunt faciundae).

The extensions, or *proiecturae*, are the "ears" that extend to either side of a lintel on some Roman doors; cf. figs. 9.11, 9.20. This passage is also considered under "miter" in "IV. Joints."

***superlimen* -inis (n.):** a lintel.

> *CIL* 11.4123 (a stone lintel): SUPERLIMINEN ⟨LA⟩PIDE⟨UM⟩

See also, in this section, *iugumentum*, *limen*, *supercilium*.

***tabulatum* -i (n.):** something built with *tabulae*, or wooden boards, and thus a "floor" or "story," including the wooden flooring of military towers. A *tabulatum* is obviously similar to a *coaxatio* (q.v. under "VII. Flooring"); the latter, however, was one element of a floor (the floorboards themselves), usually covered with a masonry paving. A *tabulatum*, however, probably refers to an exposed wooden deck. It is not surprising, then, to find that the term is also used to indicate the deck of a ship.

> Caes. *BCiv.* 1.26.1: "[Pompey] erected towers with three stories [on confiscated cargo ships in the port of Brundisium] (turres cum ternis tabulatis erigebat).
>
> Caes. *BCiv.* 2.9.9 (on the construction of a defensive tower): "in this way . . . they constructed six stories" (ita . . . sex tabulata exstruxerunt).
>
> Verg. *Aen.* 2.463–64 (Aeneas describes the prying up of a floor during the defense of Troy): "we proceeded all around with [our] sword[s], where the ends of the floorboards presented loose joints" (adgressi ferro circum, qua summa labantis / iuncturas tabulata dabant).

For commentary on *iunctura* see "IV. Joints."

> Vitr. 10.13.4 (on the mobile battle-towers of Diades): "[Diades] says, moreover, that the tower ought to incorporate ten floors" (fieri autem ait oportere eam turrem tabulatorum decem).

wattle: see, in this section, *cratis*.

window frame: see *antepagmentum* under "IX. Interior Woodwork."

VII. Flooring

***axis* -is (m.):** a single flooring plank.

> Caes. *BCiv.* 2.9.2: "they fastened these [joists] together with planks" (easque [transversas trabes] axibus religaverunt).
>
> Plin. HN 36.187 (oak *axes* unsuitable for flooring): for text and translation see *quercus* under "XII.2. Species of Trees."
>
> Vitr. 4.2.1: "in flooring [there are] joists and planks" (in contignationibus tigna et axes).
>
> Vitr. 7.1.2: "effort must be made not to mix [flooring] planks of *aesculus* oak with those of [common] oak (danda est opera, ne commisceantur axes aesculini querco).

See also *quercus* under "XII.2. Species of Trees" and *clavus* under "IV. Joints."

bearer-beam: see, in this section, *tignum (transversum)*, *trabs.*

***coaxatio* -*onis* (f.):** a floor composed of planks supported by joists. Decking.

Plin. *HN* 36.186 (regarding the subfloor of a flat roof): "it is necessary that two decks of boards should be laid across one another, and that their ends should be nailed down so that they are not twisted by warpage" (necessarium binas per diversam coaxationes substerni et capita earum praefigi, ne torqueantur).

Vitr. 5.1.2: "in the porticoes [of *fora*] . . . balconies are to be located on the upper wooden floors" (in porticibus . . . maenianaque superioribus coaxationibus conlocentur).

See also *maenianum* under "IX. Interior Woodwork."

Vitr. 6.3.9 (the floor of a balcony overlooking a reception room in a house): "above the planked floor the pavement [is to be placed]" (supra coaxationem pavimentum).

For additional comment on this passage, cf. *circumitus* under "IX. Interior Woodwork."

Vitr. 7.1.2: "when the planked floor has been made, fern (if there is any; if not, then straw) is spread around, so that the wood will be protected from the effects of lime" (coaxationibus factis, si erit, filex, si non, palea substernatur, uti materies ab calcis vitiis defendatur).

Pliny prescribes an identical flooring method in *HN* 36.187.

Vitr. 7.1.5 (floors exposed to the elements): "after laying the floorboards, a second layer of planks is to be laid above at right angles, being fixed with nails, it will furnish a double sheathing for the joists" (cum coaxatum fuerit, super altera coaxatio transversa sternatur clavisque fixa duplicem praebeat contignationi loricationem).

See also, in this section, *contignatio.*

See also *clavus* under "IV. Joints" and, in this section, *coaxo* and *axis.* Related discussion is found under "VIII. Roofing and Ceilings."

***coaxo, -are*:** to fit with floor planking.

Vitr. 2.8.17: "planked . . . with a wooden framework" (contignationibus . . . coaxatae).

For the context of this passage, cf. *contignatio*, below.

Vitr. 7.1.5: for text and translation, see, in this section, *coaxatio.*

***contabulatio* -*onis* (f.):** a floor made of boards.

Caes. *BCiv.* 2.9.1 (on building a fortification tower): "when the tower was built up to the [level of the first] planked floor" (ubi turris altitudo perducta est ad contabulationem).

Caes. *BCiv.* 2.9.4 (on building a fortification tower): "they covered the uppermost floor with bricks and clay [to make it fireproof]" (contabulationem summam lateribus lutoque constraverunt).

Vitr. 10.15.4 (description of a war machine): "it had a board floor . . . carried on joists" (habuerat . . . contabulationem supra trabiculas).

See, in this section, *tabulatum.*

***contignatio* -*onis* (f.):** a wooden floor system, including joists and floorboards. Thus, depending upon context, a "floor," or "story.".

Vitr. 2.8.17: "[mud-]brick walls a foot and a half thick—not being two or three bricks thick—cannot support more than one [upper] wooden floor" (latericii vero, nisi diplinthii aut triplinthii fuerint, sesquipedali crassitudine non possunt plus unam sustinere contignationem).

In the same passage, Vitruvius mentions "towers" (altitudines)—he probably means apartment buildings—"with walls of fired brick" (structuris testaceis) and "planked with closely spaced boarded floors (contignationibus crebris coaxatae)"; and continues, "therefore walls [are raised] to a great height by means of various stories [lit.: boarded floors]" (ergo moenibus e contignationibus variis alto spatio . . .).

Vitr. 2.9.6: for text and translation see *abies* under "XII.2. Species of Trees."

Vitr. 4.2.1: for text and translation see, in this section, *axis.*

Vitr. 5.10.3: "the vaulted ceilings [of baths] will be more functional if they are made of concrete. But if they [the ceilings] are made of timber, a revetment of tile should be placed underneath them" (concamarationes vero si ex structura factae fuerint, erunt utiliores; sin autem contignationes fuerint, figlinum opus subiciatur).

Vitr. 6.3.9: for text and translation see *circumitus* under "IX. Interior Woodwork."

Vitr. 7.1.1: "special care must be taken in the case of upper wooden flooring, lest any wall in the story below is built right up to the pavement [supported by the joists]; rather the wall is to stop short and have the joists hanging above [i.e., resting upon] it. For when the wall goes up in one piece, if the wooden decking above dries or sags [warps] as it settles, the wall, being of a rigid structure, necessarily causes cracks, right and left of it, in the pavements [above]" (in contignationibus vero diligenter est animadvertendum, ne qui paries, qui non exeat ad summum, sit extructus sub pavimentum, sed potius relaxatus supra se pendentem habeat coaxationem. Cum enim solidus exit, contignationibus arescentibus aut pandatione sidentibus, permanens structurae soliditate dextra ac sinistra secundum se facit in pavimentis necessario rimas).

Vitr. 7.1.5 (unsuitability of wooden floors for exterior applications): "wooden floors swell in dampness, shrink when drying" (contignationes umore crescentes aut siccitate decrescentes).

In the same passage, a "reinforced wooden floor" (contignationi loricationem) i.e., one created by a double layer of planks that can support a masonry pavement. See also, in this section, *coaxatio*.

See also, in this section, *contigno*.

***contigno -are*:** lit.: "to furnish with joists," i.e., for flooring (or a flat roof).

Caes. *BCiv*. 2.15.2 (a protective roof over a siege ramp): "whatever is framed [with joists] is covered with wattle; the wattle is covered with clay (quicquid est contignatum cratibus consternitur, crates luto integuntur).

Vitr. 1.5.4 (construction of a timbered catwalk between defensive towers): "the passageways [between] the interior parts of the towers are to be floored with wood, not, however, fastened with metal" (itinera sint interioribus partibus turrium contignata, neque ea ferro fixa).

Compare this passage to the construction of the Pons Sublicius, described above in chapter 5 of the text.

***cratis -is* (f.):** wattle work, employed for both floors and (more commonly) walls. For use as flooring or pavement, see especially the following passages of Julius Caesar:

Caes. *BGall*. 4.17: (Caesar's bridge over the Rhine): for text and translation see chapter 5 of the text.

Caes. *BCiv*. 2.15.2: for text and translation see, in this section, *contigno*.

See also this term under "VI. Framing."

joists: see, in this section, *contignatio*, *tignum*, *trabecula*, *trabs*.

***planities -ei* (f.):** a flat surface. Thus a *planities* can refer to a floor or a level surface defined by joists.

Vitr. 10.15.3 (the floor of Hagetor's siege machine): "above the flat surface defined by the joists, which was above the base" (ita supra transtrorum planitiem, quae supra basim fuerat).

See also, in this section, *transtrum*.

***tabulatio -onis* (f.):** a structure made of boards (*tabulae*). If in reference to flooring, a variant of *tabulatum* (q.v.).

Caes. *BCiv*. 2.9.4 (in reference to the wooden decking of a siege tower): "they placed padding on the decking, so that bolts of the artillery would not smash through the boards" (centonesque insuper iniecerunt, ne . . . tela tormentis immissa tabulationem perfringerent).

***tabulatum -i* (n.):** a floor built of wooden planks; the flooring of siege towers and storage buildings. The Latin *tablinum*, referring to the most important room in the republican house, is probably derived from *tabulinum* (*tabula* + *-inus*).

Cato *Rust.* 55: "store firewood for the owner on a wooden floor" (ligna domino in tabulato condito).

Varro *Rust.* 1.13.1: "dry provisions, such as beans and hay [should be stored] on wooden floors" (aridus, ut est faba et faenum, in tabulatis).

Varro *Rust.* 1.55.5: "those [olives], from which oil is to be made, are usually arranged in piles for a few days on a wooden floor" (haec, de qua fit oleum, congeri solet acervatim in dies singulos in tabulata).

Varro *Rust.* 1.59.3: "some think that apples remain sufficiently agreeable [stored] on boards on top of a masonry floor in the storage room" (in oporotheca mala manere putant satis commode alii in tabulis in opere marmorato).

Marmorato refers to a floor of cement or stone.

tignum -i (n.): a general term for a beam of wood. Its particular function is determined from the context of the passage. For flooring, the *tignum* usually unambiguously refers to a supporting joist.

Caes. *BCiv.* 2.9.1 (description of the joists of a wooden floor in a siege tower): "the ends of the joists" (capita tignorum).

Vitr. 4.2.1 (as a joist): for text and translation see, in this section, *axis.*

Cf. also *tignum* under "VIII. Roofing and Ceilings" and *trabs* under "XIV. Finished Lumber."

trabecula -ae (f.) (or *trabicula*): similar to *trabs* (q.v.). A beam. Vitruvius clearly uses the term to indicate a joist (10.15.4). In Cato *Rust.* 18.5, the term indicates a long (over twenty-three feet), heavy beam for an olive press. While often translated as a light joist or beam, the citations below suggest heavy, yet relatively smaller, timbers than others mentioned with them in the same passage.

Vitr. 10.15.4 (on the siege machine of Hagetor): "[the machine] had a floor in the middle supported on joists" (habuerat autem mediam contabulationem supra trabiculas).

The *mediam contabulationem* is, as translated by Granger (1983, 355), probably a central aisle.

Cato *Rust.* 18.5 (construction of an olive press): "under these beams (two by one Roman feet in dimension) . . . place a *trabecula* 1.5 feet square and 23.5 feet long—or substitute two [smaller] beams" (sub eas trabes . . . trabeculam pedum XXIII S inponito sesquipedalem, aut binas pro singulis eo supponito).

For the practice of connecting two beams to form a larger one, see *trabes conpactiles* under "VIII. Roofing and Ceilings."

trabs -bis (f.): a general term for a structural beam. Depending upon the context of the ancient passage, a *trabs* may support flooring.

Caes. *BCiv.* 2.9.5 (heavy mats hung from a siege tower to protect it from enemy missiles): "[the mats are hung from] projecting timbers" (trabes eminentiores).

Trabes eminentiores refer to the ends of the joists of the interior floors. See also, in this section, *contignatio.*

Vitr 5.1.6 (the building of the basilica at Fanum): "[the pilasters] . . . support the joists upon which the flooring of the gallery is carried" ([parastaticae] . . . quae sustinent trabes, in quibus invehuntur porticuum contignationes).

See also this term under "VIII. Roofing and Ceilings" and "XIV. Finished Lumber." See also "log-built structures" under "VI. Framing."

transtrum -i (n.), or **trastrum:** general term for a cross-beam. Used rarely to indicate a floor joist.

Vitr. 10.15.3 (description of the siege machine of Hagetor): "above the flat surface of the cross-beams" (supra transtrorum planitiem).

See, in this section, *planities.*

VIII. Roofing and Ceilings

VIII.1. GENERAL TERMS

contego -ere: to roof over; virtually identical to *tego*, in this section.

Plin. *HN* 36.186: "The Greeks invented open-air [flat roofing], and roof their houses in this way, convenient in a region with a warm climate, but unsuitable where rainfall freezes" (subdialia Graeci invenere talibus domos contegentes, facile tractu tepente, sed fallax ubicumque imbres gelant).

***contignatio -onis* (f.):** a ceiling of joists. *Contignatio* is conventionally used to describe wooden flooring. In the passage referenced here it is employed by Vitruvius to indicate a wooden framework for a ceiling in a bath building.

Vitr. 5.10.3: for text and translation see *contignatio* under "VII. Flooring."

See full discussion of this term under "VII. Flooring."

tego -ere: to cover with a roof. As a general term for roofing, it is possible only from the context of a given passage to know the nature of the roofing material. The term is synonymous with *contego*.

Caes. *BGall.* 5.43.1 (concerning the dwellings of the Gauls): "which had been, in accordance with Gallic custom, roofed with straw" (quae more Gallico stramentis erant tectae).

Liv. 27.36.8: "in that year . . . it was recorded that the *comitium* was covered for the first time" (eo anno primum . . . comitium tectum esse memoriae proditum est).

Since the *comitium* was an open-air assembly place, this statement would indicate that the *comitium* could be covered on hot days with a temporary structure or an awning.

Vitr. 4.2.1: "so that the walls are covered by the eaves [of the roof]" (uti parietes protecturis eorum tegantur).

VIII.2. TECHNICAL TERMS

***antepagmentum -i* (n.):** a covering placed over the ends of exposed joints; most commonly found on the facades of early temples.

CIL 1.577; 10.1781 (Lex operum Puteolana): for discussion see chapter 9 of the text.

Vitr. 4.7.5: for text and translation see, in this section, *traiecturae mutulorum*.

See this term also under "IX.3. Doors and Shutters."

***area -ae* (f.):** an opening—or the frame of the opening—in the roof and ceiling of a Roman atrium (i.e., the *compluvium*). Or, *area* may simply be a corruption of the term *arca* (q.v.). In Vitruvius 6.3.2, *aream* appears as *arcam* in the oldest extant manuscript (eighth century, British Museum, *Harl.* 2767).

Vitr. 6.3.2: "displuviate [atria] are those, moreover, in which the corner beams supporting the frame of the opening drop down to the eaves" (displuviata autem sunt, in quibus deliquiae aream sustinentes stillicidia reiciunt).

The *stillicidium* (q.v.) refers to the drip-line of the roof. See also, in this section, *deliquia*.

***arca -ae* (f.):** the frame of the *compluvium*, which borders the opening, *conpluvii lumen* (Vitr. 6.3.6), in the ceiling of a Roman atrium. The *arca* was formed from two *interpensiva* (q.v.) and two *tigilli* (or *trabeculae*, q.v.) which were themselves supported by the heavy *trabes* (q.v.) of the atrium.

Vitr. 6.3.2: for text and translation see, in this section, *area*.

Vitr. 6.3.4 (from the discussion of the ideal proportions of the atrium of a Roman house): "the rest [of the total height, calculated as one-quarter of the length of the atrium] is to be planned as the dimension of the coffering and the *arca* above the [main] beams [of the ceiling]" (reliquum lacunariorum et arcae supra trabes ratio habeatur).

***asser -eris* (m.):** (1) a (light) rafter of a roof. From the references made by Vitruvius in book 4, *asser* apparently represents one of the lighter set of rafters supported by purlins, *templa* (q.v.), or the principal framing rafters, the *cantherii* (q.v.). Upon the *asseres*, tiles (*tegula*, q.v.) would be laid.

Vitr. 4.2.4: "for the Greeks call the beds of beams and rafters *opae*" (opas enim Graeci tignorum cubicula et asserum appellant).

In this passage *asser* is used with *tignum* (q.v.); the latter usually refers to a horizontal beam, so the *asseres* of this passage could be in reference to rafters.

Vitr. 4.2.5: "and so in Greek buildings no one places dentils under a mutule; for *asseres* cannot be placed underneath *cantherii*" (itaque in graecis operibus nemo sub mutulo denticulos constituit; non enim possunt subtus cantherios asseres esse).

The passage is important for determining the relationship between the *cantherius* and the *asser*. For more on this passage, see also, in this section, *cantherius* and *fastigium*.

***asser* -*eris* (m.):** (2) lathwork (for plastering, especially ceilings).

Vitr. 7.3.1: "when therefore the method for curved ceilings is considered, the following should be done. Laths are placed parallel to one another, no more than two feet apart; cypress wood is preferable . . . and these lathes, when they have been arranged in the form of an arch, are to be secured by wooden ties to the floor joists or roof above, and fastened at frequent intervals with iron nails. The ties are to be made from wood that neither decay nor age nor damp can harm, such as boxwood, juniper, olive, Valonia oak, cypress, and the like, except for the common oak, which warps and causes cracks where it is used" (cum ergo camerarum postulabitur ratio, sic erit faciunda. Asseres directi disponantur inter se ne plus spatium habentes pedes binos, et hi maxime cupressei . . . hique asseres, cum ad formam circinationis fuerint distributi, catenis dispositis ad contignationes, sive tecta erunt, crebriter clavis ferreis fixi religentur. Eaeque catenae ex ea materia comparentur, cui nec caries nec vetustas nec umor possit nocere, id est e buxo, iunipero, olea, robore, cupresso ceterisque similibus praetor quercum, cum ea se torquendo rimas faciat quibus inest operibus).

Cf. also "dowels" under "IV. Joints" and *tignum* under "XIV. Finished Lumber." For the best woods to be used, see "XII.2. Species of Trees."

batten: a light timber placed on top of rafters at right angles to them. For the possible Latin equivalent, see, in this section, *templum*.

bearer-beam: see, in this section, *trabs*.

***cantherius* -*i* (m.):** a principal rafter. The pitched timbers between the ridgepole (*columen*, q.v.) and the sidewalls (or end of the eaves—*suggrundationes*, q.v.) of a building. The resultant triangular form is represented by the *fastigium* (q.v.), or pediment, at each end of the building. In traditional Etruscan (Tuscan) temples, where a triangular truss is not used, the *cantherii* probably refer to the heavy "principal" rafters, while the tile-supporting *asseres* (from *asser*, q.v.) serve as lighter rafters above the *cantherii*. (fig. 8.10)

Vitr. 4.2.1: "above the rafters are the purlins" (supra cantherios templa).

For full text and translation see chapter 8 of the text.

Vitr. 4.2.5 (provides good evidence that *cantherii* and *asseres* were placed perpendicularly to the ridgepole and were thus heavy and light rafters, respectively): for text and translation see, in this section, *fastigium*.

Vitr. 4.7.5: for text and translation see, in this section, *templum*.

For the bracing used for rafters see, in this section, *capreolus*.

***capreolus* -*i* (m.):** part of the framing of a wooden roof, perhaps the diagonal element (main rafter) of the tie-beam truss. Or, simply, any diagonal roofing element, including rafters and braces (Rowland 1999,219). Mazois (1809–11) suggests that this term (or *cantherius*) may refer to the rafters of an atrium.

Caes. *BCiv.* 2.10.3 (on the construction of a covered gallery): "these [uprights] join with low-pitched rafters where the beams were placed for the sake of roofing [the gallery]" (has (columellae) inter se capreolis molli fastigio coniungunt, ubi tigna quae musculi tegendi causa ponant collocentur).

Here there is clear distinction between the vertical framing members and the diagonal *capreoli*.

Caes. *BCiv.* 2.10.5 (continued discussion of the covered gallery): "and thus pitched . . . when beams were placed on the rafters [the roof was] covered with bricks and clay" (ita fastigato . . . ut trabes erant in capreolis collocatae, lateribus lutoque musculus).

Here *fastigato* could also mean "trussed."

Vitr. 4.2.1: "under roofs, if the spans are greater, tie-beams and rafters [are used]" (sub tectis, si maiora spatia sunt, et transtra et capreoli).

Rowland translates *capreoli* as "braces" (1999, 56); Granger: "stays" (1983, 213). For full discussion refer to chapter 8 in the text.

Vitr. 10.15.3: for the text and translation see, in this section, *laterarium.*

coffer: see, in this section, *lacunar, laqueare.*

colliciae -arum (f.) (or ***colliquae***): the valley, or the corner beam, at the juncture of two inwardly sloping roofs. Mazois (1809–11) offers the term *tigni colliquiarum* to indicate the raking beams at the four corners of an impluviate atrium's roof.

Vitr. 6.3.1 "[the Tuscan atrium has] *collicias* running down from the corners of the walls to the corners of the beams [of the frame of the *impluvium*]" (collicias ab angulis parietum ad angulos tignorum incurrentes).

See also, in this section, *interpensiva.*

columbarium: also the *cubiculum tigni*, the wall socket into which a flooring or ceiling joist is inserted.

Vitr. 4.2.4: "our [people (i.e., Romans) call] these sockets *columbaria*" (uti nostri ea cava columbaria).

Vitruvius says the space between two *columbaria* is called by the Greeks *metope;* in Latin *intertignium* (q.v.). Thus the *columbarium* may also simply be the "bed" onto which the ceiling joist is laid.

***columen -inis* (n.):** the ridgepole of a building. The *columen* runs longitudinally along the apex of a pitched roof; also the top element of a wall (e.g., Cato *Rust.* 15.1). The term may also refer to the king post of a truss (Rowland 1999, 219).

Cato *Rust.* 15.1 (a feature of a farmhouse wall): "a one-foot *columen*" (et columen P. I).

Here *columen* may refer to the coping of a masonry wall.

Vitr. 4.2.1 (prop-and-lintel vs. the timber truss): for text and translation see chapter 8 of the text.

Vitr. 4.7.5: for text and translation see, in this section, *templum.*

***compluvium -i* (n.):** the name given to the opening of the roof in the atrium of a Roman house. The *arca* (q.v.) forms the frame of the *compluvium.*

compono -ere: to construct something. In the context of the passage below *componere* refers to the act of supporting the heavy roofing beams employed in the Corinthian-style atrium upon the columns which are placed around the *impluvium.*

Vitr. 6.3.1: "in the Corinthian [atrium] the beams and *compluvium* have similar proportion, but the main beams, extending from the walls, are placed upon [*componuntur*] the columns arranged around [the *impluvium*]" (in corinthiis isdem rationibus trabes et conpluvia conlocantur, sed a parietibus trabes recedentes in circuitione circa columnas componuntur).

***contabulatio -onis* (f.):** a sheathing of boards for flooring or roofing. In siege machines, for example, a sheathing of wooden boards could be used to cover the rafters. In many permanent structures, a sheathing, or *contabulatio*, of wood was not necessary, as tiles could be laid directly upon the wooden framework of the roof. The term is discussed fully under "VII. Flooring."

See also, in this section, *operculum*, "sheathing."

contigno -are: "to furnish with joists" for flooring and/or roofing. The term is discussed fully under "VII. Flooring."

cross-beam or bearer-beam: see, in this section, *trabs, transtrum.*

deliquia (or ***delicia***) ***-ae*** **(f.)**: the corner beams of an outwardly sloping (displuviate) roof. These would correspond to the corner rafters of a modern hip (or hipped) roof.

Vitr. 6.3.2: for text and translation see, in this section, *area.*

eave: see, in this section, *protectum, suggrunda.*

fastigium -i **(n.)**: the pitched timber roof of a building. Also a pediment at the end of a pitched roof. An inclination or a slope and thus the pitch of a roof. In certain contexts, the term apparently indicates the presence of a timber truss.

Caes. *BCiv.* 2.10.3: see, in this section, *capreolus.*

Caes. *BCiv.* 2.10.5: see, in this section, *capreolus.*

Cic. *Orat.* 3.180: "it was necessity, not [the desire for] charm, that conceived the pediment of the Capitol [the temple of Jupiter Optimus Maximus] and [those pediments of] other temples; for when it was considered by what way water could flow down both sides of the roof, the majesty of the pediment of the temple resulted from the practicality [of the design]" (capitoli fastigium illud et ceterarum aedium non venustas sed necessitas ipsa fabricata est; nam cum esset habita ratio quemadmodum ex utraque tecti parte aqua delaberetur, utilitatem fastigii templi dignitas consecuta est).

Vitr. 2.1.3 (the dwellings of prehistoric, primitive men): "when the roofs could not withstand the rains of winter months, making gabled roofs [*fastigia*], and with clay smeared down the sloping roofs, they drew off the water" (posteaquam per hibernas tempestates tecta non potuerunt imbres sustinere, fastigia facientes, luto inducto proclinatis tectis, stillicidia deducebant).

Vitr. 4.2.5 (on the pediments of temples): "the ancients placed neither mutules nor dentils in pediments but only cornices, because neither *cantherii* nor *asseres* are distributed along the facades of pediments, nor can they project [from them], but are placed sloping down toward the drip-line of the eaves" (antiqui . . . neque instituerunt in fastigiis mutulos aut denticulos fieri sed puras coronas, ideo quod nec cantherii nec asseres contra fastigiorum frontes distribuuntur nec possunt prominere, sed ad stillicidia proclinati conlocantur).

Vitr. 5.1.10 (on the roof of his basilica built at Fanum, where two ridge beams intersect at right angles): "and so created from the roof is a double [i.e., intersecting] arrangement of triangular pitches" (ita fastigiorum duplex tecti nata dispositio . . .).

See also, in this section, "tie-beam truss."

gable: see, in this section, *fastigium.*

gutter: see, in this section, *colliciae.*

imbrex -icis **(f.** and **m.)**: the semicircular tile placed over the joints of pan tiles (*tegulae,* q.v.).

Plaut. *Mil.* 504: "[by running over my roof] you smashed my pan and cover tiles" (meas confregisti imbricis et tegulas).

Plaut. *Mostell.* 109 -10: "a storm comes and it cracks the pan and cover tiles" (tempestas venit / confringit tegulas imbricesque).

inassero -are: to install *asseres,* or light rafters.

CIL 1.577 (Lex operum Puteolana): "install the rafters" (TRABECULAS . . . INASSERATO).

For full discussion of this inscription, see chapter 9 of the text.

interpensiva -orum **(n. pl.)**: cross-beams or "trimmers" that, along with main beams (*trabes*), form the frame (*arca*) of the opening (*compluvium*) of the atrium in a Roman house. The *interpensiva* presumably are those timbers which run at right angles and across the *trabes.*

Vitr. 6.3.1: "Tuscan [*atria*] are those in which the main bearing beams extending across the width of the atrium have *interpensiva* and *collicias* . . ." (Tuscanica sunt, in quibus trabes in atrii latitudine traiectae habeant interpensiva et collicias . . .).

See also, in this section, *trabs* and *colliciae*.

***intertignium* -i (n.):** the space between two ceiling beams. The space lying between the ends of the beams is filled by the "metope" on the frieze level of entablatures of the Doric order.

Vitr. 4.2.4: "thus the *intertignium* between two *opae* is called the metope among them [the Greeks]" (ita quod inter duas opas est intertignium, id metope est apud eos nominata).

Opae are the beds for beams and rafters on the entablature of a Greek temple. For a similar passage (with *inter tigna*), cf. Vitr. 4.2.2.

joist: see, in this section, *tignum*, *trabecula*, *trabs*.

***lacunar* -aris (n.):** an ornamental panel covering the gaps in the framework of a ceiling. A coffer.

Petron. *Sat.* 60.1 (at Trimalchio's dinner party, a trick ceiling plays a dramatic role in the spectacles accompanying the feast): "suddenly the coffers began to creak and the whole dining room shook. I jumped up in surprise and fear that some acrobat would drop down from the ceiling. All the other guests, no less surprised than I, turned their faces upward, looking for some sign from heaven. And behold, when the paneling drew apart, a huge round ring . . . was lowered" (nam repente lacunaria sonare coeperunt totumque triclinium intremuit. Consternatus ego exsurrexi et timui, ne per tectum petauristarius aliquis descenderet. Vultus expectantes quid novi de caelo nuntiaretur. Ecce autem diductis lacunaribus subito circulus ingens . . . demittitur).

Plin. *HN* 35.124 (on the decoration of coffered ceilings): "the same [artist, Pausias,] also first introduced the painting of coffered ceilings, nor was it customary before him for vaulted ceilings to be decorated in this way" (idem et lacunaria primus pingere instituit, nec camaras ante eum taliter adornari mos fuit).

Pausias was apparently a contemporary of Apelles; both artists were taught by Pamphilus (cf. Plin. *HN* 35.123).

Sen. *Ep.* 90.9 (a wistful look backward to purer times): "for they were not readying a ceiling [or roof] for a hall about to admit a great banquet; for no such use did they carry the pine trees or the firs in a long procession of carts on the shaking streets, merely to fasten to them coffering heavy with gold" (non enim tecta cenationi epulum recepturae parabantur, nec in hunc usum pinus aut abies deferebatur longo vehiculorum ordine vicis intrementibus, ut ex illa lacunaria auro gravia penderent).

Suet. *Nero* 34.2 (plans for the murder of his mother, Agrippina): "he [Nero] tampered with the coffering which, loosened by a mechanical device, would fall upon the sleeping woman during the night" (lacunaria, quae noctu super dormientem laxata machina deciderent, paravit).

Vitruvius uses various forms of *lacunar* frequently. Interestingly, he never uses the alternate form, *laqueare* (q.v.):

Vitr. 2.9.13: "in the temple at Ephesus, the statue of Diana and the coffers of the ceiling [*lacunaria*]—both there and in other great sanctuaries—are made of [cedar]" (Ephesi in aede simulacrum Dianae ex ea, lacunaria et ibi et in ceteris nobilibus fanis).

Plin. *HN* 16.213: "it is agreed that the roof [of the Temple of Diana at Ephesus] is made of beams of cedar" (convenit tectum eius esse e cedrinis trabibus).

Vitr. 5.2.1: (on the interior finish of a Senate house, or *curia*): "but if it is longer than it is wide, let the length and breadth be added together and let half the total be given to the height under the coffered ceiling" (sin autem oblonga fuerit, longitudo et latitudo componatur, et summae compositae eius dimidia pars sub lacunaris altitudini detur).

Vitr. 6.3.4: for text and translation see, in this section, *arca*.

Vitr. 6.3.9(a) (on the decoration of a "Corinthian" *oecus*, here covered with a barrel vault): "above the cornices, curved coffering rounded to a circular section" (supra coronas curva lacunaria ad circinum delumbata).

The *oecus* was a private reception room in a house. Cf. the following passage:

Vitr. 6.3.9(b) (the "Egyptian" *oecus*, with two superimposed rows of columns which support a flat ceiling and a clerestory lighting system): "above their architraves and moldings they are adorned with paneled ceilings, and windows are placed between the upper columns. Thus they [Egyptian halls] resemble basilicas . . ." (supra earum epistylia et ornamenta lacunariis ornantur, et inter columnas superiores fenestrae conlocantur; ita basilicarum ea similitudo).
Vitr. 6.7.3 (on the sumptuous mansions of the Greeks): "the colonnades of the peristyles are fitted out with [ceilings of] stucco, plaster, and fine wooden coffering" (porticusque peristyliorum albariis et tectoriis et ex intestino opere lacunariis ornatas).

Cf., in this section, *laqueare;* and *opus intestinum* under "IX. Interior Woodwork."

***laqueare* -*is* (n.):** a coffer in a paneled ceiling. The plural form, *laquearia*, usually refers to a coffered ceiling.
Man. *Astronomica* 5.285–92 (for translation see chapter 8 of the main text):

> et, quia dispositis habitatur spica per artem
> frugibus, ac structo similis conponitur ordo,
> seminibusque suis cellas atque horrea praebet,
> sculpentem faciet sanctis laquearia templis
> condentemque novum caelum per tecta Tonantis.

Plin. *HN* 33.57 (the popularity of gilt coffered ceilings): "now, even in private houses, coffered ceilings are covered with gold; they were gilded for the first time in the Capitolium [Temple of Jupiter] after the fall of Carthage during the censorship of Lucius Mummius (146 B.C.) . . . from coffered ceilings the practice spread to vaults and even walls, which are gilded as if they are fine vessels" (laquearia, quae nunc et in privatis domibus auro teguntur, post Carthaginem eversam primo in Capitolio inaurata sunt censura L. Mummi. Inde transiere in camaras quoque et parietes, qui iam et ipsi tamquam vasa inaurantur).
Sen. *Ep.* 90.15: "[one can outfit] a dining room with movable coffering that . . . presents one pattern after another, the ceiling changing as often as the courses" (versatilia cenationum laquearia ita coagmentat, ut subinde alia facies atque alia succedat et totiens tecta quotiens fericula mutentur).
Stat. *Silv.* 4.2.31: a flattering description of a state dinner in Domitian's palace (dedicated A.D. 96); the emphasis seems to be on the height of the triclinium:

> Tectum augustum, ingens, non centum insigne columnis
> sed quantae superos caelumque Atlante remisso
> sustentare queant. Stupet hoc vicina Tonantis
> regia, teque pari laetantur sede locatum
> numina. Nec magnum properes excedere caelum:
> tanta patet moles effusaeque impetus aulae
> liberior, campi multumque amplexus operti
> aetheros, et tantum domino minor; ille penates
> implet et ingenti genio iuvat. Aemulus illic
> mons Libys Iliacusque nite⟨n⟩s ⟨et⟩ multa Syene
> et Chios et glaucae certantia Doridi saxa
> Lunaque portandis tantum suffecta columnis.
> Longa supra species: fessis vix culmina prendas
> visibus auratique putes laquearia caeli.

The Latin passage is translated in chapter 8 of the text (p. 152).

Tac. *Ann.* 4.69 (a distinction between the roof and the coffered ceiling): "[the space] between the roof and the ceiling" (tectum inter et laquearia).

From the passage we learn that such a space, the attic, is an excellent place for eavesdropping.

***laqueatus -a -um*:** the adjectival form of *laqueare*. A ceiling (note use of *tectum*) can be described as *laqueatum*:

Cic. *Verr.* 2.1.133 (a visit by Verres to the Temple of Castor): "he himself went into the Temple of Castor, he surveyed the temple, he saw the roof [ceiling] paneled most beautifully on all sides . . ." (venit ipse in aedem Castoris, considerat templum; videt undique tectum pulcherrime laqueatum).

Cic. *Tusc.* 1.85 (speculating on Priam's fate had he survived the sack of Troy; Cicero quotes Ennius): "his barbarous opulence intact / with ceilings carved and paneled" (astante ope barbarica / tectis caelatis, laqueatis).

Liv. 41.20.9: "[King Antiochus promised to build] a magnificent temple [to Jupiter Capitolinus at Antioch, Syria] of which not only the coffers, but all the walls, were to be covered with sheets of gold" (magnificum templum, non laqueatum auro tantum, sed parietibus totis lammina inauratum).

See also *lamina* under "IX. Interior Woodwork."

Luc. 10.112–13 (regarding Cleopatra's palace): "the coffered ceilings displayed wealth, and thick gold covered the beams" (laqueataque tecta ferebant / divitias crassumque trabes absconderat aurum).

Lucr. 2.23–28 (the presence of the coffered ceiling in a private house as an indicator of conspicuous consumption): "nature herself does not feel deprived . . . if the house does not gleam with silver and shine with gold, nor do the paneled golden beams reverberate from the lyre" (neque natura ipsa requirit . . . nec domus argento fulget auroque renidet / nec citharae reboant laqueata aurataque templa).

Suet. *Nero* 31.2: "there were dining rooms coffered with ivory panels that rotated" (cenationes laqueatae tabulis eburneis versatilibus).

***laqueo -are*:** "to adorn," presumably with a paneled ceiling.

Man. *Astronomica* 1.532–36: "these then are the constellations which decorate heaven with an even swath, coffering the sky with their fires in various designs. Higher than these there is nothing; they are the rooftops of the world; the public domain of nature is contented to be held by them, embracing the sea and lands lying below."

> (haec igitur texunt aequali sidera tractu
> ignibus in varias caelum laqueantia formas.
> altius his nihil est; haec sunt fastigia mundi
> publica naturae domus his contenta tenetur
> finibus, amplectens pontum terrasque iacentis).

***laterarium -i* (n.):** a horizontal roofing beam, probably running parallel to the main ridgepole (*columen*). Thus a *laterarium* could refer to a "purlin" or, for small structures, a "batten" (Rowland 1999, 131). The *laterarium* is supported by the principal rafters.

Vitr. 10.14.3 (on the roofing of a siege machine): "[The rafters (*capreoli*)] themselves are held together with *laterarii* placed on both sides and covered with planks" (ipsi autem laterariis circa fixis contineantur teganturque tabulis).

Vitr. 10.15.3 (on the roofing of a siege machine (*testudo*) of Hagetor): "above [the framed walls] rafters were raised to a height of twelve feet; above the rafters a (ridge)beam was placed that connected the joints(?) of the rafters. [These rafters] had *lateraria* attached across them, upon which planking was placed to cover the lower parts" (supra eam [compactionem] capreoli extollebantur altitudine pedum XII; supra capreolos tignum conlocatum coniungebat capreolorum compactiones. Item fixa habuerant lateraria in transverso, quibus insuper contabulatio circumdata contegebat inferiora).

Cf. also, in this section, *trabs*.

lathwork (for plastering, especially ceilings): see, in this section, *asser*.

***operculum* -i (n.):** sheathing. The *opercula* of a roof were the boards that covered the rafters. They may also be the boards that cover the back wall of an otherwise open pediment.

CIL 1.577 (Lex operum Puteolana): "place upon [the rafters] a sheathing of firwood (OPERCULAQUE ABIEGNEA INPONITO).

For full discussion of the Lex Puteolana, see chapter 9 of the text.

See also, in this section, *contabulatio.*

pediment: see, in this section, *fastigium;* see also *tympanum* under "IX.3. Doors and Shutters."

pitch: see, in this section, *fastigium, stillicidium.*

plaster, used for ceilings: see, in this section, *asser.*

prop-and-lintel: a modern term used to describe a pitched roof that is supported by vertical props (fig. 8.4).

***protectum* -i (n., also *protectus* -us, m.):** an eave; the overhang of a roof. A small shed-roof that projects from a wall. Many houses at Pompeii and Herculaneum employed a shedlike roof over their main doorways.

CIL 6.10337 (Rome, inscription referring to an overhang in front of a portico): PROTECTUM ANTE PORTICUM

Plin. *HN* 16.35: "[country people] make the '*proiecta*' of their cottages with it [bark]" (eo . . . faciunt proiecta (*sic*) tuguriorum).

The manuscript may be corrupt and therefore should read *protecta.*

See also, in this section, *protectus, suggrunda.*

purlin: see, in this section, *laterarium, templum.*

rafter: see, in this section, *cantherius, capreolus.*

ridgepole: see, in this section, *columen.*

***scandula* -ae (f.):** a wooden shingle.

Plin. *HN* 16.36: "the most suitable shingles are of Valonia oak, the next best from other acorn-bearing trees and from the beech . . . Cornelius Nepos informs us that Rome was roofed with shingles right down to the war with Pyrrhus, a period of 470 years" (scandula e robore aptissima, mox e glandiferis aliis fagoque . . . scandula contectam fuisse Romam ad Pyrrhi usque bellum annis CCCCLXX Cornelius Nepos Auctor est).

Plin. *HN* 16.42: "[the wood] of the pitch pine is suitable for shingles of split wood" (piceae ad fissiles scandulas).

Vitr. 2.1.4 (shingles and thatch as a feature of the western provinces): "to this day buildings are constructed for foreign peoples . . . such as in Gaul, Spain, Portugual, and Aquitaine, with oak shingles or thatch" (ad hunc diem nationibus exteris . . . aedificia constituantur, uti Gallia, Hispania, Lusitania, Aquitania scandalis robusteis aut stramentis).

See also under "XII.2. Species of Trees," *fagus, pinus, quercus.* Cf. *scandularius* under "II. Areas of Specialization."

***scandularis* -is -e:** characterized by the use of wooden shingles.

Apul. *Met.* 3.17: "a shingled roof" (tectum scandulare).

sheathing: light boarding used to cover the structural members of a roof.

Vitr. 10.15.3: for text and translation see, in this section, *laterarium.*

See also, in this section, *contabulatio, operculum.*

shingle: see, in this section, *scandula.*

slot ceiling: a type of "beam ceiling." Ceiling joists are laid together closely, leaving narrow slots in between them; these are covered by boards laid on top. Cf. Hodge 1960, 101–5.

socket: a cutting, often in a masonry wall of ashlar or concrete, which receives the butt end of a joist (e.g., upper floor or roof beams).

See also *columbarium* under "VI. Framing."

***stillicidium* -*i* (n.):** used to denote liquid falling in drops. The term has been interpreted to indicate the angle of a roof, or pitch, or, simply, "eaves" (Andrén 1940, lxiv).

> Vitr. 4.7.5: "[in the Tuscan temple,] the ridge-beam, rafters, and purlins are to be so installed that the *stillicidium tecti* corresponds to one-third of the whole" (columen, cantherii, templa ita sunt conlocanda, ut stillicidium tecti absoluti tertiario respondeat).

Note Rowland 1999, 61: "one-third of its entire run."

See also, in this section, *protectus* and *fastigium* (especially Vitr. 4.2.5).

subgrunda = *suggrunda*.

suggrunda* -*ae* (f.)** (also ***subgrunda or ***suggrundatio*):** the overhang of a roof beyond the load-bearing exterior wall(?) The eave.

> Vitr. 4.2.1: (the manner in which a rafter [*cantherius*] extends from the ridgepole *ad extremam suggrundationem*): for commentary see chapter 8 of the text.
>
> Vitr. 10.15.1 (the siege machine of Hagetor): "above [the wooden merlons it has] a sloping overhang" (superne [pinnas ex tabulis] subgrundas proclinatas).

See also, in this section, *protectum*.

***suggrundatio* -*onis* (f.):** see, in this section, *suggrunda*.

***tectum* -*i* (n.):** the usual meaning is "roof," but often *tectum* refers to a "ceiling," as indicated in the references. *Tectum proclinatum*: "a pitched roof" (cf. Vitr. 2.1.3; the passage is included in this section under *fastigium*). *Sub tecto*: "indoors"; *sub uno tecto*: "under one roof," i.e., in reference to the same building.

> Cic. *Verr.* 1.133: (coffered ceilings of the Temple of Castor in the Roman forum): for text and translation see, in this section, *laqueatus*.
>
> Cic. *Tusc.* 1.85 (quoting Ennius): for text and translation see, in this section, *laqueatus*.
>
> Vitr. 4.2.1: "under roofs, if the spans are great, both tie-beams and rafters" (*sub tectis, si maiora spatia sunt, et transtra et capreoli*).

Here *tectum* appears to refer to the roof, but not its system of support. Thus *sub tecto* refers to the framing system of cross-beams, braces, and rafters which support the *tectum*. For additional discussion see chapter 8 of the text. See also, in this section, *transtrum*, *capreolus*.

> Vitr. 7.5.6: "tiled roofs" (tegularum tecta).

***tectus* -*a* -*um*:** "roofed," from the verb *tego* (q.v., under "VIII.1. General Terms").

> *CIL* 1.1633.4 (the *odeon* at Pompeii, a roofed theater): THEATRUM TECTUM.

***tegula* -*ae* (f.):** a roof tile. The most common form is a flat terra-cotta "pan" tile with a flange on either side. The seam between two *tegulae* is covered with an *imbrex* (q.v.).

> Vitr. 7.5.6: for text and translation see, in this section, *tectum*.

***templum* -*i* (n.):** as a roofing term, possibly a purlin.

> Fest. 505L: "the word *templum* means both the structure that is sacred to a god, and the beam of wood which is placed crosswise on the building" (templum significat et aedificium deo sacratum, et tignum, quod in aedificio transversum ponitur).

Here Festus may mean that the beam is "crosswise," lit. *tranversum*, in relation to the main rafters.

> Lucr. 2.28 (ceiling beams in a house): for text and translation see, in this section, *laqueatus*.
>
> Vitr. 4.2.1: "the *templa* are above the rafters" (supra cantherios templa).
>
> Vitr. 4.2.5: cf. comments in chap. 8, p. 137.
>
> Vitr. 4.7.5 (Greek temple roofing): "above the pediment, the ridge-beam, the rafters, and

the *templa* are placed" (supraque eum fastigium, columen, cantherii, templa ita sunt conlocanda).

***testudo* -inis (f.):** a tortoise or the shell of a tortoise, thus, as an architectural term, a roof. In military usage, the *testudo* indicates a protective covering formed by shields or other (often timber-framed) construction. Its employment to describe the coverings of temples and basilicas suggests that a timber frame is meant, and specifically the "wooden vault," or tie-beam truss (q.v.) used to span wide spaces.

Varro *Ling.* 5.161: "the *cavum aedium* [*atrium*] is said to be the roofed place which was left accessible within the walls of the house . . . if in this place there was nothing left [i.e., no part of the roof] which was under the open sky, it was said to be *testudo* [English: "testudinate"] from the similarity to the *testudo* as it is in the general's headquarters and in the military camp (cavum aedium dictum qui locus tectus intra parietes relinquebatur patulus . . . in hoc locus si nullus relictus erat, sub divo qui esset, dicebatur testudo ab testudinis similitudine, ut est in praetorio et castris).

Vitr. 5.1.6 (roof over the nave of the basilica at Fanum): "in the middle, between the columns, is a *testudo* 120 feet long and 60 wide" (mediana testudo inter columnas est longa pedes CXX, lata pedes LX).

The full passage is considered in chapter 8 of the text.

Vitr. 6.3.2 (houses with roofed halls, or *atria*): "testudinate [*atria*] are employed when the spans are not great, and they provide roomy living rooms upon the flooring above" (testudinata vero ibi fiunt, ubi non sunt impetus magni et in contignationibus supra spatiosae redduntur habitationes).

Thus, in this situation, a small atrium can be covered with a joist-based superstructure which creates attic rooms above the main hall of the house.

Verg. *Aen.* 1.505: "[Dido sat] at the doors [of the temple] of the goddess, in [under] the middle *testudo* of the temple" (foribus divae, media testudine templi).

Since the doorways to the cult rooms of Greek and Roman temples were preceded by a deep porch (covered by a timber-framed roof), the *testudo* of this passage probably refers to the roof over the porch.

See also, in this section, "tie-beam truss."

thatch: sheathing of straw or grass. Thatch was used throughout Roman and postclassical times as a roofing material for farm buildings.

Caes. *BGall.*5.43.1 (used by Gauls): "houses that were roofed with straw" ([casae] stramentis erant tectae).

See also, in this section, *tego* under "VIII.1. General Terms."

tie-beam: see, in this section, *trabs.*

tie-beam truss: a triangular configuration of two rafters and a cross-beam, connected at each corner, to create a frame for a roof. *Fastigium* (q.v.) or *testudo* (q.v.) may have sufficed to indicate the presence of a triangular truss.

***tignum* -i (n.):** a beam of wood. The term is paired with *asser* in Vitr. 4.2.4, apparently to indicate "joist" and "rafter," respectively. *Tignum* is usually translated as "beam" (or "joist"). In the context of roofing, the *tignum* would thus refer to a ceiling joist.

Vitr. 4.2.4: for text, translation, and commentary see, in this section, *asser*, *columbarium*, and *intertignium*.

See also the term *tignum* under "XIV. Finished Lumber." For related discussion, see, in this section, *trabs.*

tile, roofing: see, in this section, *imbrex*, *tegula*.

***trabes conpactiles*:** beams attached to each other along their lengths. The result is a thicker and stronger composite beam.

Vitr. 5.1.8 (regarding the basilica at Fanum): "above the columns, placed all around, are *trabes* made from three two-foot beams put together" (supra columnas ex tribus tignis bipedalibus conpactis trabes sunt circa conlocatae).

For full discussion of the basilica, see chapter 8 of the text.

Vitr. 4.7.4 (composite beams for the architraves of temples): for text and translation see, under "IV.I Joints," *securicula.*

Vitr. 10.14.2 (composite beams in vertical applications): for text and translation see *postis* under "VI. Framing."

trabs -bis (f.): a heavy timber, presumably squared, often employed horizontally as a load-bearing beam. An architrave beam. *Trabes* formed the frame for both flat ceilings and roofs, but also were used for floor joists (q.v., under "VII. Flooring"). *Trabes* are the main beam(s) holding up the roof of an atrium. If part of a triangular roofing truss, the *trabs* formed the bottom horizontal tie-beam.

Plin. HN 16.213 (regarding use in the Temple of Artemis at Ephesus): for text and translation see, in this section, *lacunar.*

Plin. HN 16.216 (regarding use in the Temple of Apollo at Utica): "beams of Numidian cedar" (cedro Numidica trabes); "[at the Temple of Diana at Saguntum (Spain)], the beams were of juniper, and even now still exist" (iuniperi trabibus etiam nunc durantibus).

Plin. HN 16.223: "walnut [or 'chestnut'] bends easily, for beams are made out of it; it warns that it is about to break by creaking, which in fact happened at Antandro when the bathers, frightened by the sound, fled from the public baths" (facile pandatur iuglans, fiunt enim et ex ea trabes; frangi se praenuntiat crepitu, quod et in Antandro accidit, cum e balineis territi sono profugerunt).

Vitr. 6.3.6 (on the proportions of the *tablinum* in a Roman house): "the height of the tablinum to the beam [*trabem*] is to be one-eighth more than its breadth" (altitudo tablini ad trabem adiecta latitudinis octava constituatur).

Here, the *trabs* may refer to the beam, in fact a lintel, which would have spanned the broad doorway between the *tablinum* and the main area of the atrium.

Cf. additional discussion under "VII. Flooring" and "XIV. Finished Lumber."

For the suitability of certain types of wood to serve as *trabes* see, under "XII.2. Species of Trees," especially *abies* and *larix.*

traiecturae mutulorum: lit. "the projecting part of the mutules." I.e., beams that rest upon the lateral architraves, support the rafters, and are cantilevered over the facade of a Tuscan temple.

Vitr. 4.7.5: "above the [architrave] beams and above the [cella] walls, the projecting parts of the mutules [*traiecturae mutulorum*] should extend [beyond the facade of the temple] one-quarter the height of the columns; and on the front [or ends] of these [mutules] [*antepagmenta*] should be attached" (supra trabes et supra parietes traiecturae mutulorum parte IIII altitudinis columnae proiciantur; item in eorum frontibus antepagmenta figantur).

transtrum -i (n.) or **trastrum:** a cross-beam. In Vitr. 4.2.1., the term may refer specifically to the horizontal tie-beam of the tie-beam truss (Andrén 1940, lxii).

Caes. *BGall.* 3.13.3 (on the boats made by the Gauls): "the crosspieces [were made] from beams a foot thick" (transtra ex pedalibus in altitudinem trabibus).

Vitr. 4.2.1: for text and translation, see, in this section, *tectum.*

See also, in this section, comments under "VIII.1. General Terms" and *transtrum* under "VII. Flooring."

trastrum: see, in this section, *transtrum.*

truss, timber: see, in this section, *testudo,* "tie-beam truss."

IX. Interior Woodwork

IX.1. DEFINITION OF *OPUS INTESTINUM*

Plin. HN 16.225: for text and translation see *ars fabrica* under "I. General Woodworking Terms."

Vitr. 2.9.7: "[lumber from] the lowest part [of the tree] . . . is used for interior work (ima autem . . . ad intestina opera conparatur).

Vitr. 4.4.1 (in reference to the inner porch, or *pronaos*, of a temple): "the three intercolumniations which will be between the *antae* and the columns should be closed off with a fence of marble or woodwork" (item intercolumnia tria quae erunt inter antas et columnas, pluteis marmoreis sive ex intestino opere factis intercludantur).

Vitr. 5.2.2: "the interior walls are to be skirted halfway up by cornices of fine woodwork or plaster" (praecingendi sunt parietes medii coronis ex intestino opere aut albario).

Vitr. 6.3.2: "[rainfall dribbling down the walls from a displuviate roof can] damage the walls and fine woodwork in these kinds of buildings [houses]" (et intestinum et parietes in eis generibus aedificiorum corrumpunt).

Vitr. 6.3.9 "[above the columns, Corinthian-style reception rooms] have architraves and cornices made either from fine woodwork or from plaster" (habeant epistylia et coronas aut ex intestino opere aut albario).

See also *intestinarius* under "II. Areas of Specialization."

IX.2. TECHNICAL TERMS

architrave: see, in this section, *epistylium*.

balcony: see, in this section, *circumitus*, *maenianum*.

***circumitus* -*us* (m.):** a balcony.

Vitr. 6.3.9 (a description of an "Egyptian"-style *oecus* in a Roman house): "upon the floorboards there is a pavement, so that there can be a *circumitus* in the open" (supra coaxationem pavimentum, subdiu ut sit circumitus).

The floorboards in question are supported by the architrave of a colonnade below, and thus the *circumitus* is an elevated balcony.

For a description of *coaxatio*, cf. under "VII. Flooring."

See also, in this section, *maenianum*.

***contignatio*:** see this term under "VII. Flooring" and "VIII. Roofing and Ceilings."

cornice: see, in this section, *corona*.

***corona* -*ae* (f.):** a cornice, which can be made of wood, plaster, or stone. The passages cited here refer to examples in wood or plaster.

Vitr. 5.2.2 (cornices to aid acoustics): for text and translation see IX.1

Vitr. 6.3.9 (the Corinthian *oecus*): for text and translation see the introduction to this section.

door: doors of all types are considered later in this section under IX.3.

***epistylium* -*ii* (n.):** an architrave. Architraves can be made of woodwork in a Roman house.

Vitr. 6.3.9: for text and translation see IX.1.

***intestinarius*:** a specialist of *opus intestinum*. See "II. Areas of Specialization."

lamina* -*ae (f.), also ***lammina*** and ***lamna*:** veneer. A thin facing of high-quality wood is glued to backing wood. Thin sheets used for bentwood containers.

Plin. HN 13.94 (a table with a veneer): for text and translation see *citrus*, under "XII.2 Species of Trees."

Plin. HN 16.226: "for this use [veneer] they approve of a threadlike grain, and they call it fennel-pattern for the resemblance" (stamineam in hoc usu probant venam, et vocant ferulaceam argumento similitudinis).

For peacock (or bird's-eye) grain, cf. Mart. 14.85.1 under *lectus* in "XI.2. Beds and Couches."

Plin. *HN* 16.229: "beech also is easy [to work], although brittle and soft; also sawn into thin layers of veneer it is flexible and is uniquely suitable for boxes and [round] document cases" (facilis et fagus, quamquam fragilis et tenera; eadem sectilibus laminis in tenui flexilis capsisque ac scrineis sola utilis).

Plin. *HN* 16.231: "the preferred woods that can be cut into sheets to use as a veneer to cover other woods are citrus, turpentine-tree, varieties of maple, box, palm, holly, holm oak, elder root, and poplar" (quae in lamnas secentur quorumque operimento vestiatur alia materies, praecipua sunt citrum, terebinthus, aceris genera, buxum, palma, aquifolium, ilex, sabuci radix, populus).

Additional information on each species of tree and its wood can be found under "XII.2. Species of Trees."

Plin. *HN* 16.231 -2: "the middle part of trees is more wavy [in grain] and closer to the root smaller and more curly are the markings . . . so that a single tree can be sold over and over, thin sheets [*bratteae*] have been invented" (media pars arborum crispior, et quo proprior radici minoribus magisque flexilibus maculis . . . ut una arbor saepius veniret, excogitatae sunt et ligni bratteae).

"*Brattea*" is usually used to indicate an extremely thin sheet of metal (like gold leaf). It is used here to emphasis the thinness of the veneer.

See also *lamina* under "III. Tools."

intestinarius: a specialist of *opus intestinum*. See "II. Areas of Specialization."

***maenianum -i* (n.):** a balcony. The term can also refer to a projecting upper story. Balconies were built upon joists (*tigna*) cantilevered out from walls. Projecting balconies were also a feature of (comic) stage scenery, which was designed "in imitation of ordinary buildings" (imitatione communium aedificiorum) (Vitr. 5.6.9). (fig. 6.14)

Vitr. 5.1.2: for text and translation see *coaxatio* under "VII. Flooring."

See also, in this section, *circumitus*.

veneer: see, in this section, *lamina*.

IX.3. DOORS AND SHUTTERS

***antepagmentum -i* (n.):** door and window frames, including the frames of temple doorways. The exterior moldings of these frames could be highly decorative. As a trim element for a roof, see this term under "VIII. Roofing and Ceilings."

Cato *Rust.* 14.4: "[after the foundation of a farmhouse has been laid] add the lintels and door [and/or window] frames that are necessary" (iugumenta et antepagmenta quae opus erunt indito).

Vitr. 4.6.2: for text and translation see *supercilium* under "VI. Framing."

Vitr. 4.6.3 (Ionic doorways): "the width of the door frame is to be one-fourteenth of the height of the opening in front" (crassitudo antepagmentorum ⟨ex⟩ altitudine luminis in fronte XIIII parte).

Vitr. 4.6.1: "these are the rules of the door frames in temples" (antepagmentorum in aedibus hae sunt rationes).

Vitr. 4.6.6.: for text and translation see, in this section, *corsa*.

See also, in this section, *coassamentum*.

***armilla -ae* (f.):** the socket, often of bronze, into which a pivot (*cnodax*, q.v.) fits.

Vitr. 10.2.12: see, in this section, *cnodax*. The passage makes clear the relationship between the *armilla* and the *cnodax*.

See also, in this section, *cardo*, *foramen*.

biforis -*is* -*e* (adj.): having a double door, such as might act like a window shutter. The literary record often associates double doors with prestigious buildings, but in fact they were common in houses as well. Vitruvius distinguishes between *biforis* and *valvata* (q.v.) in 4.6.6., but the distinction is not always clear in other texts.

> Ov. *Met.* 2.4: "the folding double doors shone with the gleam of silver" (argenti bifores radiabant lumine valvae).

The implication is that each of the two doors is composed of folding leaves.

> Ov. *Pont.* 3.3.5: "double-shuttered windows" (bifores fenestras).
>
> Vitr. 4.6.6: "they [Attic doors] are to be without grates and are not *bifores* but of hinged leaves [*valvata*] and open outward" (ipsaque non fiunt clathrata neque bifora sed valvata, et aperturas habent in exteriores partes).

See also, in this section, *quadriforis*.

cardo -*inis* (f.): a hinge or pivot on which a door turns. From the passages cited below, the term must refer to the pin attached to the bottom of the hinge stile of the door that rests in the socket (*armilla*, q.v.) embedded in the threshold. For multileaved doors, the *cardo* may refer to the hinges which allowed the leaves to fold.

> Plin. *HN* 16.210: "elm keeps its stiffness most steadfastly, on account of which it is most useful for the pivots and the frameworks of doors" (rigorem fortissime servat ulmus, ob id cardinibus coassamentisque portarum utilissima).
>
> Pliny goes on to say that the plank should be inverted so that the wood which came from the top of the tree is directed down toward the pivot. This would suggest that he is referring to the "hinge stile," or *scapus cardinalis* (q.v.). Cf. also, in this section, *coassamentum*.
>
> Plin. *HN* 16.230: "pivots made of olive, the hardest wood, left unmoved in doorways for too long have sprouted like a plant" (quippe cum ex olea, durissimo ligno, cardines in foribus diutius immoti plantae modo germinaverint).
>
> Verg. *Aen.* 1.449: "the hinge shrieked from the bronze doors" (foribus cardo stridebat aenis).

See also, in this section, *cnodax*.

clatratus -*a* -*um* (or *clathrata*): used to describe a door, window, or passageway with a grate. Some doors employed a latticework, presumably of bronze, but perhaps of wood as well.

> *CIL* 1.577 (*Lex operum Puteolana*): "grated doors" (FORES CLATRATAS).

See additional discussion of this passage above in chapter 9 of the text.

> Vitr. 4.6.6: for text and translation see, in this section, *biforis*.

See also the relevant passage in Vitr. 4.4.1 under "IX.1. Definition of *opus intestinum*."

clatri -*orum* (m. pl.): a grate(s) or latticework used to cover windows on a structure. Those found *in situ* at sites such as Herculaneum are of iron. Wooden *clatri* were also used on the doors of *armaria*, and, presumably, some houses (fig. 4.10).

> Cato *Rust.* 14.2 (on the fittings for a farmhouse): "ten two-foot *clatri* [are required] for the larger windows (clatros in fenestras maioris bipedalis X).

The meaning of "two-foot lattices" (Hooper 1993, 29) is not clear. Possibly this means the lower two feet of the window were to be protected by such a grate. Or the standard size for a larger window was two feet across.

cnodax -*acis* (m.): a pin or pivot. The point of the pivot rested in a socket placed in the threshold of the doorway (see, in this section, *armilla*). See also, in this section, *cardo*, which apparently has a similar, if not identical, meaning.

> Vitr. 10.2.11 (concerning iron pivots, *cnodaces ferrei*, placed in the ends of a column so that it can be "wheeled" along for transport): "he [Chersiphron] inserted sockets in the wood [frame] to receive the pivots" (armillas in materia ad cnodacas circumdandos infixit).

See also, in this section, *armilla*.

***coassamentum* -i (n.):** the frame of the doorway or the structural frame of rails and stiles for the door itself.

Plin. *HN* 16.210: for text and translation see, in this section, *cardo*.

Cf. also *ulmus* under "XII.2. Species of Trees," and, in this section, *antepagmentum*.

***corsa* -ae (f.):** the face (also known as *fascia*) of the architrave that runs vertically and horizontally along a doorjamb or lintel.

Vitr. 4.6.6: "the *corsae* [of Attic doors] are carried around under the cymations on the door frames" (corsae sub cymatiis in antepagmentis circumdantur).

***cyma (reversa)*:** see, in this section, *cymatium*.

***cymatium* -i (n.):** the Roman ogee molding bordering the rectangular panel in a door, or as a decorative molding on the frame of the door. The molding is mentioned in the Lex operum Puteolana as *cumatium*; see chapter 9 of the text.

Vitr. 4.6.4: for text and translation see chapter 9 of the text.

door(way): see, in this section, *biforis*, *foris*, *ianua*, *lumen*, *ostium*, *quadriforis*, *thyromum*, *valvae*.

***fenestra* -ae (f.):** a general term for a window; possibly a window shutter (cf. Juvenal 9).

Cato *Rust.* 14.2 (farmhouse windows): for the text and translation of this passage, see, in this section, *clatri*.

Ov. *Pont.* 3.3.5: for text and translation see, in this section, *biforis*.

Juv. 9.104–05: "let him close the shutters [or simply: 'close the windows (with shutters)'], let the curtains cover the chinks, latch the doors" (claude fenestras, / vela tegant rimas, iunge ostia).

***foramen* -inis (n.):** a socket, a mortise, a hole, or a metal bushing in a wooden workpiece. Perhaps another term for the socket into which a pivot fit.

Cato *Rust.* 19.2 (wine press construction): "make the first [of six] *foramen* . . . half a foot from the joint" (foramen quod primum . . . semipedem ab cardine facito).

Vitr. 10.6.3 (on the water screw): "in these [wooden crosspieces] iron sockets are inserted (in his foramina ferrea sunt inclusa).

See also, in this section, *armilla*.

***foris* -is (f.):** a gate, door, or doorway. Used to describe the doors of temples and palaces. When used in the plural, the term can refer to a folding door.

CIL 1.577: see, in this section, *clatratus*.

Tac. *Hist.* 1.43: "Piso was slain by the door of the temple [of Vesta]" (Piso in foribus templi trucidatur).

In this passage, the *fores* could be in reference to either the doors of the temple or those that entered into the precinct of Vesta.

Verg. *Aen.* 1.449: (the doors [or gates?] of Dido's Temple of Juno): for text and translation see, in this section, *cardo*.

Vitr. 4.4.1 (access to the pronaos of a temple): "so they [the intercolumniations] have *fores*, through which there is a passage to the pronaos" (ita uti fores habeant, per quas itinera pronao fiant).

Granger (1983, 229, n. 2) takes special pains to translate *fores* here as "gates," not "doors." This passage is also considered under the introduction to "IX. Interior Woodwork."

Vitr. 4.6.4 (assembly of doors): "doors are thusly joined together" (fores ita compingantur).

Cf. also *compages* under "IV.1 Joints."

gate: see, in this section, *foris*, *ianua*.

grates, grating, for doors and windows: see in this section, *clatri*, *clatratus*.

ianua -ae (f.): a door, or simply an "entrance" to a house or other building. Generally a term for the means of access to a place (cf. the god Janus, divinity of passageways). Cato (*Rust.* 14.2) refers to the main door of a farmhouse as the *ianua maxima*. Boëthius (1978) defines *ianua* as the outer door to a house; the term could be modified to refer to the inner door (*ianua interior*) of a house (cf. Vitr. 6.7.1). See also, in this section, *lumen*.

Cato *Rust.* 14.2: "[for the farmhouse the builder will furnish] one main entrance" (ianuam maximam).

Catul. 61.76: "release the bolts of the door, the bride is coming" (claustra pandite ianuae, / virgo adest).

Cic. *Nat. D.* 2.67: "thus the name [*iani* for archways and] *ianuae* for doors at the thresholds [i.e., the "front doors"] of secular buildings" (fores . . . in liminibus profanarum aedium ianuae nominantur).

Cic. *Verr.* 1.66: "Rubrius ordered [the slaves] to shut the outer door and to stand by at the entrance" (Rubrius ut ianuam clauderent et ipsi ad foris adsisterent imperat).

Vitr. 6.7.1 (the entrance to a Greek house): "and these [porter's room] immediately flank the inner doorways (statimque ianuae interiores finiuntur).

Vitr. 6.7.3: "moreover, these [Greek] houses [with a block of rooms around the peristyle] have distinguished vestibules and individual, dignified doorways" (habent autem eae domus vestibula egregia et ianuas proprias cum dignitate).

Vitr. 6.7.4 (guest rooms of a Greek villa): "having their own entrances" (habentes proprias ianuas).

impages -is (f.) = inpages: the horizontal framing pieces, or rails, of the door. Thus, *medii inpages*: the "middle rails" (Vitr. 4.6.5). (fig. 9.12)

Vitr. 4.6.5 (ideal proportions of temple doors): "for the rails, let the positioning be such that . . ." (inpagibus distributiones ita fient . . .).

inpages: see, in this section, *impages*.

lattice: see, in this section, *clatri*, *clatratus*.

leaf (of a folding door): see, in this section, *pagina*.

lumen -inis (n.): a general term for an opening, thus a door, a window (or its shutters), a *compluvium*. Generally any aperture that admits light.

Cato *Rust.* 14.2 (on requirements for a farmhouse): "six [window] shutters" (luminaria VI).

In the list of features for the farmhouse, Cato places *luminaria* after *fenestras*, hence Hooper (1993, 29): "shutters."

Vitr. 4.6.3: "the height of the opening [of a temple door]: (⟨ex⟩ altitudine luminis).

For a related use of *lumen* (an opening in a roof), see *arca* under "VIII. Roofing and Ceilings."

ogee: see, in this section, *cymatium*.

ostium -i (n.): a doorway, an opening for the door. The *ostium* is framed by the *antepagmentum* (q.v.).

Vitr. 4.6.1: "the [proportional] rules for doorways to temples" (ostiorum . . . in aedibus . . . rationes).

Vitr. 6.3.6: "[the relation between] widths to height of doors" (latitudines ostiorum ad altitudinem).

pagina -ae (f.): a panel of a multileaved door; one leaf of a folding door (cf. *valvae*).

Plin. *HN* 16.225: for text and translation, see, in this section, *valvae*.

quadriforis -is -e (adj.): a door with four folds or leaves. Or a pair of doors, each with two leaves. Cf. also *biforis*.

Vitr. 4.6.5: for text and translation see, in this section, *valvae*.

rail: a horizontal framing piece in a door. See, in this section, *impages*.

replum -i (n.): a covering molding. In the case of double doors, the molding conceals the gap between the two leaves at the center and acts as a stop for one of the pair. (fig. 9.12)

Vitr. 4.6.5: "the breadth of the inner stiles is to be one-half that of the *impages*; the center molding [*replum*] is two-thirds the width of the *impages*" (scaporum latitudines inpagis dimidia parte, item replum de inpage dimidia et sexta parte).

scapus -i (m.): part of the vertical frame of a door. The *scapus cardinalis* is the main stile (or "hinge stile") that serves to hold the hinges or pivots of the door. Cf. also *cardo*. (fig. 9.12)

Vitr. 4.6.5: for text and translation see, in this section, *replum*.

shutter, for a window: see, in this section, *fenestra*, *lumen*.

stile: see, in this section, *scapus*.

thyroma -atis (n.): a door(way). When Vitruvius uses this term, he refers specifically to doors built with Greek proportional rules.

Vitr. 6.3.6: "[the proportions of Doric and Ionic style doors should be] in the same way as for [Attic] doorways" (quemadmodum de thyromatis).

tympanum -i (n.): one of the rectangular sunken panels placed between the stiles and rails of a door. The term can also be used to indicate the triangular field enclosed by a pediment. (fig. 9.12)

Vitr. 4.6.4 (concerning ideal proportions of temple doors): "the *tympana* between the two stiles are three parts out of twelve in width" (inter duos scapos tympana ex XII partibus habeant ternas partes).

Vitr. 4.6.5 (context same as previous): "let the height of the rail be one third that of the *tympanum*" (altitudo inpagis fiat tympani tertia parte).

valvae -arum (f. plur.): multileaved folding doors or the panels thereof. In some contexts *valvae* may be the same as simple double doors (*bifores*, q.v.). The most complicated appear to have been multileaved doors that could be folded open. In most literary passages it is difficult to distinguish between the different types.

Apul. Met. 1.15: "Open the gates of the inn!" (valvas stabuli absolue).

The same doors are described as *ianuae* earlier in the same passage.

Cic. *Verr.* 2.61: "[you cannot prove that you bought] those two most beautiful statues which now stand alongside the *impluvium* in your house and which for many years stood in front of the *valvae* of [the Temple of] Juno on Samos" (ne haec quidem duo signa pulcherrima quae nunc ad impluvium tuum stant quae multos annos ad valvas Iunonis Samiae steterunt).

Juv. 4.63 (a fisherman with an enormous turbot arrives at the emperor's residence at Alba Longa): "the doors swung open on their smooth hinges" (facili patuerunt cardine valvae).

Ov. Met. 1.172 (the dwelling places of gods are imagined): "to the right and left the atria of the noble gods, with their *valvae* open, are crowded with guests" (dextra laevaque deorum atria nobilium valvis celebrantur apertis).

Ov. Met. 2.4: "the palace of the sun was high with lofty columns, bright with gleaming gold and bronze—like fire—whose polished ivory covered the high gables, the double folding doors [*bifores valvae*] shone with the gleam of silver. And the worksmanship [*opus*] surpassed the raw material[s] [*materiam*]" (regia Solis erat subliminibus alta columnis / clara micante auro flammaque imitante pyropo, / cuius ebur nitidum fastigia summa tegebat, / argenti bifores radiabat lumine valvae./ materiam superabat opus . . .).

Bifores valvae may be another way of describing the *quadriforis* (q.v.).

Petron. 65.3: "meanwhile a lictor banged on the folding doors of the *triclinium*" (inter haec triclinii valvas lictor percussit).

Plin. Ep. 2.17.5 (the *triclinium* of Pliny's seaside villa): "on all sides it has folding doors or windows as big as folding doors" (undique valvas aut fenestras non minores valvis habet).

Plin. Ep. 2.17.20 (a bedroom in the villa of the previous entry): "a bedroom (looks through) folding doors to the cryptoporticus; a window opens on to the sea" (cubiculum . . . valvis cryptoporticum, fenestra prospicit mare).

Plin. *HN* 16.215: "the folding doors [of the Temple of Artemis at Ephesus] are made of cypress, and all the wood is still like new after four hundred years" (valvas esse e cupresso et iam CCCC prope annis durare materiem omnem novae similem).

See also *cupressus* under "XII.2. Species of Trees."

Plin. *HN* 16.225: "fir wood is most suitable . . . for the panels of folding doors" (abies . . . paginis valvarum . . . aptissima).

For full text see *ars fabrica* under "I. General Woodworking Terms."

Vitr. 3.2.8 (*valvae* as a feature of hypaethral—partially unroofed—temples): "entrances of folding doors are in both the pronaos and in the *posticum*" (aditus valvarum et utraque parte in pronao et postico).

The *posticum* refers to the back of the temple.

Vitr. 3.3.3 (commenting on the fault of intercolumniations on the facade of a temple being too closely spaced): "the view of the double doors is concealed by the closeness of the columns" (valvarum adspectus abstruditur columnarum crebritate . . .).

Vitr. 4.6.1: "the opening of the doorway is to be so determined that the height of the temple from the pavement to the coffered ceiling is to be divided into 3.5 parts, and of these 2.5 in height are to be fixed for the opening of the folding doors" (lumen autem hypaethri constituatur sic, uti quae altitudo aedis a pavimento ad lacunaria fuerit, dividatur in partes tres semis et ex eis duae partes ⟨semis⟩ lumini valvarum altitudine constituantur).

Vitr. 4.6.5 (on dimensions of doors): "if the doors are *valvatae*, the heights remain the same; but let the width be increased by the width of a doorway. If the doors are quadri-paneled, let the height be increased" (sin autem valvatae erunt, altitudines ita manebunt, in latitudinem adiciatur amplius foris latitudo. Si quadriforis futura est, altitudo adiciatur).

The point being made is that for a four-leaved door, the overall height of the doorway needs to be increased in order to preserve harmonious proportions.

Vitr. 4.8.2 (on the arrangement of circular temples): "in the middle a place for the folding doors for the access is to be left" (*medioque valvarum locus ad aditus relinquatur*).

Vitr. 5.5.7 (the acoustics of the theater): "when [performers] wish to sing with a louder tone, they turn themselves to the double doors of the scene building, and thus from the help of the doors gain resonance for their voice" (superiore tono cum volunt canere, avertunt se ad scaenae valvas et ita recipiunt ab earum auxilio consonantiam vocis).

***valvatus -a -um* (adj.):** characterized by double or folding doors or similarly shuttered windows.

Varro *Ling.* 8.29: "and so we do not make winter dining rooms and summer dining rooms with the same [types of] folding doors and windows" (hiberna triclinia et aestiva non item valvata ac fenestrata facimus).

Varro may be referring to the relative sizes of the doors.

Vitr. 4.6.6: for text and translation see, in this section, *biforis*.

Vitr. 6.3.10 (on the layout of a "Cyzicene" hall in a Roman house, a kind of fancy dining room which looks out over a peristyle garden): "they have shuttered openings for windows on the right and the left" (habentque dextra ac sinistra lumina fenestrarum valvata).

window: see, in this section, *fenestra*.

window, framing of: see *antepagmentum* under "VI. Framing."

X. Wheels

***radius -i* (m.):** the spoke of a wheel.

Ov. *Met.* 2.107–08 (the chariot of Apollo): "the axle was of gold, the yoke-beam of gold, the rim of the wheel was golden, and a row of silver spokes" (aureus axis erat, temo aureus, aurea summae / curvatura rotae, radiorum argenteus ordo).

Plin. *HN* 2.64 (discussing the arcs of celestial bodies): "lines [projected] from the top of the arc to the center necessarily converge like the spokes of wheels" (ab summa apside lineas coarctari ad centrum necesse est sicut in rotis radios).

Verg. *Georg.* 2.444: for text and translation see, in this section, *rota.*

rim: the circular element of a wheel that is connected to the hub, or nave, by spokes. In Latin, perhaps *curvatura rotae.* See the passage in Ovid (*Met.* 2.107) from the previous entry.

***rota* -*ae* (f.):** a wheel for a vehicle or a wheel for a mechanical device.

Cato *Rust.* 11.3 (furnishing for a farmstead): "one waterwheel" (rotam aquariam I).

Lucr. 6.551: "[the road surface] rattles the iron rims of the wheels on both sides" (*ferratos utrimque rotarum succutit orbes*).

Ov. *Met.* 2.107: for text and translation see, in this section, *radius.*

Plin. *HN.* 7.199: "a vehicle with four wheels [was invented by] the Phrygians" (vehiculum cum quattuor rotis Phryges).

Verg. *Georg.* 2.444: "[from these trees] farmers turn spokes for their wheels, or disks for their wagons" (hinc radios trivere rotis, hinc tympana plaustris agricolae).

spoke: see, in this section, *radius.*

***tympanum* -i (n.):** a solid disc used as a wheel. A drum used in a waterwheel or an olive press.

Verg. *Georg.* 2.444: for text and translation see, in this section, *rota.*

wheel: see, in this section, *rota.*

XI. Furniture

XI.1 GENERAL COMMENTS: THE TERM FURNITURE

***supellex* -*ectilis* (or *suppellex*, f.):** a general term for furniture and household furnishings.

Cato *Rust.* 98.2 (the rubbing of furniture with *amurca*): for text and translation see "Treatment by Oiling" under "XII.3. Drying, Preservatives, and Conditioners."

Plaut. *Per.* 732: "both my house and my furnishings are caked with dirt" (mihi supellex squalet atque aedes meae).

Suet. *Aug.* 73: "the frugality of his [Augustus's] furniture and household fittings is apparent in the couches and tables still in existence" (instrumenti eius et supellectilis parsimonia apparet etiam nunc residuis lectis atque mensis).

Varro *Ling.* 9.47: "and it is possible to observe this [the attraction for matching objects] from the same [type] of furniture; for no one makes the [three] beds of the triclinium unless they match in wood, height, and form" (itaque ex eadem supellectili licet videre: nam nemo facit triclinii lectos nisi paris et materia et altitudine et figura).

XI.2 BEDS AND COUCHES

bed: see, in this subsection, *lectus.*

bier: see, in this subsection, *lectus.*

couch: see, in this subsection, *lectus.*

***lectus* -i (m.):** a bed or couch, most commonly with turned legs. The *lectus* was used for sleeping, eating, and studying (reading). A bier. Related terms: *lectulus* -i: couch or bed; *lectica*, -*ae*: a litter; *lecticula* -*ae*: a small litter or chair for studying.

Cato *Rust.* 10.5 (the proper equipment for a farmstead): "one bed in the bedroom, four beds/couches with taut [leather?] straps, three beds/couches" (lectum in cubiculo I, lectos loris subtentos IIII et lectos III).

Mart. 66.5–6: "jewelled couches gleam with first-class tortoiseshell, and Moorish citrus tables, massive and choice" (gemmantes prima fulgent testudine lecti / et Maurusiaci pondera rara citri).

Mart. 14.85.1: "a peacock couch: a most beautiful bird with painted feathers gives the name to the frame of a bed" (lectus pavoninus. nomina dat spondae pictis pulcherrima pinnis . . . avis).

Probably referred to as such because of the intricate grain of the wood.

Ov. *Met.* 7.710: "I was talking about the first union [in marriage] on my now abandoned bed" (primaque deserti referebam foedera lecti).

Ov. *Met.* 8.656: for text and translation see *salix* under "XII.2. Species of Trees."

Plin. *Epist.* 2.17.21: for text and translation, see, in this section, *cathedra.*

Plin. *Epist.* 5.6.38 (concerning Pliny's country villa): "there is a bed here [in the alcove], and windows on all sides" (lectus hic et undique fenestrae).

Suet. *Aug.* 73: for text and translation see, in the introduction to this section, "*supellex.*"

Varro *Ling.* 9.4.7: for text and translation see *supellex* under "XI.1. General Comments."

See also, in this subsection, *pluteus, sponda.*

***pluteus* -i (m.):** The side and back walls of a couch. Specifically the side near one's head (see the reference to Suetonius here).

Mart. 3.91.10: "[the old man] was lying on the *sponda* part [i.e., on the outer side] while the boy was safe by the barrier of the *pluteus* [i.e., against the back wall]" (spondae qui parte iacebat / namque puer pluteo vindice tutus erat).

Suet. Calig. 26.2: "the head [of the couch] versus the foot [of the couch]" (ad pluteum . . . ad pedes).

***sponda* -ae (f.):** the frame of a bed or a couch; also the open side not protected by the *plutei.*

Mart. 3. 91.10: for text and translation see, in this section (XI.2), *pluteus.*

Mart. 14.85.1: for text and translation see, in this section (XI.2), *lectus.*

Ov. *Met.* 8.656: for text and translation see *salix* under "XII.2. Species of Trees."

XI.3. BOXES, CHESTS, CUPBOARDS, CONTAINERS

***arca* -ae** (or ***arcula* -ae,** both **f.**): a chest or large box, used for a great variety of household goods. The *arcula* is presumably a smaller and portable version.

Cato *Rust.* 11.3 (furnishings recommended for a farmstead): "a clothes chest" (arcam vestiariam); thus nominative form: *arca vestaria.*

Cato *Rust.* 98.2: for text and translation see "Treatment by Oiling" under "XII.3. Drying, Preservatives, and Conditioners."

Plaut. *Aul.* 823: "where is the gold? In a chest at home" (ubi id est aurum? in arca apud me).

Varro *Ling.* 5.128: "an *arca* [is named such] because robbers were kept from it when it was locked" (arca, quod arcebantur fures ab ea clausa).

Varro *Ling.* 5.140: "because the wagon had been made of boards like an *arca* it was called *arcera*" (quod ex tabulis vehiculum erat factum ut arca, arcera dictum).

***armarium* -i (n.):** a cupboard which could be used for books or other household items, including clothing. As a bookcase, the *armarium* was a rare furnishing in a private house.

Cato *Rust.* 11.3 (list of the equipment necessary for a vineyard): "one storage cupboard" (armarium promptarium I).

An "armarium promptarium" is literally an *armarium* which stores things that are ready to be used. It could in this context refer to a clothes cupboard.

Plin. *HN* 35.6 (on the display of the images of ancestors in the atrium of one's house): "impressions of faces in wax were organized in individual *armarii*" (expressi cera vultus singulis disponebantur armariis).

If the translation of "singulis . . . armariis" is correct, the use of a separate cupboard for each image suggests that the *armarium* could range in size from quite small to large.

bookcase: see, in this subsection, *armarium, pegma.*

box: see, in this subsection, *arca.*

***capsa* -*ae* (f.):** a case (perhaps cylindrical) for holding books or other objects.

Mart. 11.8.3: "[the scent of] apples ripening in their winter *capsa*" (poma quod hiberna maturescentia capsa).

Plin. HN 16.229 (*capsae* made from beechwood): for text and translation see *lamina* under "IX. Interior Woodwork."

The context of this passage suggests that boxes were made by cylinders of very thin beech laminate. See also, in this section, *scrinium.*

chest: see, in this subsection, *arca (arcula), cista, capsa, loculus,* and *scrinium.*

***cista* -*ae* (f.):** a small box or chest.

Cic. *Verr.* 3.197: "[the four sesterces] I will keep and transfer from the state treasury to [my own money] box" (ego habebo et in cistam transferam de fisco).

cupboard: see, in this subsection, *armarium.*

***loculamentum* -i (n.):** a receptacle for holding things, a case. Perhaps also a shelf. A dovecote.

Columella *Rust.* 8.8.3 (discussion of pigeons): "boxes, in which the birds may nest" (loculamenta, quibus nidificent aves).

Vitr. 10.10.3 (as part of a catapult): for text and translation see *securicula* under "IV. Joints."

***loculus* -i** (or ***locellus,* m.):** a chest or box, perhaps with interior divisions for holding things. In plural form, a (portable) cashbox or a small case for holding writing instruments.

Mart. 14.13 (title of a couplet): "wooden cashbox" (loculi lignei).

Suet. *Gal.* 12.3: "he gave him five denarii which he took from his own personal *loculus* with his own hand" (denarios quinque donasse prolatos manu sua e peculiaribus loculis suis).

The plural form of *loculus* in this passage suggests a small box with multiple interior compartments.

***scrinium* -i (n.):** a small box, a case for holding writing materials. The *scrinium* may have been fashioned from thin sheets of wood bent into the form of a cylindrical container.

Mart. 1.2.4 (on the portability of Martial's "complete works," which could be contained within one volume): "give book boxes to the great, one hand holds me [i.e., writings of Martial]" (scrinia da magnis, me manus una capit).

Plin. HN 16.229 (fashioned from beechwood): for text and translation see *lamina* under "IX. Interior Woodwork."

Round containers made from thin sheets of beech are implied in this passage.

Suet. *Nero* 47.2: "afterwards a speech composed for this purpose was found in his *scrinium*" (inventus est postea in scrinio eius hac de re sermo formatus).

Rolfe (1979, 79) translates *scrinium* as "writing desk."

XI.4. CHAIRS, BENCHES, FOOTSTOOLS

GENERAL INTRODUCTION

Varro *Ling.* 5.128: "From the verb 'to sit' [*sedere*] were named *sedes* 'seat,' *sedile* 'chair,' *solium* 'throne,' *sellae* 'stools,' *siliquastrum* 'wicker chair,' then from these *subsellium* 'bench' . . . where two had room on a seat of this sort, it was called *bisellium* 'double seat' (ab sedendo appellatae sedes, sedile, solium, sellae, siliquastum; deinde ab his subsellium . . . ubi in eiusmodi duo, bisellium dictum).

SPECIFIC EXAMPLES

bench: see, in this subsection, *sigma, stibadium, subsellium.*

***cathedra* -*ae* (f.):** a chair with a back, an armchair.

Ambrosiaster, *Epistolas ad Corinthios* 1.14.31 (the passage indicates the significance of the *cathedra* in a Christian context): "the elders on *cathedrae* according to dignity, the followers on benches, and the novices on floor mats" (seniores dignitate in cathedris, sequentes in subsellis, novissimi in pavimento super mat[t]as).

Hor. *Sat.* 1.10.90–91 (the poet insults two of his contemporary critics, suggesting they direct their attentions to spoiled girls in school): "but you, Demetrius, and you, Tigellius, I bid you [go] whine amidst the *cathedrae* of your skirted students" (Demetri, teque, Tigelli, / discipularum inter iubeo plorare cathedras).

On the improbability of students being seated in *cathedrae*, at least in the seventh century, see the following passage:

Isid. *Orig.* 20.11.9 (use of the *cathedra* by those of higher rank): "the benches are for the others, the *cathedrae* for teachers" (subsellia vero ceterorum, cathedrae doctorum).

Plin. *Epist.* 2.17.21 (appointments of Pliny's country villa): "[the alcove is large enough] to hold a *lectus* ["couch"] and two *cathedrae*" (lectum et duas cathedras capit)."

Plin. *HN* 16.174: for text and translation see *salix* under "XII.2. Species of Trees."

chair (with a back): see, in this subsection, *cathedra.*

scabellum* -*i (or ***scabillum***, both **n.**)**:** a (folding) stool.

Cato *Rust.* 10.4 (furnishings for a farm): "three stools" (scabilla III).

Varro *Ling.* 5.168 (as a footstool for getting into bed): "for a bed not too high, a *scabellum*" (in lectum non altum, scabellum).

For additional comment on this passage, see, in this section, *scamnum.*

scamnum* -*i **(n.):** a stool, a bench; an aid for climbing into a high bed.

Cat. *Rust.* 10.4 (furnishings for a farm): "three large *scamna* and one *scamnum* in the bedroom" (scamna magna III, scamnum in cubiculo I).

Perhaps the "scamna magna" signify benches or simple chairs, while the "scamnum" alone refers to a (foot)-stool.

Varro 5.168: "[to climb into] a higher [bed, use] a *scamnum*" (in altiorem [lectum] scamnum).

sella* -*ae **(f.):** a simple seat for one person, a stool. Attached to poles and carried by slaves, a *sella* could also be a means of conveyance.

Cato *Rust.* 14.2 (furnishings for a farm): "five stools" (sellas V).

sigma* -*atis **(n.):** a bench. A curved or semicircular bench used at a table for dining. Perhaps also the type found in gardens.

Mart. 10.48.6 (in reference to a dinner party): ". . . Flaccus, are you coming? The *sigma* holds seven" (. . . Flacce, venitis? septem sigma capit).

Mart. 14.87.1 (from a couplet titled "Stibadia," q.v.): "accept a crescent *sigma* inlaid with tortoiseshell" (accipe lunata scriptum testudine sigma).

solium* -*i **(n.):** a chair with backrest, armrest, and footstool. Similar to *cathedra.*

Cato *Rust.* 10.5 (furnishings for a farm): "two *solia*" (solia II).

Cic. *de Orat.* 3.133 (praise for the versatility of the great orators of the republic): "in the old days individuals sought out these men . . . seated in their *solia* at home, not only to consult them on a legal issue, but also about arranging a daughter's marriage, buying a farm, tilling their land, and finally about every kind of responsibility or business" (ad quos olim et ita ambulantes et in solio sedentes domi sic adibatur non solum ut de iure civili ad eos verum etiam de filia collocanda, de fundo emendo, de agro colendo, de omni denique aut officio aut negotio referretur).

Isid. *Orig.* 20.11.9: "the Greeks say *thronum* for what we [Latin speakers] call *solium*" (item thronum Graeci dicunt; nos solium).

Ovid. *Fast.* 3.358–60 (an account of the legendary King Numa of early Rome): "the people assembled at the doorstep of their king. He came forth and sat down in their midst on a *solium* of maple wood; unnumbered men were standing around him silent" (ante sui populus limina regis adest / prodit et in solio medius consedit acerno / innumeri circa stantque silentque viri).

Suet. *Aug.* 82.2: "[when Augustus enjoyed a hot salt bath] he was content sitting on [in?] a wooden *solium*, which he called by the Spanish name *dureta*" (contentus hoc erat ut insidens ligneo solio, quod ipse Hispanico verbo duretam vocabat).

The *solium* in question is either a wooden bathtub or a seat within the bathtub.

Suet. *Cal.* 57.3: "[Caligula] dreamt that he stood in heaven next to the *solium* of Jupiter" (somniavit consistere se in caelo iuxta solium Iovis).

Verg. *Aen.* 8.178: "[King Evander] invites Aeneas [to sit on] a maple wood *solium*" (Aenean solioque invitat acerno).

***stibadium* -ii (n.):** a semicircular bench, of the type often found in gardens.

Mart. 14.87: *Stibadia* is used as the heading for this couplet. For the text and translation of this passage see, in this section, *sigma*.

stool: see, in this subsection, *sella*.

stool, footstool: see, in this subsection, *scabellum*, *scamnum*.

stool, folding: see, in this subsection, *scabellum* = *scabillum*.

***subsellium* -i (n.):** simple chair or bench, supported by four straight legs, without back- or armrests.

Ambrosiaster, *Epistulas ad Corinthios* 1.14.31: for text and translation see, in this section, *cathedra*.

Cic. *Brut.* 84.290: "this is what I wish for the orator: when it is reported that he is about to speak let every place on the benches be taken, the [judge's] tribunal full" (volo hoc oratori contingat, ut cum auditum sit eum esse dicturum, locus in subselliis occupetur, compleatur tribunal).

Cic. *Cat.* 1.16 (regarding senatorial animosity toward the conspirator Cataline when he arrived at the Senate House): "at your arrival the benches around you were vacated . . . as soon as you sat down they left that section of benches bare and unoccupied" (adventu tuo ista subsellia vacuefacta sunt . . . simul atque adsedisti, partem istam subselliorum nudam atque inanem reliquerunt).

Isid. *Orig.* 20.11.9: for text and translation, see, in this section, *cathedra*.

Plaut. *Poen.* 5 (addressing the audience in a theater): "sit on your benches with good cheer" (bonoque ut animo sedeate in subsellis).

Plaut. *Stich.* 93 (Antipho motions to some seats in his home): "not I; you two sit there; I'll sit on this bench myself" (non sedeo istic; vos sedete; ego sedero in subsellio).

Suet. *Caes.* 84.3 (objects burned in Julius Caesar's funeral pyre in the forum during a riot): "the [wooden] judges' platforms with their benches" (cum subselliis tribunalia).

Suet. *Claud.* 41.1 (the young prince Claudius offers a reading to an audience without much success): "for at the beginning of the reading the splintering of several benches by a fat man [raised a laugh]" (nam cum initio recitationis defractis compluribus subselliis obesitate).

Suet *Nero* 26.2 "[brawling in the theater] settled with stones and broken benches" (lapidibusque et subselliorum fragminibus decerneretur).

***subsellium cathedrarium* -i (n.):** a simple seat or bench with a backrest.

Paul. *Dig.* 33.10.5: "[rugs] with which *subsellia cathedraria* are commonly covered" (subsellia cathedraria quibus insterni solent).

throne: see, in this subsection, *solium*.

XI.5. TABLES

***abacus* -i (m.):** a table with a rectangular slab top and three or more legs. A sideboard used to display fine vessels (cf. Varro *Ling.* 9.46).

Cato *Rust.* 11.3 (furnishings for a farm): "one table" (abacum I).

Immediately before this passage, Cato mentions the need for two *mensae* (*mensas* II); thus the terms are not interchangeable to indicate "table." Hooper (1993, 27) suggests the abacus may be a "kneading-trough."

Plin. HN 34.14 (introduction of the *abacus* to Italy): "according to Lucius Piso Gnaeus Manlius first introduced dining couches fitted with bronze and *abaci* and one-legged tables at his triumph after the defeat of Asia" (nam triclinia aerata abacosque et monopodia Cn. Manlium Asia devicta primum invexisse triumpho suo . . . L. Piso auctor est)."

The event is dated to 187 B.C.

Cic. *Verr.* 4.35: "from him [Verres] swept clean all the table's [silver] vessels just as they stood there (ab hoc abaci vasa omnia, ut exposita fuerunt abstulit).

Varro *Ling.* 9.46: "just as the *abacus* is embellished with silver . . . [with silver vessels that match and others that do not] . . . so also is speech adorned" (sicut abacum argento ornari . . . sic orationem).

See also, in this subsection, *cartibulum*, *urnarium*.

***cartibulum* -*a* (n.):** rectangular table carried on a single pedestal. Perhaps derived from wooden prototypes; examples known are of stone.

Varro *Ling.* 5.125: "a second kind of table for vessels was an oblong stone rectangle with one pedestal; it was called a *cartibulum*" (altera vasaria mensa erat lapidea quadrata oblonga una columella; vocabatur cartibulum).

See also, in this section (XI.5), *urnarium*.

***cilliba* -*ae* (f.):** a form of round table. See, in this section (XI.5), *mensa*.

desk: see, in this section (XI.5), *scrinium*.

***mensa* -*ae* (f.):** a table. Often with round tabletops (*orbes*) and three legs (*pedes*); thus the tripodlike "Delphic table" (*mensa delphica*). A one-legged version can be described simply as *monopodium*. A round table for serving wine may have also been called *cilliba*. *Citrus* alone can refer to a round table made of citrus wood.

Cato *Rust.* 11.3 (furnishings for a farm): "two tables" (mensas II); see also, in this section (XI.5), *abacus*.

Hor. *Sat.* 2.8.10–11: "a high-girt boy [slave] with a purple napkin wiped well the maple wood table" (sublatis puer alte cinctus acernam gausape purpureo mensam pertersit).

Isid. *Orig.* 20.1.1: "Daedalus first made a table and a stool" (primus Daedalus mensam et sellam fecit).

Mart. 10.98.6: "[your] antique citrus table" (citrum vetus).

Mart. 12.66.7: "a Delphic table [i.e., three-legged] of complicated design bears silver and gold" (argentum atque aurum non simplex Delphica portat).

Ov. *Met.* 8.661 -2 (inside the rustic hut of Baucis and Philemon): "the third leg of the table was too short: she made it level with a potsherd" (mensae sed erat pes tertius inpar:/ testa parem fecit).

Ovid's reference to a "third leg" suggests a *mensa delphica*.

Mart. 14.89.2 (on the value of citrus tables, or *mensa citrea*): "whoever gives gifts of gold gives less" (aurea qui dederit dona, minora dabit).

Plin. HN 12.5: "the legs of tables" (mensarum pedes).

Plin. HN 13.94 (a citrus table owned by the emperor Tiberius): for text and translation see *citrus*, under "XII.2. Species of Trees."

Plin. HN 13.98 (lack of interesting grain): "the faults of a table are 'woodiness,' that is the name given to a dull and uniform plainness to the wood" (mensae vitia lignum—ita vocatur materiae surda et indigesta simplicitas).

Plin. HN 34.14 (one-legged table): for text and translation see, in this section (XI.5), *abacus*.

Varro *Ling.* 5.121: "a round wine-table used to be called *cilliba*, as even now [it is] in the military camp" (mensa vinaria rotunda nominabatur ci(l)liba (a)nte ut etiam nunc in castris).

Note that Varro uses the same term to describe a square eating table employed in military camps (*Ling.* 5.118).

***scrinium* -i (n.):** a small writing desk(?). This definition is hard to verify.

Suet. *Nero* 47.2: for text and translation see *scrinium* in this section, under the subheading "Boxes, Chests."

sideboard: see, in this subsection, *abacus.*

table: see, in this subsection, *abacus, mensa.*

***urnarium* -i (n.):** rectangular table for kitchen or bath to hold water vessels.

Varro *Ling.* 5.126: "there was a third kind of rectangular table for vessels, like the second one [cf. *cartibulum*]; it was called an *urnarium*" (erat tertium genus mensae it(em) quadratae vasorum; voca[ba]tur urnarium).

See, in this subsection, *cartibulum.*

XI.6. MISCELLANEOUS HOUSEHOLD FURNISHINGS

***candelabrum* -i (n.):** candelabra, for holding candles or lamps, were usually made of metal, but Martial lists a wooden candelabrum as one of his gifts.

Mart. 14.44: "wooden candelabrum. You see that it is wooden" (candelabrum ligneum. Esse vides lignum).

The point Martial is making is that a wooden *candelabrum* is dangerous.

***pegma* -atis (n.):** shelf, bookcase, or similar fixture.

Cic. *Att.* 4.8.2: "nothing could be more charming than those *pegmata* of yours now that the books are adorned with title-slips" (nihil venustius quam illa tua pegmata, postquam sillybis [or *sittybae*] libros illustrarunt).

shelf: see, in this subsection, *pegma, and loculamentum* (XI.3).

XII. Trees and Timber

XII.1. PARTS OF TREES, TYPES OF TIMBER

***arbor* -oris (f.):** a generic term for a tree or its trunk.

Caes. *BCiv.* 2.15.1: "since all the trees far and wide within Massilian borders had been cut down" (omnibus arboribus longe lateque in finibus Massiliensium excisis).

Cat. *Rust.* 133.1: "for the propagation of fruit and other trees" (propagatio pomorum ceterarumque arborum).

Plin. *HN* 16.48 (in reference to the "female" [*femina*] of the silver fir as superior [*prolixior*] to the male): "with softer wood more easily worked and a tree [or trunk] more rounded" (materie mollior utiliorque, arbore rotundior).

The phrase "arbore rotundior" presumably refers to the profile of the tree.

***arbuscula* -ae (f.):** an immature tree, sapling, or shrub.

Varro *Rust.* 1.23.6: "thus in new orchards, when the cuttings have been planted and the saplings set out in rows" (nam et in recentibus pomariis dissitis seminibus in ordinemque arbusculis positis).

See this term also under "XIV. Finished Lumber."

***arbusta* -ae (f.):** cultivated trees, a plantation.

bark: see, in this section, *cortex, liber.*

***bruscum* -i (n.):** a burl or growth on a tree (maple in Plin. *HN* 16.68) that is prized for its intricate veining.

Plin. *HN* 16.68 (a feature of the maple): "most beautiful is [its] *bruscum*, even more remarkable by far is the *molluscum*, both are burls of this tree, the [grain of the] *bruscum* curled more tightly, the *molluscum* veined less elaborately, and if it reached the diameter of table(top)s, there's no doubt it would be preferred to citrus; now, however, it is seen in rare use for writing tablets, marquetry for beds, [or veneers] (pulcherrimum vero est bruscum, multoque excellentius etiamnum molluscum; tuber utrumque arboris eius, bruscum intortius crispum, molluscum simplicius sparsum, et si magnitudinem mensarum caperet, haut dubie praeferretur citro; nunc intra pugillares lectorum silicios [aut lamnas] raro usu spectatur).

See also *acer* under "XII.2. Species of Trees," and, in this section, *molluscum*.

cambrium: see, in this section, *torulus*.

***centrum* -*i* (n.):** a hard spot in a tree, where knots or clumps of wood fibers make cutting the wood difficult.
Plin. *HN* 16.198: "in some trees *centra* are found, just as in marble, that is, hard points similar to a nail, detrimental to saws" (inveniuntur in quibusdam sicut in marmore centra, id est duritia clavo similis, inimica serris).

***cortex* -*icis* (m.):** the bark of a tree; the bark of some species was used for roofing.
Plin. *HN* 16.35: for the text and translation see *abies* under "XII.2. Species of Trees."
Cf. also, in this section, *liber*.

club-wood: see, in this section, *fusterna*.

***fusterna* -*ae* (f.):** knotwood. The upper portion of a tree, particularly fir, where the branches spring, creating a hard, knotty wood. Club-wood (note that *fustis* refers to a club).
Plin. *HN* 16. 196: "the upper part [of the tree], which is knotted and harder, is called *fusterna*" (superior pars nodosa duriorque fusterna).
Vitr. 2.9.7: "the upper part of the tree . . . is called *fusterna* because of the hardness of the knots" (quae vero est superior . . . propter nodationis duritiem dicitur esse fusterna).

grain (of wood): see, in this section, *materia spissa*, *vena*.

green wood: see, in this section, *materia humida*.

knotwood: see, in this section, *fusterna*.

***liber* -*bri* (m.):** the bark of a tree.
Plin. *HN* 12.1: for text and translation see this passage above in the introduction to the text (p. 1).

See also, in this section, *cortex*.

***materia humida*:** literally, "damp wood." Julius Caesar uses this term to mean "green" or unseasoned wood.
Caes. *BCiv*. 1.58: "for, [the ships] having been made in haste from green wood, they did not exhibit the same capability of speed" (factae enim subito ex humida materia non eundem usum celeritatis habebant).

***materia spissa*:** literally, "packed" or "dense" wood. Thus, probably meaning "close-grained."
Plin. *HN* 16.204: for text and translation see, under "XII.2. Species of Trees," *buxus*.

For "peacock" grain, cf. Mart. 14.85.1 under "*lectus*" in "XI.2. Beds and Couches."

See also, in this section, *vena*.

***medulla* -*ae* (f.):** the pith of a tree, that is, the living tissue (sapwood) that lies between the bark and the heartwood.
Plin. *HN* 16.192 (girdling a tree): For text and translation see *circumcisura* under "XIII. Harvesting of Trees."
Vitr. 2.9.3 (initial cutting of a tree for timber): "therefore [the tree] should be cut so that the thickness of the tree is cut to the middle of the pith, and then left, so that the sap may drain

from it by dripping" (caedi autem ita oportet, uti incidatur arboris crassitudo ad mediam medullam, et reliquatur, uti per eam exsiccescat stillando sucus).

See also, in this section, *torulus*.

molluscum -i (n.): a burl or growth on the trunk of a tree valued for its intricate grain and used for veneer work and small implements like bowls.

Plin. HN 16.68: for text and translation see, in this section, *bruscum*.

See also, in this section, *tuber*.

nodus -i (m.): a knot. Knot-free (or clear) timber is described as *enodus*.

Plin. HN 16.196: "the part of the fir tree that was near the ground is knot-free" (abietis quae pars a terra fuit enodis est).

pando -are (vb.): to warp or sag, or to cause to warp.

Vitr. 2.9.11 (elm and ash beams in construction): for text and translation see *coagmentum* under "IV. Joints."

Vitr. 6.8.2: for the text and translation of this passage see "*postis*" under "VI. Framing."

roundwood: wood that has been used as found. Usually in prehistoric applications, but also in construction where the woodwork will be hidden.

sappineus -a -um (adj.): adjective referring to the choicest clear-grained wood from a tree.

Vitr. 2.9.7 (used for interior woodwork): for text and translation see *opus intestinum* under section IX.1.

stipes -itis (m.): the trunk of a tree or its lower part (sometimes merely the stump).

Verg. *Aen.* 4.444: "and the high leafy boughs blanket the ground when the trunk is buffeted" (et altae / consternunt terram concusso stipite frondes).

See also this term under "VI. Framing."

torqueo -ere: to warp.

Vitr. 2.9.8: "[the oak] shrinking from moisture . . . becomes warped ([quercus] fugiens ab umore . . . torquetur).

torulus -i (m.): the moist layer of outer wood in the trunk of a tree, which surrounds the heartwood. Vitruvius uses both the terms *medulla* (q.v.) and *torulus* to refer to what is often called sapwood. Possibly the latter refers to the outer layer of wood which contains knots.

Vitr. 2.9.7: "the *torulus* is removed . . . from the tree" (eiecto torulo . . . arbore).

At issue here is the use of clear heartwood *sappinea* (q.v.) for inside (fine) woodwork. Cf. also Vitr. 2.9.3.

truncus -i (m.): the trunk of a tree.

Caes. *BGall.* 4.17 (concerning Caesar's bridge over the Rhine): "trunks of trees" (arborum trunci).

For full text and translation see chapter 5 of the text.

Varro *Rust.* 1.14.2 (on the construction of a primitive fence): "tree trunks placed one after another, [with the trimmed branches] driven into the ground" (ex arboribus truncis demissis in terram deinceps constitutis).

trunk: see, in this section, *arbor*, *stipes*, *truncus*.

tuber -eris (n.): a swelling or growth. On a tree, a burl, of which specific types are mentioned: e.g., *bruscum* (q.v.), *molluscum* (q.v.).

Plin. HN 16.68: for text and translation see, in this section, *bruscum*.

vena -ae (f.): a vein or fiber in wood making up the grain.

Plin. HN 226: for text and translation see *lamina* under "IX. Interior Woodwork."

knot: see, in this section, *nodus.*

warp: see, in this section, *pando, torqueo.*

XII.2. SPECIES OF TREES USED BY ROMAN WOODWORKERS

***abies -etis* (f.):** fir. The most common species in Italy is *Abies alba,* or silver fir.

Plin. *HN* 16.35: "the bark of the beech, the linden, the fir, and the pitch pine is commonly used by country folk. With it they fashion baskets and wider vessels for carrying crops and grapes and for the shingling of country cottages" (cortex et fagi, tiliae, abietis, piceae, in magno usu agrestium. vasa eo corbesque ac patentiora quaedam messibus convehendis vindemiisque faciunt atque proiecta tuguriorum).

Plin. *HN* 16.42: "a first rate timber [for shipbuilding] . . . beams, and many household objects" (materies vero praecipua . . . trabibus et plurimus vitae operibus).

Plin. *HN* 16.48: (qualities of "male" vs. "female" firs): for text and translation see *arbor* in "XII.1. Parts of Trees."

Plin. *HN* 16.195(1): "[along with the larch] the tallest and straightest of all trees" (omnium arborum altissimae ac rectissimae).

Plin. *HN* 16.195(2): "for the masts and spars of ships fir is preferred because of its light weight" (navium malis antemnisque propter levitem praefertur abies).

Plin. *HN* 16.195(3): for text and translation see *nodus* under "XII.1. Parts of Trees."

Plin. *HN* 16.222: "fir and larch are strong load-bearers, even when placed horizontally" (pondus sustinere validae abies, larix, etiam in traversum positae).

Plin. *HN* 16.225: "fir is strongest in a vertical position; it is suitable for the panels of folding doors and any kinds of inlay work . . . of all sorts of wood it is most adapted for being glued together, so much so that it will split at a solid place before it parts at a join" (firmissima in rectum abies, eadem valvarum paginis et ad quaecumque libeat intestina opera aptissima . . . eadem e cunctis maxime sociabilis glutino, in tantum ut findatur ante qua solida est).

Sen. *Ep.* 90.9 (transport of firs for building): for text and translation see *lacunar* under "VIII. Roofing and Ceilings."

Vitr. 2.9.6: "[fir] remains straight (i.e., without warping) in flooring" (directa permanet in contignatione).

See also *contignatio* under "VII. Flooring."

***acer -ris* (n.):** maple.

Plin. *HN* 16.66: "the maple . . . is second only to the *citrus* for woodcraft in terms of its elegance and fine grain. It is of several types: the white maple, for its notable light-hued wood, is called Gallic maple . . . the second kind, with markings running in wavy lines, . . . has received the name 'peacock (maple)'" (acer . . . operum elegantia ac subtilitate citro secundum. plura eius genera: album, quod praecipui candoris, vocatur Gallicum . . . alterum genus crispo macularum discursu . . . pavonum nomen accepit).

Plin. *HN* 16.68 (the burls on the maple): for text and translation see *bruscum* under "XII.1. Parts of Trees."

Plin. *HN* 16.231 (veneer): for text and translation see *lamina* under "IX.2 Interior Woodwork."

See also, in this section, *zygia* and *carpinus.*

***acernus -a -um* (adj.):** belonging to the maple tree, made of maple wood.

Hor. *Sat.* 2.8.10: for text and translation, see, *mensa* under "XI.5. Tables."

Ovid *Ars Am.* I.325: "the cow [made for Queen Pasiphae by Daedalus of Crete was made] of maple wood (vacca . . . acerna).

Prop. 4.4.7 (Propertius imagines that the camp of the Sabine king Tatius was protected): "by a maple palisade" (vallo . . . acerno).

Verg. *Aen.* 8.178: for text and translation see *solium* under "XI.4. Chairs, Benches, Footstools."

alder: see, in this section, *alnus.*

***alnus* -i (f.):** alder.

Plin. HN 16.218: "larch and black alder do best in wet conditions" (larix in umore praecipua et alnus nigra).

Plin. HN 16.219: "the alder if driven into the ground in boggy places lasts forever and carries a load of any amount" (adacta in terram in palustribus alnus aeterna onerisque quantilibet patiens).

Plin. HN 16.224 (pipes): for text and translation see, in this section, *pinus.*

Vitr. 2.9.10 (alder piling used to shore up foundations): "lasts unperished for eternity, supports the crushing load of walls, and preserves them without failing" (permanet inmortalis ad aeternitatem et sustinet inmania pondera structurae et sine vitiis conservat).

Vitr. 2.9.11: for text and translation see *palus* under "V. Foundations."

Plin. HN 16.68: "[the burl occasionally found on the alder] is as inferior [to the other kinds of burls] as the alder itself is to the maple" (tanto deterius quantum ab acere alnus ipsa distat).

***aquifolium* -i** (also *aquifolia -ae*) **(f.):** holly

Plin. HN 16.231 (veneer): for text and translation see *lamina* under "IX.2 Interior Woodwork."

ash: see, in this section, *fraxinus, farnus.*

beech: see, in this section, *fagus.*

***betulla* -ae (f.):** the (Gallic) white birch.

Plin. HN 16.74-75: "this Gallic tree, with its wonderful whiteness and slender form, frightening because [it supplies] the [whipping] rods of the magistrates, easily bent for hoops, also for the ribs of baskets" (Gallica haec arbor mirabili candore atque tenuitate, terribilis magistratuum virgis, eadem circulis flexilis, item corbium costis).

birch: see, in this section, *betulla.*

box: see, in this section, *buxus.*

***buxus* -i (f.):** the box tree and its wood (the latter is neuter: *buxum, -i*).

Ovid *Met.* 4.30 (reference to a flute): "box [drilled] with holes" (foramine buxus).

Mart. 14.25: "[a comb of] boxwood with multifissured teeth" (multifido buxus . . . dente). Boxwood for combs is also mentioned by Ovid in the *Fasti* (6.229) and by Juvenal (14.194).

Plin. HN 16.70: "a wood rated among the best is that from the box tree, rarely with uneven grain except at the root, the rest clear-grained; the tree's wood is valued for a certain toughness and hardness and light color, the tree itself for ornamental gardens" (in primis vero materies honorata buxo est raro crispanti nec nisi radice, de cetero levi, cuius materia est lentitie quadam et duritie ac pallore commendabilis, ipsa vero arbor et topiario opere).

Plin. HN 16.73: "the holly, the box, the holm oak, the juniper, the turpentine-tree, the poplar, the mountain ash and the hornbeam love the mountains" (montes amant . . . aquifolia, buxus, ilex, iuniperus, terebinthus, populus, ornus, carpinus).

Plin. HN 16.204: "the densest grain of all timber—and thus the heaviest wood—is considered the box and ebony; the [trees are] slender in form. Neither floats in water" (spississima ex omni materie, ideo et gravissima, iudicatur hebenus et buxus, graciles natura. neutra in aquis fluvitat).

Plin. HN 16.212 (a: resistance to rot and age): for text and translation see, in this section, *cupressus.*

Plin. HN 16.212 (b: resistance to checking, splitting): for text and translation see, in this section, *cedrus.*

Plin. *HN* 16.226 (problems with gluing): for text and translation see, in this section, *cornus*. See also *gluten* under "III. Tools."

Plin. *HN* 16.230: "there are in fact minor applications [of wood] in the operations of carpentry; it is significant to relate that the most serviceable handles for augers are made from wild olive, box, holm oak, elm, and ash, and from these same woods mallets, and larger mallets from pine and holm oak" (sunt vero et parvi usus fabrilium ministeriorum, insigneque proditum terebris vaginas ex oleastro, buxo, ilice, ulmo, fraxino utilissimas fieri, ex iisdem malleos, maioresque e pinu et ilice).

Plin. *HN* 16.231 (veneer): for text and translation see *lamina* under "IX. Interior Woodwork."

Plin. *HN* 35.77 (educational innovations of classical Greece): "freeborn children . . . were taught drawing on boxwood" (pueri ingenui . . . graphicen in buxo, docerentur).

Pliny's reference is to flat tablets, presumably covered with wax, upon which sketches could be made.

Verg. *Aen.* 10.136–37: "as ivory shines when it has been enclosed by boxwood or Orician terebinth" (inclusum buxo aut Oricia terebintho / lucet ebur).

Virgil's reference is to inlay work.

Verg. *Georg.* 2.449: "boxwood, polished smooth [or "shaped"] on the lathe" (torno rasile buxum).

Vitr. 7.3.1 (interior ties of a hanging Roman vault): for text and translation see *asser*, under "VIII.2 Roofing and Ceilings."

***caprificus* -*i* (f.):** the wild fig tree.

Plin. *HN* 16.227: "all the woods we have described as pliant are bendable for every application, so too are the mulberry and wild fig" (cuicumque operi flexilia omnia quae lenta diximus, praeterque morus et caprificus).

***carpinus* -*i* (f.):** hornbeam. Hornbeam was valued as a hard, tough, close-grained wood.

Cato *Rust.* 31.1 (olive press): "make the pressing-beam ideally from black hornbeam" (prelum ex carpino potissimum facito).

Vitr. 2.9.12: "the hornbeam . . . is not fragile but exhibits a most useful flexibility [or is 'most easily handled']. Therefore the Greeks, since they make the yokes for draft animals from this wood, and since among them yokes are called *zyga*, they also call this wood '*zygia*'" (carpinus . . . non est fragilis, sed habet utilissimam tractabilitatem. Itaque Graeci, quod ex ea materia iuga iumentis conparant, quod apud eos iuga "zyga" vocitantur, item "zygian" eam appellant).

***castanea* -*ae* (f.):** the chestnut tree. Pliny tends to confuse this tree with walnut (*iuglans*).

Plin. *HN* 16.223 (roofing beams; Pliny uses *iuglans* but intends to refer to chestnut): for text and translation see *trabs* under "VIII.2 Roofing and Ceilings."

cedar: see, in this section, *cedrus*.

***cedrus* -*i* (f.):** cedar.

Plin. *HN* 16.73: "the cedar, the larch, the torch-pine and the other trees from which resin is obtained love the mountains" (montes amant cedrus, larix, taeda et ceterae e quibus resina gignitur).

Plin. *HN* 16.187: "the cedar, the larch, and the juniper are red" (cedrus et larix et iuniperus rubent).

Plin. *HN* 16. 203: "but in Egypt and Syria for want of fir the kings are said to have used cedar for their fleets; the largest cedar is reported to have been acquired in Cyprus" (at in Aegypto ac Syria reges inopia abietis cedro ad classes feruntur usi; maxima in Cypro traditur).

Plin. *HN* 16.207: "cedar cannot hold a nail" (clavum non tenet cedrus).

Plin. *HN* 16.212(a) (resistance to rot and age): for text and translation see, in this section, *cupressus*.

Plin. *HN* 16.212(b): "the cedar, cypress, olive, and box do not split or check by themselves" (rimam fissuramque non capit sponte cedrus, cupressus, olea, buxum).

Plin. *HN* 16.213 (durability of the wood): for text and translation see, in this section, *hebenus*. As roofing material for the Temple of Artemis at Ephesus see *lacunar* under "VIII.2 Roofing and Ceilings."

Plin. *HN* 16.216 (beams of Numidian cedar): for text and translation see *trabs* under "VIII.2 Roofing and Ceilings."

Vitr. 2.9.13: "from cedar is derived an oil called '*cedrium*.' When other things are soaked with this, for example books, they are not damaged by worms or rot" (ex cedro oleum quod cedrium dicitur, nascitur. quo reliquae res cum sunt unctae, uti etiam libri, a tineis et carie non laeduntur).

***cerrus*:** see, in this section, *quercus*.

chestnut: see, in this section, *castanea*.

***citrus* -i (f.):** the citrus or thuja tree. The neuter form, *citrum*, is used to refer specifically to the wood of the *citrus*. *Citrum* can also refer to a table made of citrus wood; see *mensa* under "XI.5. Tables."

Luc. 9.429–30: "our axes have invaded the unknown forest [of Mauretania], we have sought tables and delicacies from the ends of the earth" (in nemus ignotum nostrae venere secures, / extremoque epulas mensasque petimus ab orbe).

Plin. *HN* 13.91 (an intense desire for citrus): "table mania" (mensarum insania).

Plin. *HN* 13.93 (construction of citrus tabletops): for text and translation see *iunctura* under "IV. Joints."

Plin. *HN* 13.94: "under this topic it seems proper not to omit a table that belonged to the emperor Tiberius which was four feet two and a quarter inches across and one and one-half inches thick all over, but it was only covered with a veneer [of citrus wood]" (qua in re non omittendum videtur Tiberio principi mensam quattuor pedes sextante et sicilico excedentum, tota vero crassitudine sescunciali, operimento lamnae vestitam fuisse).

Plin. *HN* 13.97: "the highest of all [its merits] is its color, the color of mead is the most pleasing" (summa vero omnium in colore: hic maxime mulsi placet).

Plin. *HN* 16.66 (along with maple a superior wood): see, in this section, *acer*.

Plin. *HN* 16.231 (veneer): for text and translation see *lamina* under "IX. Interior Woodwork."

Stat. *Silv.* 3.3.94 (part of the contents of the imperial treasury): "Massylian citrus wood" (Massylaque robora).

Lit: "Massylian oaks," here the hard wood of the citrus is referenced poetically as "oak."

See also *lamina* under "IX.2 Interior Woodwork"; *acer* under "XII.2. Species of Trees"; "XII.3. Drying, Preservatives and Conditioners"; and *mensa* under "XI.5. Tables."

***citrum* -i (n.):** the wood of the citrus. See, in this section, *citrus*.

cornel: see, in this section, *cornus*.

***cornus* -i (f.):** the cornel. A member of the dogwood family, having extremely hard wood.

Cato *Rust.* 18.9 (dowels or tenons of cornel to fasten boards in an olive press): for text and translation see *occludo* under "IV. Joints."

Plin. *HN* 16.187: "cornus wood shines bright yellow in hunting spears notched with incisions for decoration" (fulva cornus in venabulis nitet incisuris nodata propter decorem).

Plin. *HN* 16.206: "[cornel] wood is hardly useful for anything else except the spokes of wheels or in case something has to be wedged in wood or fixed with dowels [made of it], which are like iron" (sed lignum non alio paene quam ad radios rotarum utile aut si quid cuneandum sit in ligno clavisve figendum ceu ferreis).

Plin. *HN* 16.226 (difficulty in gluing cornel to other woods): "the [wood of the] service-tree, hornbeam, box, and to a lesser degree linden, have a strong dislike for cornel" (cornum maxime odit sorbus, carpinus, buxus, postea tilia).

See also *gluten* under "III. Tools."

Verg. *Aen.* 12.267 -8 (a launched spear of cornel): "the cornel [shaft] slices through the air, high pitched and unerring" (stridula cornus et auras certa secat).

Verg. *Georg.* 457 -8: "the cornel, with its sturdy spearshafts, good for war" (validis hastilibus et bona bello cornus).

***corylus* -i (f.):** hazel (or filbert). Since the hazel grows only to a small size, it was used only for small implements (Cato 18.9, however, suggests its use in the making of a disk for an olive press).

Cato *Rust.* 18.9 (a disk for an olive press): "fashion the disk from elm or hazel, if you have both, arrange them alternately" (orbem ex ulmo aut ex corylo facito: si utrumque habebis, alternas indito).

***cupressus* -i (f.):** cypress.

Plin. *HN* 16.141: "both [the male and female trees], having their branches pruned off, are rendered into poles or beams, which [having grown] twelve years sell for a denarius apiece . . . the native country of this tree is Crete" (utraque autem immittitur in perticas asseresve amputatione ramorum, qui XIII anno denariis singulis veneunt . . . huic patria insula Creta).

Plin. *HN* 16.212(a): "the cypress, cedar, ebony, nettle-tree, box, yew, juniper, wild olive and olive do not suffer from age or rot" (cariem vetustatemque non sentiunt cupressus, cedrus, hebenus, lotus, buxum, taxus, iuniperus, oleaster, olea).

Plin. *HN* 16.212(b): resistance to checking, splitting. For text and translation see, in this section, *cedrus*.

Plin. *HN* 16.213 (durability): for text and translation see, in this section, *hebenus*.

Plin. *HN* 16.215: doors of Artemis at Ephesus; best of woods for retaining its polish (see also *valvae* under "IX.3. Doors and Shutters").

Plin. *HN* 16.216: "the statue of Veiovis on the citadel [of Rome] of cypress has lasted . . ." (simulacrum Veiovis in arce e cupresso durat . . .).

Plin. *HN* 16.223: "pine and cypress are the most resistant against rot and wood-worms" (pinus et cupressus adversus cariem tiniasque firmissimae).

Vitr. 2.9.12 (cypress and pine): "because of their abundance of moisture, in structural applications they often warp [or sag], but they last for a long time without degrading" (propter umoris satietatem in operibus solent esse pandae, sed in vetustatem sine vitiis conservantur).

Vitr. 2.9.13: "resin [comes from] cypress and pine" (ex cupressu et pinu resina).

Vitr. 7.3.1 (interior lath- and plasterwork): for text and translation see *asser* under "VIII. Roofing and Ceilings."

cypress: see, in this section, *cupressus*.

dogwood: see, in this section, *cornus*.

ebony: see, in this section, *hebenus*.

elm: see, in this section, *ulmus*.

***fagus* -i (f.):** beech. Note: in the passages below, Pliny can confuse *fagus* with the Greek *phegos* (used by Theophrastus, meaning "oak"). Thus some uses of the beech listed in this section by Pliny are in fact more appropriate for the *quercus* (cf. discussion in Meiggs 1982, 25). The statements below are contradictory as far as the relationship between beech and water goes. In fact, if left submerged, beech lasts well and was used in shipbuilding.

Ov. *Met.* 8.669–70 (description of the hut of Baucis and Philemon): "hollowed goblets of beechwood, coated on the inside with yellow wax" (fago / pocula, qua cava sunt, flaventibus inlita ceris).

Plin. *HN* 16.35 (use of bark): for text and translation see, in this section, *abies*.

Later in this passage Pliny says that fresh bark was "used for ritual purposes" (in . . . usu sacrorum religiosus est)"; this practice of stripping the fresh bark killed the tree.

Plin. *HN* 16.36: good for roofing shingles. For text and translation see *scandula* under "VIII. Roofing and Ceilings."

Plin. *HN* 16.218: "both beech and walnut are well-judged [for use in] water, these indeed are best among those [woods] which are placed underground, and juniper, too [which is also most suitable for open-air structures]" (non inprobatur et fagus in aqua et iuglans, hae quidem in iis quae defodiuntur vel principales, item iuniperus [eadem et subdialibus aptissima]).

Plin. *HN* 16.229 (suitable for boxes and document cases): for text and translation see *lamina* under "IX. Interior Woodwork."

Vitr. 2.9.9: "the turkey oak and the beech, because they have a similar mixture of water and earth and fire, and most of all air, being furnished with open pores admit moisture and quickly lose their integrity" (cerrus quercus fagus, quod pariter habent mixtionem umoris et ignis et terreni, aeris plurimum, provisa raritates umoris penitus recipiendo celeriter marcescunt).

Vitr. 7.1.2: "no [flooring board] of turkey oak, beech, or ash can last for a long time" (namque de cerro (quercus cerris) aut fago seu farno nullus ad vestutatem potest permanere).

***farnus* -i (f.):** the ash or its wood.

Vitr. 7.1.2: for text and translation see, in this section, *fagus*.

fig, wild: see, in this section, *caprificus*.

fir: see, in this section, *abies*.

***fraxinus* -i (f.):** ash.

Plin. *HN* 16.219 (curing ash by girdling): see *circumcisura* under "XIII. Harvesting of Trees."

Plin. *HN* 16.228: "ash is the most compliant wood for whatever application, it is better than hazelwood for spears, lighter than dogwood, and more flexible than the mountain ash; indeed the Gallic ash has the lightness and suppleness for chariots" (oboedientissima quocumque in opere fraxinus, eademque hastis corylo melior, cornu levior, sorbo lentior; Gallica vero etiam ad currus flexili levitate).

Plin. *HN* 16.230 (tool handles and mallets): for text and translation see, in this section, *buxus*.

Vitr. 2.9.11 (tendency to warp or sag when not cured): for text and translation see *coagmentum* under "IV. Joints."

hazel: see, in this section, *corylus*.

***hebenus* -i (m. and f.):** ebony.

Luc. 10.117: "Mareotican [Egyptian] ebony" (hebenus Mareotica).

Plin. *HN* 212 (resistance to rot and age): for text and translation see, in this section, *cupressus*.

Plin. *HN* 16.213: "it is believed that ebony lasts forever, and also cypress and cedar" (maxime aeternam putant hebenum, et cupressum cedrumque).

Verg. *Georg.* 2.116–17: "India alone produces the black ebony" (sola India nigrum / fert hebenum).

holly: and the wood of the holly: see, in this section, *aquifolium*.

hornbeam: see, in this section, *carpinus*.

***ilex* -icis (f.):** the holm oak, or ilex. An evergreen species of oak, a member of the holly family, with rounded or pointed leaves.

Cato *Rust.* 18.9 (useful for tenons or dowels): for text and translation see *subscus* under "IV. Joints."

Plin. *HN* 16.19: "there are two species of ilex. One of these from Italy [with rounded leaf, the other type from] the provinces with pointed leaves" (ilicis duo genera. ex his in Italia folio . . . in provinciis aquifoliae sunt ilices).
Plin. HN 16.73 (thrives in high altitudes): for text and translation see, in this section, *buxus.*
Plin. HN 16.230 (tool handles and mallets): for text and translation see, in this section, *buxus.*
Plin. HN 16.231 (veneer): for text and translation see *lamina* under "IX.2 Interior Woodwork."
Plin. HN 16.229: "the ilex also can be cut into extremely thin layers, and also has a color not unpleasing" (secatur in lamnas praetenues et ilex, colore quoque non ingrata).

***iuglans -ndis* (f.):** walnut.
Plin.HN 16.218 (resistance to moisture): for text and translation see, in this section, *fagus.*
Plin. HN 16.223 (probably in reference to chestnut): for text and translation see *trabs* under "VIII. Roofing and Ceilings."

***iuniperus -i* (f.):** juniper.
Plin. HN 16.73 (thrives in high altitudes): for text and translation see in this section, *buxus.*
Plin.HN 16.187 (red color of wood): for text and translation see, in this section, *cedrus.*
Plin. HN 16.212 (resistance to rot and age): for text and translation see, in this section, *cupressus.*
Plin. HN 16.216 (used for the beams of the Temple of Diana at Saguntum in Spain): for text and translation see *trabs* under "VIII. Roofing and Ceilings."
Plin. HN 16.218 (resistance to moisture): for text and translation see, in this section, *fagus.*
Vitr. 7.3.1 (interior ties, or *catenae,* of a hanging Roman vault): for text and translation see *asser* under "VIII. Roofing and Ceilings."

juniper: see, in this section, *iuniperus.*

larch: see, in this section, *larix.*

***larix -icis* (f.):** larch.
Plin. HN 16.45: "[resinous trees burn] except the larch which neither burns nor makes charcoal nor is consumed by the action of fire any more than stones" (excepta larice quae nec ardet nec carbonem facit nec alio modo ignis vi consumitur quam lapides).
Plin. HN 16.73 (thrives in high altitudes): for text and translation see in this section, *cedrus.*
Plin. HN 16.187 (a: red color of wood): for text and translation see, in this section, *cedrus.*
Plin. HN 16.187 (b): "the female larch has [wood] that the Greeks call '*aegis*,' this wood has been found to last forever without splitting [when used for] the panels of paintings" (larix femina habet quam Graeci vocant aegida . . . inventum pictorum tabellis inmortale nullisque fissile rimis hoc lignum).
Plin. HN 16.200 (a beam of record size): for text and translation see chapter 12 of the text.
Plin. HN 16.218 (suitability for wet conditions): for text and translation see, in this section, *alnus.*
Plin. HN 16.222 (used to form strong beams): for text and translation see, in this section, *abies.*
Vitr. 2.9.14: "the larch is known only to locals on the banks of the river Po and the shores of the Adriatic Sea. Because of the fierce bitterness of its sap, it is not harmed by rot or worms . . . nor can it burn by itself . . . because of its weight it does not float on water" (larix vero, qui non est notus nisi is municipalibus qui sunt circa ripam fluminis Padi et litora maris Hadriani, non solum ab suco vehementi amaritate ab carie aut tinea non nocetur, . . . nec ipse per se potest ardere . . . propterque pondus ab aqua non sustinetur).
Vitr. 2.9.16 (discovery of this wood by Julius Caesar): "and so the fort was called 'Larignum' in the same way the wood [used to build its fortifications] was called larch" (et ideo id castellum Larignum, item materies larigna est appellata).

Vitr. 2.9.17: "[larch is] suitable for interior woodwork not less than clear fir" (tractabilis ad intestinum opus non minus sappinea).

See also *opus intestinum* under "IX. Interior Woodwork."

lime: see, in this section, *tilia.*

linden: see, in this section, *tilia.*

***molluscum*:** see, in this section, *acer.*

***morus* -i (f.):** the (black) mulberry.

Plin. HN 16.218 (praises the mulberry for its durability): "it blackens with age" (quae vetustate etiam nigrescit).

Plin. HN 16.227 (flexibility): for text and translation see, in this section, *caprificus.*

maple: see, in this section, *acer.*

mulberry: see, in this section, *morus.*

oak: see, in this section, *quercus, robur.*

***olea* -ae (f.):** olive. *Oleaster* refers to the wild olive.

Plin. HN 16.212 (a: resistance to rot and age): for text and translation see, in this section, *cupressus.*

Plin. HN 16.212 (b: resistance to checking, splitting): for text and translation see, in this section, *cedrus.*

Plin. HN 16.222 (not suitable for beams): "oak and olive bend and yield under weight" (robur, olea incurvantur ceduntque ponderi).

Plin. HN 16.230 (a: hinges of olive): for text and translation see *cardo* under "IX.3. Doors and Shutters."

Plin. HN 16.230 (b: tool handles and mallets): for text and translation see, in this section, *buxus.*

Vitr. 1.5.3 (used to reinforce the walls of fortifications): "charred olivewood beams" (tabulae oleagineae ustilatae).

The use of *tabulae* for "beams" is unusual; perhaps Vitruvius means "planks."

Vitr. 7.3.1 (interior ties of a hanging or false vault): for text and translation see *asser* under "VIII. Roofing and Ceilings."

olive: see, in this section, *olea.*

***palma* -ae (f.):** the palm tree.

Plin. HN 16.231 (veneer): for text and translation see *lamina* under "IX.2 Interior Woodwork."

***picea* -ae (f.):** pitch pine. The *picea* was associated with death; it was grown on graves or placed at the doorways of houses where a family member was being mourned. It was also an important source of pine resin. As far as woodworking is concerned, Pliny considers it a tree of few practical functions.

Plin. HN 16.40: "[it is] a tree associated with funerals" (feralis arbor).

Plin. HN 16.42 (uses of pitch pine): "vats . . . split shingles . . . pieces of joinery" (cupae . . . fissiles scandulae . . . secamenta).

Plin. HN 16.35 (use of bark): for text and translation see, in this section, *abies.*

Plin. HN 16.224 (pipes): for text and translation see, in this section, *pinus.*

pine: see, in this section, *pinus.*

***pinus* -us (or -i) (f.):** pine.

Cato Rust. 18.8: "make the anchor posts (for an olive press) of oak or pine" (arbores stipites robustas facito aut pineas).

Plin. *HN* 16.36 (suitability for roofing shingles): for text and translation see *scandula* under "VIII. Roofing and Ceilings."

Plin. *HN* 16.223: "pine and cypress are the strongest to resist rot and destructive worms" (pinus et cupressus adversus cariem tiniasque firmissimae).

Plin. *HN* 16.224: "pines, pitch pines, and alders are hollowed to form pipes for carrying water" (pinus, piceae, alni ad aquarum ductus in tubos cavantur).

Plin. *HN* 16.230 (large mallets): for text and translation see, in this section, *buxus.*

Sen. *Ep.* 90. 9 (transport of pines for building): for text and translation see *lacunar* under "VIII. Roofing and Ceilings."

Verg. *Aen.* 11.136: "they fell pines that had soared as high as the stars" (evertunt actas ad sidera pinus).

Vitr. 2.9.12 (warpage, resistance to decay): for text and translation see, in this section, *cupressus.*

Vitr. 2.9.13 (source of resin): for text and translation see, in this section, *cupressus.*

pine, pitch: see, in this section, *picea.*

pistachio: see, in this section, *staphylodendron.*

poplar: see, in this section, *populus.*

***populus* -i (f.)** (or *poplus*): the poplar tree, sacred to Hercules. Pliny and Vitruvius mention the white (*alba*) and black (*nigra*) varieties, Pliny the *libyca* as well, an unidentified species.

Cato *Rust.* 6.3: "around the perimeters [of the farm] and along the streets plant elms and some poplars . . . so that you will have [leaves for fodder] and timber, if there is a job to be done, will be ready" (circum coronas et circum vias ulmos serito et partim populos, uti . . . habeas et materies, siquo opus sit, parata erit).

Plin. *HN* 16.73 (found in high altitudes): for text and translation see in this section, *buxus.*

Plin. *HN* 16.85: "[there are] three kinds of poplar: the white, the black and the one called the Libyan" (populi tria genera, alba ac nigra et quae Libyca appellatur).

Plin. *HN* 16.231 (veneer): for text and translation see *lamina* under "IX. Interior Woodwork."

Vitr. 2.9.9 (poplar, willow, and linden): "on account of their porosity [the wood] is white and for carving is convenient to handle" (propter raritatem sunt candida et in sculpturis commodam praestant tractabilitatem).

***quercus* -us (f.):** oak; *quercus robur* (English or Valonia oak), *cerrus* (turkey or scrub oak). The evergreen oak (*ilex*) is considered separately. (fig. 12.3)

Caes. *BGall.* 3.13.3: "the ships [of the Gauls] were made entirely of *robur*, to endure any violence and buffeting" (naves totae factae ex robore ad quamvis vim et contumeliam perferendam).

Cato *Rust.* 17.1: "oak wood . . . is always ready [for cutting] at the winter solstice" (robus materies . . . ubi solstitium fuerit ad brumam semper tempestiva est).

Cato *Rust.* 18.8 (use in an olive press): for text and translation see, in this section, *pinus.*

Plin. *HN* 16.36 (*robur* good for roofing shingles): for text and translation see *scandula* under "VIII. Roofing and Ceilings."

Plin. *HN* 16.207 (hardness of *robur*): "[it is] so hard it cannot be bored unless it has been soaked in water and even then when a nail has been hammered into it it cannot be pulled out" (tanta duritia ut terebrari nisi madefactum non queat et ne sic quidem adactus avelli clavus).

Plin. *HN* 16.218: "elm endures in the open air, [*quercus*] *robur* underground and *quercus* submerged in water; the same [*quercus* employed] above ground makes masonry crack by its warping" (ulmus in perflatu firma, robur defossum et in aquis quercus obruta; eadem supra terram rimosa facit opera torquendo sese).

Note the differentiation between *robur* (Valonia oak) and *quercus* (oak).

Plin. *HN* 16.222 (beams of *robur* bend under a load): for text and translation see, in this section, *olea*.

Plin. *HN* 16.226 (forms a poor bond when glued): see *gluten*, under "III. Tools."

Plin. *HN* 36.187: "it is considered counterproductive to sheathe the wood floor with oak planks, because they warp" (quernis axibus contabulari, quia torquentur, inutile putant).

Vitr. 2.9.8 (foundations): "when it [*robur*] is buried in underground works [i.e., foundations] it has indefinite durability" (cum in terrenis operibus obruitur, infinitam habet aeternitatem).

Vitr. 2.9.9: "*aesculus* . . . has great applications in building" (aesculus . . . habet in aedificiis magnas utlitates).

For commentary on the identification of *aesculus* see chapter 12 of the text, n. 8.

Vitr. 2.9.9 (turkey oak prone to decay): for text and translation, see, in this section, *fagus*.

Vitr. 7.1.2: for text and translation see, in this section, *fagus*.

Later in the same passage, Vitruvius warns against mixing planks of "winter oak" (*quercus aesculus*) with common oak (*quercus robur*), since the latter tends to warp when moist.

Vitr. 7.3.1 (concerning interior ties of a hanging Roman vault): for text and translation see *asser* under "VIII. Roofing and Ceilings."

See also, in this section, comments under *fagus*.

robur, -oris (n.): oak (common oak, English oak, Valonia oak). See, in this section, *quercus*.

salix -icis (f.): willow. A grove of willows (or osier bed): *salictum*.

Cato *Rust.* 1.7 (ranking the most important features of a farm): "a vineyard is first . . . in second place is an irrigated garden, third a willow grove . . ." (vinea est prima . . . secundo loco hortus inriguus, tertio salictum).

Cato *Rust.* 20.1: "[stabilize an iron bar in an olive press] with wedges of willow wood" (cuneis salignis).

Ov. *Met.* 8.655–56: "a cushion of soft rushes was placed on the couch with a frame and feet of willow wood" (torum de molli fluminis ulva / inpositum lecto sponda pedibusque salignis).

Plin. *HN* 16.174: "the uses of the willow take many forms . . . vine trellises . . . basketry . . . [the branches] most suitable for luxurious armchairs" (salicis utilitatum plura genera . . . iugis vinearum . . . corbibus . . . in delicias cathedrarum aptissimae).

Verg. *Georg.* 2.446: "the willows' fruitfulness is in their branches" (viminibus salices fecundae).

Vitr. 2.9.9 (suitability for carving): for text and translation see, in this section, *populus*.

sorbus, -i (f.): a genus of trees, the European service tree (*Sorbus domestica*), a member of the rose family (*Rosaceae*), has small edible fruit.

Plin. *HN* 74: "the sorbus thrives in cold places" (gaudet frigidis sorbus).

Plin. *HN* 16.226 (poor bond when glued to *cornus*): for text and translation see, in this section, *cornus*. See also "*gluten*" under "III. Tools."

staphylodendron -i (n.): the (wild) pistachio.

Plin. *HN* 16.69: "across [i.e., north of] the Alps is a tree with wood very similar to the white maple (=ash?) which is called staphylodendron" (est trans Alpes arbor simillima aceri albo materie quae vocatur staphylodendron).

taxus -i (f.): the yew tree.

Verg. *Georg.* 2.458: "yews are bent into Ituraean bows" (Ituraeos taxi torquentur in arcus).

The Ituraeans were an Arab people of the eastern empire famed for archery.

terebinthus -i (f.): the terebinth or turpentine-tree. A wood valued for its grain, it was used for furniture and fine wooden utensils.

Plin. *HN* 16.73 (found at high altitudes): for text and translation see in this section, *buxus*.

Plin. *HN* 16.205 (goblets of *terebinth*): for text and translation see *tornus* under "III. Tools."
Plin. *HN* 16.231 (veneer): for text and translation see *lamina* under "IX. Interior Woodwork."
Verg. *Aen.* 10.136–37: "as ivory shines when it has been enclosed by boxwood or Orician terebinth" (inclusum buxo aut Oricia terebintho / lucet ebur).

***tilia* -*ae* (f.):** linden (or lime). The wood of the linden was relatively soft; thus it was good for small items and could be easily turned on a lathe (see *tornus* under "III. Tools"). *Tilia* is related to the American basswood, also known for its suitability for carving.

Plin. *HN* 16.35 (use of bark): for text and translation see, in this section, *abies.*
Plin. *HN* 16.207: "linden is the softest [of woods] and it seems to be the hottest as well: the proof of this is evident since it blunts adzes so quickly" (mollissima tilia; eadem videtur et caldissima: argumentum adferunt quod citissime ascias retundat).
Plin. *HN* 16.226 (poor bond when glued to *cornus*): for text and translation see, in this section, *cornus.* See also *gluten* under "III. Tools."
Vitr. 2.9.9 (suitability for carving): for text and translation see, in this section, *populus.*

turkey oak: see, in this section, *quercus.*

turpentine-tree: see, in this section, *terebrinthus.*

***ulmus* -*i* (f.):** elm.

Cato *Rust.* 6.3 (used on farms as a source of timber): for text and translation see, in this section, *populus.*
Cato *Rust.* 18.9 (disk for an olive press): for text and translation see, in this section, *corylus.*
Plin. *HN* 16.72: mentions four varieties in Italy: *Atinia, Gallica, Italica, Silvestris.*
Plin. *HN* 16.210: (suitability for hinges and door frames): for text and translation see *cardo* under "IX.3. Doors and Shutters."
Plin. *HN* 16.218 (durability of elm in open air): for text and translation see, in this section, *quercus.*
Plin.*HN* 16.219 (curing elm by girdling): for text and translation see *circumcisura* under "XIII. Harvesting of Trees."
Plin. *HN* 16.229: "the elm would compete [with ash in terms of flexibility] if its weight did not count against it" (aemularetur ulmus ni pondus esset in culpa).
Plin. *HN* 16.230 (tool handles and mallets): for text and translation see, in this section, *buxus.*
Verg. Georg. 2.446: "the [value of] elms is in the leaves" (frondibus ulmi).
Vitr. 2.9.11 (warps when uncured, good for joinery): for text and translation see *coagmentum* under "IV. Joints."

walnut: see, in this section, *iuglans.*

willow: see, in this section, *salix.*

***zygia*:** perhaps Norway maple (*Acer platanoides*)?

Plin. *HN* 16.67: "easy to split" (fissili ligno).

XII.3. DRYING, PRESERVATIVES, AND CONDITIONERS

DRYING BY APPLICATION OF HEAT

Columella *Rust.* 1.6.19: "the smoke room too, in which timber, if it has not been cut for a long time, may be dried quickly, can be built in the part of the country villa next to the rural baths" (fumarium quoque, quo materia, si non sit iam pridem caesa, festinato siccetur, in parte rusticae villae fieri potest iunctum rusticis balneis).

TREATMENT BY CHARRING

Vitr. 1.5.3 (on binding and reinforcing the outer and inner walls of a defensive circuit): "timbers of charred olive wood should be frequently placed [transversely] in order that both facades of the wall as if connected by pins . . . may have everlasting strength. For such timber cannot be injured by decay or weather or age" (tabulae oleagineae ustilatae quam creberrime instruantur, uti utraeque muri frontes inter se, quemadmodum fibulis . . . conligatae . . . aeternam habeant firmitatem; namque ei materiae nec caries nec tempestates nec vetustas . . . potest nocere).

TREATMENT BY OILING

The use of cedar oil on wood:

Plin. *HN* 16.198: "timber daubed with cedar oil suffers from neither worms nor decay" (cedri oleo peruncta materies nec tiniam nec cariem sentit).

The use of *amurca* (the by-product of olive oil production):

Cato *Rust.* 98.2: "lest moths ruin clothing . . . rub the exterior of the bottom, the feet, and the corners of the chest with [*amurca*] . . . If you rub wooden furniture all over [with *amurca*] it will not decay, and so treated will shine more brightly" (vestimenta ne tiniae tangant . . . ea (amurca) unguito fundam arcae et extrinsecus et pedes et angulos . . . et item (amurca) ligneam supellectilem omnem si ungues, non putescet, et cum ea terseris spendidior fiet).

TREATMENT BY SMEARING WITH OR IMMERSING IN DUNG

Cato *Rust.* 31.1: for text and translation see *fibula* under "IV.2 Joints."

Plin. *HN* 16.222: "we have in our country some timbers that check [split] by themselves, on account of which architects prescribe that they should be smeared with dung and then dried, so that [drying] winds will not harm them (apud nos materiae finduntur aliquae sponte; ob id architecti eas fimo inlitas siccari iubent ut adflatus non noceant).

TREATMENT BY IMMERSING WOOD IN SALTWATER

Plin. *HN* 13.99: "the shattered remains of ships have recently shown that this timber, dried by the action of seawater, is solidified with a hardness that resists decay more strongly than any other method" (naufragia docuere nuper hanc quoque materiam siccatam mari duritie incorrupta cospissari non ullo modo vehementius).

TREATMENT BY DRYING WOOD IN PILES OF GRAIN (AND BY BURYING IT IN EARTH): (IN TEXT)

Plin. *HN* 13.99: "carpenters lay [citrus wood] in heaps of grain for periods of a week with intervals of a week between, and it is surprising how much its weight is reduced by the process" (artifices vero frumenti acervis inponunt septenis diebus totidem intermissis, mirumque ponderi quantum ita detrahatur).

In the same passage Pliny reports that some *barbari* (probably the locals of north African Mauretania where the citrus wood was harvested) treat the raw timber by coating it with wax and burying it in the ground.

XIII. Harvesting of Trees

XIII.1. INTRODUCTION: BEST CONDITIONS FOR FELLING

Cato. *Rust.* 31.1–2 (cutting in late autumn): "when the moon is waning, in the afternoon and without a southerly wind, take out the elm, pine, nut and all the other timber. It will be the right time when their seed is ripe, and take care you neither cut it nor haul it in wet conditions" (ulmeam, pineam, nuceam, hanc atque aliam materiem omnem . . . luna decrescente eximito post meridiem sine vento austro. Tum erit tempestiva, cum semen suum maturum erit, cavetoque per rorem trahas aut doles).

Cato *Rust.* 37.4: "take care that you neither chop timber nor fell it nor even touch it, if you can [avoid it] unless it is dry and neither frosty nor covered with dew" (caveto nequam materiem doles neu caedas neu tangas, si potes, nisi siccam neu gelidam neu rorulentam).

Plin. *HN* 16.188: "[if tree trunks] are to be stripped of their bark and, being well turned, used for temples and other applications calling for columns, the favorable time for felling is when they bud [i.e., early spring, so that the bark can be removed easily] . . . [however,] for beams and those [timbers] for which the axe removes the bark [the favorable time for cutting] is from the winter solstice to the rise of the westerly wind [or, Pliny adds, even sooner] . . . generally people think it is enough to take care that [no tree] is felled for hewing before its fruits have been borne" (caedi tempestivum quae decorticentur, ut teretes ad templa ceteraque usus rotundi, cum germinant . . . tigna et quibus auferat securis corticem a bruma ad favonium . . . vulgo satis putant observare ne qua dedolanda sternatur ante editos suos fructus).

Plin. *HN* 16.190: "the emperor Tiberius ordained that the larches be cut in Raetia [during the full moon] for the reconstruction of the deck used for the mock naval battles that had been destroyed by fire" (Tiberius Caesar concremato ponte naumachiario larices ad restituendum caedi in Raetia praefinivit).

Plin. *HN* 16.191–92: "some add [say] the rising of the dog [star is also a good time for cutting] and thus timber was felled for the Forum of Augustus" (quidam et canis ortum addunt et sic caesas materias in forum Augustum).

The dog star, Sirius, rises with the sun on July 3.

Plin. *HN* 16.192: "[trees that are] neither immature nor old are the most useful for timber" (nec novellae autem ad materiem nec veteres utilissimae).

Plin. *HN* 16.193–94: Pliny cites Cato (*Rust.* 31.1ff., cited above, XIII.1).

Vitr. 2.9.1–2: "wood should be cut from the beginning of autumn to the time before the west wind [*favonius*] starts to blow. . . the force of the winter wind compresses and consolidates them [trees and thus their wood]" (materies caedenda est a primo autumno ad id tempus, quod erit antequam flare incipiat favonius . . . aeris hiberni vis conprimit et consolidat eas).

XIII.2. FELLING AND ROUGH CUTTING

caedo, -ere: to cut down, hew

Cato *Rust.* 37.4: for text and translation see above under XIII.1.

***circumcisura -ae* (f.):** the practice of removing a strip of bark from around the trunk of a living tree before it is cut down, to aid in removing excess moisture.

Plin. *HN* 16.192 (girdling): "also, some people purposely leave [trees] cut all around as far as the *medulla*, so that all the moisture [sap] may drain out of them while they are still standing" (circumcisas quoque in medullam aliqui non inutiliter relinquunt, ut omnis umor stantibus defluat).

Plin. *HN* 16.219: "elm and ash are . . . more reliable if the trees are left standing and dried by girdling" (ulmus et fraxinus . . . stantesque ac circumcisura siccatae fideliores).

Note Vitruvius (2.9.11) also recommends that these same species be allowed to dry out "in the field" (in agro perfecto).

Vitr. 2.9.3: for text and translation see *medulla* under "XII.1. Parts of Trees."

decido -ere: to cut or saw.

Sen. *Ep.* 90.9: "to cut timbers square" (in quadratum tigna decidere).

Cf. also "*serra*" under "III. Tools," and "*tignum*"under "XIV. Finished Lumber."

***decorticatio -onis* (f.):** the act of stripping bark (*cortex*) off of timber; *decorto, -are:* to strip the bark off of timber.

Plin. *HN* 17.236: "if the stripping of the bark is only narrow [or shallow], it does not harm [the oak]" (si angusta decorticatio fuit, nihil nocet).

discindo: see, in this section, *scindo.*

dolo -are: to hew or chop wood.

Cato *Rust.* 14.3 (preparation of lumber for building): for text and translation see *materia* under "I. General Woodworking Terms."

Vitr. 2.10.1: "when they are cut down and hewed" (cum autem excisae et dolatae).

The combination of *excido* (q.v.) and *dolo* is interesting. The former may refer specifically to the act of felling, while the latter refers to the initial dressing (limb removal, debarking) of the wood.

See also, in this section, *seco*.

***dolamen -inis*:** the act of hewing wood.

Apul. *Fl.* 1: "a trunk carved by hewing" (truncus dolamine effigiatus).

***excido -dere*:** to remove by cutting. Thus, to cut down a tree.

Caes. *BCiv.* 2.15.1: "since all the trees far and wide within Massilian borders had been cut down" (omnibus arboribus longe lateque in finibus Massiliensium excisis).

Enn. *Ann.* 188: "the ilex is cut down" (exciditur ilex).

Vitr. 2.10.1: for text and translation see, in this section, *dolo*.

girdling: see, in this section, *circumcisura*.

***lignum -i* (n.):** the most general term for wood. While *materia* usually refers to wood for finished work, *lignum* may refer to wood used for fuel or tree-length wood that has not yet been processed (Meiggs 1982, 359; Mols 1999, 75).

Plin. *HN* 16.186: "*lignum* [i.e., a log] floats horizontally on water" (lignum in longitudinem fluitat).

Cf. also *tignum* under "XIV. Finished Lumber."

***perdolo -are*:** to hew fallen trees into usable timber.

Vitr. 2.10.2 (trees harvested from sunny locations): "such [trees], when they are cut into lumber, offer great advantages in terms of durability" (quae, cum in materiem perdolantur, reddunt magnas utilitates ad vetustatem).

See also, in this section, *dolo*.

***scindo -ere*:** to split wood with a blow. Also *discindo*. Seneca uses the verb to indicate both ripping with a saw and splitting with wedges.

Cato *Rust.* 40.2 (on grafting fruit trees): "split apart a Greek willow" (salicem Graecam discindito).

Sen. *Ep.* 90.9 (cites Verg. *Georg.* 1.144): "the first [men] split fissile wood with wedges" (nam primi cuneis scindebant fissile lignum).

***seco -are*:** to cut.

Cato *Rust.* 14.3: (for the passage see *materia* under "I. General Woodworking Terms").

Cato discusses the preparation of lumber for building.

Cf. also, in this section, *caedo*, *dolo*.

***sterno -ere*:** to cast down and thus fell a tree.

Plin. *HN.* 16.188: for text and translation see the introduction to this section.

splitting wood: see, in this section, *scindo* and *cuneus* under "III. Tools."

transport on roads: smaller pieces of lumber could be carried on carts. Long beams could be drawn along roads by attaching wheels to each end of the beam.

Sen. *Ep.* 90. 9: (long lines of carts used to transport unfinished beams along the roads): for text and translation see *lacunar* under "VIII. Roofing and Ceilings."

Juv. 3.254–56: "a long fir rattles on a passing wagon, and yet another cart hauls a pine, they sway from on high and threaten the people" (longa coruscat / serraco veniente abies, atque altera pinum / plaustra vehunt, nutant alte populoque minantur).

XIV. Finished Lumber: General Terms for Planks and Boards

***arbuscula* -*ae* (f.):** wood bored to receive an axle, thus possibly the nave of a wheel.

Vitr. 10.14.1 (on the construction of the siege machine): "timbers . . . in which the axles of wheels rotate, these are sheathed with iron plates" (arbusculae . . . in quibus versantur rotarum axes conclusi lamnis ferreis).

***asser* -*eris* (m.):** a beam of wood, a wooden post or pole. The function of an *asser* ranges from an element in roofing to a weapon to a pole upon which a vine is trained.

Caes. *BCiv.* 2.2.2: "poles 12 feet long, fixed with points, and launched from the greatest catapults" (asseres enim pedum XII cuspidibus praefixi atque hi maximis ballistis missi).
Plin. *HN* 16.141: [young cypresses are stripped of their branches for] poles or props" (perticas asseresve).

A *pertica* refers to a rod or pole.

Varro *Ling.* 5.140: for text and translation see, in this section, *tignum*.

See additional commentary on *asser* under "VIII. Roofing and Ceilings."

***assis*:** see *axis*.

***axis* -*is* (m.):** A variation of *assis*, a plank or board.

Caes. *BCiv.* 2.9.2 (description of the construction of an upper floor of a siege tower): "they laid cross-joists and connected them with planks" (transversas trabes iniecerunt easque axibus religaverunt).
Vitr. 4.2.1: "in [decked] floors there are joists and planks" (in contignationibus tigna et axes).

See also, in this section, *asser*.

board: see, in this section, *axis*. See also *coaxatio*, *contabulatio*, and *tabulatio* under "VII. Flooring," and *tabula* under "IX.3. Doors and Shutters."

***postis* -*is* (m.):** a vertical structural beam. The *postis* was used to frame windows and doors. In Livy 2.8.7 the term may refer to the doorpost (or one of the columns?) of the temple.

Cat. *Rust.* 14.1: for text and translation of this passage, see chapter 5 of the text.
Liv. 2.8.7: "keeping his hand on the *postis* he finished his prayer and dedicated the temple [of Jupiter Optimus Maximus]" (tenens postem precationem peragit et dedicat templum).
Vitr. 6.8.2: for text and translation see *postis* under "VI. Framing."
Vitr. 10.14.2: for text and translation see *compactio* and *postis* under "VI. Framing."

***regula* -*ae* (f.):** a straight piece of wood; a ruler.

Varro *Ling.* 5.135 (upright handle of a plow): "the *regula* that stands above this [the actual plow assembly] is called *stiva*" (super id regula quae stat, stiva).
Varro *Ling.* 5.136: "*irpices* [are made from] a *regula* with many teeth" (irpices regula compluribus dentibus).

The *irpices* here refer to harrows pulled by oxen.

***scapus* -*i* (m.):** applied to various upright framing pieces, such as the stiles of a door, the sides of a ladder, the shaft of a windlass. See also *postis*.

See discussion of this term under "IX.3. Doors and Shutters."

***tabula* -*ae* (f.):** a general term in reference to a flat piece of wood.

See *contabulatio* and *tabulatio* under "VII. Flooring," and "*tabula*" under "IX.3. Doors and Shutters."

***tigillum* -*i* (n.):** a small plank or beam.

Plaut. *Aul.* 301 (reference to a miser who kindles a tiny fire): "[a puff of] smoke from his *tigillum* escapes outdoors" (de suo tigillo fumus . . . exit foras).

tignum -i (n.): the general term for a piece of wood, timber, beam.

Caes. *BGal.* 4.17.3: for text and translation, see chapter 5 of the text.

Ov. *Met.* 8.648 (old Philemon's hut): "[bacon] hanging from a blackened beam" (nigro pendentia tigno).

Sen. *Ep.* 90.8–9: "to cut timbers square" (in quadratum tigna decidere).

Varro *Ling.* 5.140 (the cargo of open carts, *plaustri*): "stones, *asseres, tignum*" (lapides, asseres, tignum).

Varro's use of two terms for lumber here is notable. The *asseres* perhaps refer to finished boards, the *tignum* to either a beam (note the use of the singular noun) or, generally, unfinished(?) timber.

Vitr. 6.3.1 (on the roof of an atrium): "[the framing has] valleys running down from the walls to the corners of the beams" (collicias ab angulis parietum ad angulos tignorum incurrentes).

trabs (or **trabes**) **-is (f.):** heavier support plank or beam. Commonly used to refer to a support beam placed in a horizontal position.

Caes. *BGall.* 4.17 (horizontal beams for trestles built to support a bridge): for text and translation see chapter 5 of the text.

Sen. *Ep.* 90.9 (ripping large beams with a saw): for text and translation see *serra* under "III. Tools."

See also *trabs* under "VII. Flooring" and "VIII. Roofing and Ceilings."

XV. Roman Linear Measurements used by Woodworkers and Architects

Foot measurements given in the text are based upon a Roman foot (*pes*) of 29.6 cm.

From smaller to larger measurements:

digitus, -i: a "finger"; one-sixteenth of a foot (18.5 mm).

Thus the adjective *digitalis*: measuring a finger's width. Variations include *semidigitale* (half-digit), *sesquidigitale* (one and a half digits).

uncia, -ae: one-twelfth of a foot (24.6 mm, based upon a foot of 296 mm). The measurement is represented in inscriptions with a short horizontal line. One-half of an *uncia* is a *semuncia* (12.3 mm).

1 **palm** (**palma,** also **palmus**) = 4 *digiti* = 3 *unciae* = 7.4 cm.

1 **foot** (**pes, pedis**) = 16 fingers = 12 unciae = 4 palmae = 29.60 cm. Subdivisions: *semipes* (half-foot); *quadrantalis* (quarter-foot).

1 **cubit** (**cubitum** or **cubitus**) = 6 palms = 1.5 Roman feet = 44.4 cm (distance from elbow to end of middle finger).

bipedalis = two feet. This was a standard size for some building material, including tiles.

Appendix
Archaeological Evidence of Roman Woodworking Tools

A Guide to Select Examples

The following should be considered a representative list that includes most of the best-known examples of Roman woodworking tools. Figure numbers at the end of individual listings refer to images in the text.

Adze

ACTUAL EXAMPLES

Adze blades

Blade fragment, iron, Gorhambury, St. Albans, U.K. Flaring blade, 8 cm cutting edge, socket not preserved. (Neal 1990, 138, fig. 131. 364)

Adze-head, iron, Silchester, Reading Museum, U.K. Socket diameter ca. 1.5 cm, spoon-shaped blade length ca. 11.5 cm, cutting edge, gougelike, 4.5 cm across. (Evans 1894, 148, fig. 14; Goodman 1964, 25, fig. 15)

Adze-head, iron, Hod Hill, Dorset, British Museum inv. 1892.9–1.1257. Socket collared with a round eye, length 15.7 cm, splayed blade with curved sides. The form is known from Iron Age and Roman contexts in Britain; this example possibly first century. (Manning 1985, 16, B7, pl. 9)

Adze-head, iron, Hod Hill, Dorset, British Museum inv. 1892.9–1.1256. Socket collared with oval eye, length 15.9 cm, flat, triangular blade, attached to socket at a sharp angle, similar to blades on the adze-hammers from Silchester and London described below. Mid-first century. (Manning 1985, 17, B10, pl. 8)

Adze-Axes

Adze-axe, iron, Pompeii, Naples Museum, inv. s. n. 307.

Adze-axe, iron with bronze blade guard from Vindonissia, Landesmuseum, Zurich. (Goodman 1964, 25, fig. 14a)

Adze-Hammers

Adze-hammer, iron, Pompeii, Naples Museum, inv. 71952. Collared, 21 cm; the hammerhead is roughly square in section; flat, triangular blade attached to the collar at a sharp angle. First century. (Ciarallo and De Carolis 1999, 129)

Adze-hammer, iron, Silchester, Reading Museum, U.K. Socket diameter ca. 3.5 cm, hammerhead roughly square in section, ca. 6 cm long, adze blade ca. 11 cm long, cutting edge 6.5 cm wide. Blade angle ca. 40 degrees. (Evans 1894, 148, fig. 13; Goodman 1964, 25, fig. 15)

Adze-hammer, iron, London, British Museum inv. 1956.4–3.1. Oval socket with collar, length 22 cm, hammerhead roughly square in section, adze blade at a sharp angle similar to the Silchester example; fragment of wood left in socket identified as whitebeam (*Sorbus* sp.) (Manning 1985, 18, B16, pl. 9; Painter 1961, 116, pl. 49, C)
(fig. 3.3)

Adze-hammer, iron, Pakenham, Suffolk, British Museum inv. 1867.7–11.7. Circular socket with extended conical collar, length 18.3 cm, hammerhead rectangular in section, gently curving, splayed blade; in profile the tool looks much like a modern hammer. (Manning 1985, 18, B17, pl. 9)
(fig. 3.3)

RELIEF SCULPTURE

Adze in use on a funerary relief of P. Longidienus Cam(ilia), *faber navalis*, Ravenna, Museo Nazionale, inv. 7. A man in a tunic shapes a curved (framing?) plank; a ship under construction rests in the background. First century (Bianchi-Bandinelli 1970, 88, fig. 96; Blümner 1875 (II), 341, fig. 55; Casson 1991, fig. 46; *CIL* 11.139; Gummerus 1913, 92, figs. 14–15; Kampen 1981, fig. 68; Mansuelli 1967, 126; Squarciapino 1941, 47; Zimmer 1985, pl. 7)
(fig. 3.9)

Adze in use, depiction of a seated Daedalus, ash urn of C. Volcacius Artemidorus, found in the vicinity of Tivoli, Museo Nazionale, Rome. Early second century. (Gummerus 1913, 88, fig. 13; *NSc* 1898, 456)

Adze in use, sarcophagus, Vatican, Museo Gregoriano Profano, inv. 3262. Depiction of a seated woodworker carving a chair leg with a short-handled adze. Mid–third century. (Amelung 1903, 864, no. 162; Blümner 1875 (II), 343, fig. 57; Liversidge 1955, fig. 64; Mols 1999, fig. 25; Richter 1966, fig. 614; Zimmer 1982, 139, no. 57)
(fig. 3.8)

Adze, possible adze-hammer, altar to Minerva dedicated by the priests of the *collegium fabrum tignuariorum*, Rome (S. Giorgio in Velabro), Capitoline Museums, inv. 1909. Early imperial (Augustan). (*CIL* 6.30982; Fava 1969, 35; Felletti Maj 1977, 324, pl. 63, fig. 155; Gaitzsch 1980, 381; Goodman 1964, 121, fig. 126; Gummerus 1913, 101, fig. 20; Ryberg 1955, 87, fig. 40; Zimmer 1982, 162, no. 84)
(fig. 2.2)

Adze, possibly adze-hammer, funerary relief in Augsburg (Germany), Römisches Museum. (Kellner 1971, pl. 51; Zimmer 1985, pl. 6, n. 3)

Adze, funerary relief, Autun, Musée Rollin d'Autun. Depicted with a bucksaw; short handled, lugs above and below blade, hammer(?) opposite sharply angled blade. (*CIL* 13.2721; Gaitzsch 1980, 380, pl. 64, fig. 306)

Adze, funerary relief from Bordeaux, Musée d'Aquitaine, Bordeaux, inv. 60.1.83. Known as the "Bearded Carpenter of Bordeaux," the craftsman holds a short-handled adze-hammer in his right hand and a ruler in his left. (Adam 1994, 98, fig. 223; *CIL* 13.644; Espérandieu 1908, 1117)

Adze-Plane ("Ascia-Hobel" variants)

Adze-plane, funerary relief from Isola Sacra (Ostia). A hand-adze ("ascia-Hobel" variant) is shown on a plaque depicting a knife sharpener at work. Isola Sacra (Ostia), tomb 29. Mid–second century. (Becatti 1951, pl. 68; Gaitzsch 1985, 190, fig. 1b; Squarciapino 1941, 14; Zimmer 1982, 183, no. 117; *id.* 1985, pl. 4.2)
(fig. 3.5)

Adze-plane, funerary relief of P. Ferrarius Hermes, from Bientina (near Pisa), Italy, Archaeological Museum, Florence, inv. 1914. The tool is shown with a ruler, plumb line, and square. (*CIL* XI 1471; Felletti Maj 1977, 350, pl. 74, fig. 180; Gaitzsch 1980, 379, pl. 61.299; Gummerus 1913, nr. 84, fig. 31; Goodman 1964, 189; Matthäus 1984, 91, fig. 15; Zimmer 1982, 166, no. 90)
(fig. 3.4)

Adze-plane, funerary relief of P. Beitenos Hermes, a bed maker, Louvre, inv. MA 934. The tool is shown with calipers and a square. From Greece (province unknown), Roman imperial period. (Burford 1973, 182, fig. 8; Matthäus 1984, 99, fig. 23; Richter 1966, 127, fig. 612; Strong 1976, fig. 271)
(fig. 3.6)

OTHER RENDITIONS

Adze-hammer with straight handle, among miniature models of tools in bronze from the British Museum, acquired in 1854. (Goodman 1964, 24, fig. 13)

Small adze, gold on glass, from the Catacombs of Saturninus, Rome, Museo Biblioteca Apostolica Vaticana, inv. 60788. A shipwright uses a short-handled adze to shape a plank. Early fourth century. (Blümner 1875 (II) 343, fig. 58; Gaitzsch 1980, 384, pl. 73, fig. 320c; Gaitzsch and Matthäus 1981, 213, fig. 11; Garrucci 1864, 171; Goodman 1964, 119, fig. 123; Liversidge 1976, 158, fig. 264; Mercer 1960, 152; Morey 1959, pl. 16, no. 96; Neuburger 1919, 77, fig. 117; Richter 1966, 128, fig. 613; Squarciapino 1941, 44)
(fig. 3.23)

Auger (bits)

ACTUAL EXAMPLES

Spoon bit, iron, Museum of London, inv. 1529. Squared tang, 14 cm long.

Spoon bit, iron, Naples Museum, inv. 120737. Shaft square in section, 18.3 cm long. (Ciarello and De Carolis 1999, 130)

Twisted bit, iron, with spiral cutting point, Aquileia, Museo Nazionale. 12 cm long by 1 cm diameter. (Gaitzsch 1985, fig. 5.f)
(fig. 3.11)

Twisted bit, iron, with spiral cutting point, Aquileia, Museo Nazionale, inv. 19721. 15.2 cm long by 1 cm diameter. (Gaitzsch 1985, fig. 5.g)
(fig. 3.11)

Axe and Hatchet

ACTUAL EXAMPLES

Double-bladed axes

Double-bladed axehead, iron, Pompeii, Naples Museum, inv. 286771. 28 cm long, blade 9.7 cm wide. First century. (Ciarello and De Carolis 1999, 123.78; Petrie 1917, 13, pl. 12.40)
(fig. 3.12)

Single, curving blade (Manning 1985, 16; Type 4),

Axehead, beaten copper, Monte Rovello (Rome), Italy. Curved edge, oval eye flared on top and bottom. (Petrie 1917, 11. pl. 0.4)

Axehead, iron, Coldham Common, U.K., British Museum inv. 1870.12–8.45. 17.2 cm long, oval eye. (Manning 1985, 16, B4)

Axeheads, iron, five of similar design, Silchester, Reading Museum, U.K. All have gently curving blades, flat butts, oval eyes, lengths 19–20 cm, slightly curved cutting edges 7.5–8.5 cm long, 1.3–1.5 kg. (Evans 1894, 147, fig. 12; Goodman 1964, 23, fig. 12)

Axehead, iron, Pompeii, Naples Museum, inv. 286770. 19.2 cm long, blade width 7.3 cm (Ciarello and De Carolis 1999, 122.77)

Single, curving blade with lugs on top and bottom

Axehead, iron, Silchester, Reading Museum, U.K. 25.5 cm long, slightly curved cutting edge 7 cm wide, 1.65 kg.

Hatchet, iron, Gorhambury, St. Albans, U.K. Triangular in longitudinal section, 14 cm long, lugged oval eye, 2.9 cm long, poll has two vertical notches on back face. (Neal 1990, 138, fig. 131.363)

Single blade, straight or gently curving, cutting edge roughly same width as poll (Manning 1985, 15, Type 1)

Axehead, iron, London, British Museum inv. 1917.12–3.1. 22.8 cm long, round eye. (Manning 1985, 15, B1)

Axehead, iron, Verulamium, 19.1 cm long, 1.8 kg, A.D. 155–60. (Frere and Manning 1972, 165, no. 7)
(fig. 3.12)

Single blade, straight top with lower curve to produce a broad cutting edge

Heavy felling axe (Manning 1985, 15 Type 2)

Axehead, iron, Combend, Gloucestershire, British Museum inv. 1810.2–10.5/6. 23 cm long, heavy poll with round eye. (Manning 1985, 15, B2)

Axehead, iron, London, British Museum inv. 1856.7–1.1420. 21.4 cm long, oval eye (Manning 1985, 15, B3)

Axehead, iron, Pompeii, Naples Museum, inv. 286769. 21.5 cm long, blade 18 cm wide; the butt end of the axe is thick (5.5 cm) and therefore heavy. First century. (Ciarello and De Carolis 1999, 122.75).

Single blade, both top and bottom faces flare to produce a broad cutting edge; smaller blades for lighter carpentry work (Manning 1985, 16, Type 3)

Axehead from Sandy, Bedfordshire, British Museum inv. 1915.12–8.337. 15.0 cm long, oval eye, curved blade. (Manning 1985, 16, B4)

Dolabra, with blade and opposed pick

Five axeheads, iron, Newstead, Scotland, Museum of Scotland, Edinburgh. The smallest measures 36 cm long with a blade 9 cm wide weighing 1.2 kg. The largest is 46 cm long with a blade 10 cm wide weighing 2.72 kg. Second century. (Curle 1911, pl. 57.1–5)
(fig. 3.14)

RELIEF SCULPTURE

Double-bladed axe, funerary relief of Eutyches from Priolo, Sicily, Museo Nazionale in Syracuse. 1.07 x 0.46 x 0.39 m. Third–fourth century. (Adam 1994, 99, fig. 228; Colini 1958, 618, no. 57; Gaitzsch and Matthäus 1981, 210, fig. 5; Goodman 1964, 46, fig. 43; Gummerus 1913, 90; Orsi 1891, 359; Zimmer 1982, 142, no. 61)
(fig. 3.15)

Double-bladed axe and pickaxe, terra-cotta relief from tomb 29, Isola Sacra (Ostia). The tools are shown on a plaque depicting a knife sharpener at work. (References under *Adze-plane*)
(fig. 3.5)

Trajan's column, multiple renditions. Refer to fig. 3.13.

Axe-Adzes

See Adze-Axes, above.

Axe-Hammer

Axe-hammer, iron, Augst, Landesmuseum, Zurich. (Goodman 1964, 25, fig. 14b)

Chisels and Gouges

ACTUAL EXAMPLES

Mortising chisels, socketed handle

Chisel, iron, Museum of London, inv. 23317. 26.3 cm long, blade width 2.2 cm; socketed handle with a flange (3.5 cm diameter) that has cracked from hammering.

Chisel, iron, Museum of London, inv. 19159. 29.6 cm long (exactly one Roman foot), blade width ca. 1.5 cm; heavy construction, for timber framing, socketed with opening 3 cm in diameter and 9 cm deep.

Chisels, group of four socketed iron tools found in 1890 at Silchester. Two of these are 25.5 cm long; socket opening ca. 2.5 cm; the third ca. 21.5 cm long, the fourth 17.5 cm long, blade widths between 1.3 and 2 cm. (Evans 1894, 149, fig. 15)

Chisel, iron, socketed, Augst, inv. 64.11454. 24.6 cm long, blade width 1.05 cm, socket opening 2 cm, heavy construction with sharply tapered cutting edge. (Mutz 1980, 127, fig. 17)

Chisel, iron, socketed, Augst, inv. 64.11455. 25.1 cm long, blade width 1.10 cm, socket opening 2 cm, heavy construction with sharply tapered cutting edge. (Mutz 1980, 127, fig. 18)

Mortising chisels, tanged handle

Chisel, iron, tanged, Caerleon. 11.8 cm long, rectangular cross section, slightly waisted, cutting edge damaged. (Evans and Scott 2000, 393, cat. 27)

Chisel, iron, tanged, Augst, inv. 64.11462. 19.1 cm preserved length, blade width 6.3 mm, tang not fully preserved, thick shaft with sharply tapered cutting edge. (Mutz 1980, 127, fig. 15)

Chisel, iron, tanged, Augst, inv. 64.11463. 15.8 cm preserved length, blade width 9 mm, only base of tang preserved as well as iron reinforcing ring for wooden handle, thick shaft with sharply tapered cutting edge. (Mutz 1980, 127, fig. 16)

Mortising chisels, solid, one-piece handle or fragmentary

Chisel, iron, Caerleon. Rectangular cross section, angled cutting edge, no evidence of method of handle attachment, 9.2 cm preserved length. (Evans and Scott 2000, 393, cat. 26)

Chisel, iron, Silchester, 21.5 cm long. (Evans 1894, 149, fig. 15)

Chisel, iron, Museum of London, inv. 1893. 17.5 cm long, narrow blade width (.07 cm) suggests fine cabinetry work; evidence of hammer marks on the butt end. (Wheeler 1930, pl. 32.11)
(fig. 3.10)

Chisel, iron, Gorhambury. Fragment 5 cm long, includes preserved sharply splayed blade, ca. 2 cm wide, with bevel on one side; stem is square in section. (Neal 1990, 140, fig. 131.377)

Paring chisels (thin blade with bilateral taper), socketed handle

Paring chisel, iron, socketed, Augst, inv. 64.11450. 28.8 cm long, blade width 3.10 cm, socketed opening ca. 2 cm, gently flaring blade. (Mutz 1980, 129, fig. 5.19)

Paring chisel, iron, socketed, Augst, inv. 64.11452. 22.7 cm long, blade width 2.08 cm, socketed opening ca. 2 cm. (Mutz 1980, 129, fig. 5.20)

Paring chisel, iron, socketed, Augst, inv. 64.11452. 28.4 cm long, blade width 9.6 mm, socketed opening ca. 2.25 cm; wooden handle partially preserved with iron reinforcing ring. (Mutz 1980, 129, fig. 5.21)

Paring chisels, tanged handle

Paring chisel, iron with wooden handle, tanged, from Aquileia, inv. 80697. Length, including handle, 28.6 cm; wooden handle was turned on a lathe, with a mushroom-shaped butt, reinforced with a bronze collar at attachment point; slightly waisted blade, thin cutting edge. (Gaitzsch 1980, 363, pl. 38, no. 181)
(fig. 3.16)

Paring chisel, iron, with tanged handle, Camulodunum. (Hawkes and Hull 1947, 343, pl. 105, 6)

Paring chisel, iron, tanged, Museum of London, inv. 19763. 17.5 cm long, slightly waisted, from 11 to 17 mm; blade width 17 mm, tang 5 cm long

Paring chisels, solid handle

Paring chisel, iron, Verulamium. Solid handle 21.1 cm long, A.D. 155–60.
(Frere and Manning 1972, 164, no. 10)

Paring chisel, iron, Verulamium. Solid handle and partial blade, 16 cm long, A.D. 270–75. (Frere and Manning 1972, 164, no. 11)

Gouges, socketed handle

Gouges, group of five socketed iron tools found in 1890 at Silchester:

a. 31.5 cm long, stem octagonal, 1.4 cm channel, ground on outside, possible turning tool for use on a lathe;

b. 30.5 cm long, octagonal stem, 2 cm channel, ground on inside;

c. 25.8 cm long, round stem, 2.5 cm semicircular cutting channel;

d. 27 cm long, nearly square stem, 1.3 cm channel, ground on both sides;

e. 26.1 cm long, round stem, 1.3 cm channel, ground on both sides.

(Evans 1894, 150, fig. 16)
(fig. 3.18)

Gouge, socketed gouge, iron, Caerleon, Wales, inv. 65.170A. 28.1 cm long, round stem, with evidence of hardening on the inside. Dated to second–third century. (Tylecote in Zienkiewicz 1986, 195, fig. 68.23)

Gouges, tanged handle

Gouge, iron, Museum of London, inv. 16267. Tanged handle, 14.7 cm long; possibly used as an auger blade(?).

Gouges, solid handle

Gouge, iron, Museum of London, inv. 19891. Nearly cylindrical in form with solid handle, 12 cm long, cutting tip 6 mm wide.

Gouge, iron, Museum of London, inv. 16267. Tanged handle, 14.7 cm long; possibly used as an auger blade(?).

RELIEF SCULPTURE

Paring chisel, tanged, without handle, Frascati (Italy), British Museum inv. 1954.12–14.1. On right margin of relief of Publius Licinius Philonic(os) and P. Licinius Demetrius. Early imperial period. (Goodman 1964, 162, fig. 161; Kleiner 1977, 196, no. 3; Manning 1985, P1.1; Zimmer 1982, 191, cat. 128)
(fig. 3.20)

Paring chisel, broad flaring cutting edge, Vatican Museums, inv. 9277. Depicted hanging on the wall of a knife seller's shop. First century. (Blümner 1884.4, 371, fig. 60; Burford 1972, fig. 36; Zimmer 1982, 180, no. 114)
(fig. 3.25)

Paring chisels and tanged chisel, funerary relief from tomb 29 at Isola Sacra (Ostia). Three broad-bladed paring chisels and two narrower chisels (one tanged) are shown on a plaque depicting a knife sharpener at work. Second century. (References under *Adze-plane*)
(fig. 3.5)

PAINTING

House of the Vettii, Ixion Room, Pompeii (VI, 15,1). Ca. A.D. 70, Icarus and Daedalus, a tanged chisel is being used by Icarus to cut mortises. (Clarke 1991, 224, fig. 132; Gaitzsch 1980, pl. 22, fig. 117)
(fig. 3.17)

OTHER RENDITIONS

Chisel being struck by a mallet, gold leaf on glass, from the Catacombs of Saturninus, Rome, Museo Biblioteca Apostolica Vaticana, inv. 60788. The broad blade of the tool suggests a paring chisel, but it is being used as if cutting mortises. Early fourth century. (References under *Adze*)
(fig. 3.23)

Drawknife

ACTUAL EXAMPLES

Drawknife, iron, Königsforst, Rheinisches Landesmuseum, Bonn, inv. 75.1708. Blade length 15.5 cm, width at center 2.8 cm, tangs attached at ninety degrees to curved cutting edge. (Gaitzsch 1980, 376; pl. 56.280)

Drawknife, iron, London, Museum of London, inv. 19167. Blade length 11.8 cm, width at center 2.90 cm, tangs concealed by modern restored handles, attached thirty degrees from straight cutting edge.
(fig. 3.10)

Drawknife, iron, Pompeii (I, 15), Naples Museum, inv. 286780. Total length, 60 cm, blade length 46 cm, gently arched, width at center 5 cm, tang length 7 cm. Handles are not inclined toward the direction of the blade. (Ciarallo and De Carolis 1999, 128, fig. 97; Gaitzsch 1980, pl. 31, fig. 132)

Drill Bits

ACTUAL EXAMPLES

Drill bit, iron, pyramidal tang, Gorhambury, St. Albans, U.K. 11.3 cm long, simple pointed bit, tang 5 cm long. (Neal 1990, 138, fig. 131.371)

Drill bit, iron, pyramidal tang, Museum of London, inv. 13667. 16 cm long, tang 7 cm long.

Drill bit, iron, fragment, Caerleon. 12.6 cm long, shaft is circular in cross section, with pyramidal point. (Evans and Scott 2000, 393, cat. 28)

Drill bit, iron, pyramidal tang, square in section, Augst, inv. 64.11465. 14.1 cm long, tang 5.6 cm long, terminates in a simple point. (Mutz 1980, 128, fig. 3.14)

Drill (spoon) bit, iron, pyramidal tang, square in section, Augst, inv. 64.11464. 19.3 cm long, tang 6.3 cm long, width of spoon 1.10 cm. (Mutz 1980, 126, fig. 3.13)

RELIEF SCULPTURE, SEE "DRILL, BOW TYPE," "DRILL, STRAP TYPE"
PAINTING, SEE "DRILL, BOW TYPE," "DRILL, STRAP TYPE"
OTHER RENDITIONS, SEE "DRILL, BOW TYPE"

Drill, Bow Type

ACTUAL EXAMPLES

Bit (iron), wooden stock, and cap, London, University College, inv. 27979. From Hawara, Egypt (Roman period), total length, 55 cm. (Petrie 1917, 39, pl. 51, M 15)
(fig. 3.19)

RELIEF SCULPTURE

Bit, drill stock, and bow, British Museum inv. 1954.12–14.1. Funerary relief of Publius Licinius Philonic(os) and P. Licinius Demetrius, Frascati (Italy). Early imperial period. (References under *Paring chisels*)
(fig. 3.20)

Bow and arrows, funerary relief of Monimus, Mainz, Landesmuseum, inv. S166. The objects are held in the hands of the deceased. First half of the first century A.D. (*CIL* 13.7041; Selzer 1988, 153, cat. 80; Stribrny 1987, no. 8). The image is included for comparative purposes.
(fig. 3.22)

Bit, drill stock, and bow depicted on funerary altar of Eutyches, Priolo, Sicily, Syracuse, Museo Nazionale. The handle is shown with compound, turned moldings, terminating in a nearly spherical butt. (Adam 1994, 99, fig. 228; Gaitzsch 1980, pl. 67, fig. 303; and see additional references under *Axe, relief sculpture*)
(fig. 3.21)

PAINTING

House of the Vettii, Ixion Room, Pompeii. Icarus and Daedalus, a bow and drill lie on the floor under the carpenter's bench depicted in the foreground. Mid-first century A.D. (References under *Chisels and Gouges, painting*)
(fig. 3.17)

OTHER RENDITIONS

Bow and drill being used by a standing craftsman, gold leaf on glass, from the Catacombs of Saturninus, Rome, Museo Biblioteca Apostolica Vaticana, inv. 60788. Note that the craftsman's left hand (holding the drill) should be placed on the top of the handle, not in the center. Early fourth century. (References under *Adze, other renditions*)
(fig. 3.23)

Drill, Strap Type

RELIEF SCULPTURE

Two men operate the strap drill while carving a sarcophagus, Urbino, Museo Archeologico Lapidario. Sarcophagus of Eutropos, found on the Via Labicana (Rome). Mid-fourth century. (Adam 1994, 39, fig. 74; Blümner 1875 (III), 220, fig. 28; *CIG* 4.9598; Colini 1958, 613, no. 27; Gaitzsch 1980, 383, pl. 72, fig. 318; Gummerus 1913, 95; Strong and Brown 1976, 199, fig. 327)
(fig. 3.24)

Files and Rasps

ACTUAL EXAMPLES

File, iron with tang, Naples Museum, inv. 286779. Ca. 23 cm long, 4 cm wide, tanged, handle missing, corroded surface. Before A.D. 79. (Ciarallo and De Carolis 1999, 127, fig. 95)

File, iron with "cranked" tang (bent like the handle of a trowel), Silchester. Cutting face 19 cm long, 2.3 cm maximum width, filed teeth on flat surface, ca. 5 rows per 2.5 cm, in parallel rows but at a slight bias to the sides of the file; thus probably used for working wood. (Evans 1894, 152, fig. 19)

File, iron, London, Museum of London, inv. 19863. Tapering blade 18 cm long, tang concealed by modern restored handle, average blade width 2 cm, triangular in cross section, with a flat bottom and two shorter sloping faces, teeth, finely spaced on all sides. Presumably used for sharpening metal cutting edges.
Rasps (distinguished by a convex cutting face)

Rasp, iron, tanged, Augst, inv. 64.11466. Total length 28.3 cm, width 2.2–0.70 cm, tang reinforced by an iron rim (wood handle now missing). (Mutz 1980, 123, fig. 2.6)

Hammers and Mallets

ACTUAL EXAMPLES

Hammer, iron, with tang, Naples Museum, inv. 7188. Head width 10 cm, overall length 20 cm. The corroded head probably included a claw. A second, similar hammer measuring 27 cm long is also numbered as 7188. First century.

Hammer, iron, with tang, Naples Museum, inv. 411. Head width 6.5 cm, overall length 13.3 cm. First century.

Hammer, iron, with long handle, forged from one piece, Naples Museum, inv. 428. Clawed head 11 cm long, overall length 27.5 cm. First century.

Hammer, iron, head, socketed, with nail in socket, Verulamium Museum, St. Albans, U.K., inv. 856.1501. 8 cm long.

Mallet, wooden, Comacchio Museum, inv. 56394. Head width 21.6 cm, overall length 37.2 cm. The handle is inserted so that the head strikes on the end-grain. The mallet was made of ash (*fraxinus*) and holm oak (*ilex*). First century B.C. (Desantis in Berti 1990, 281, no. 292)

Mallet, wooden, Comacchio Museum, inv. 55077. Single-piece construction with grain of handle and head aligned, head length 22.4 cm; with handle total length 34.4 cm. (Desantis in Berti 1990, 285 no. 295; similar examples are illustrated in this catalog)

RELIEF SCULPTURE

Mallet, waisted, from a funerary relief found in Italy. (Blümner 1884, 92, fig. 2)

Mallet, waisted, on a dedicatory relief in the Antiquarium del Celio, Rome
(fig. 3.45)

Mallet, waisted, used by a Roman soldier to pound a wooden pile, Trajan's column, Rome, A.D. 113.
(fig. 3.42)

Mallet, waisted, used by a Roman soldier to hit a heavy chisel, Trajan's column, Rome, A.D. 113 (Gaitzsch 1980, pl. 64, fig. 305)
(fig. 3.41)

PAINTING

House of the Vettii, Ixion Room, Pompeii. Icarus and Daedalus. Icarus holds a socketed mallet to strike a tanged chisel. Ca. A.D. 70. (References under *Chisels and Gouges, painting*)

Lathe

RELIEF SCULPTURE

Depiction of a lathe operated by a wheel (?), part of a scene of a furniture maker's shop, Museo Capitolini (Montemartini), inv. 2743; Museo della Civiltà Romana, inv. 3408. (Colini 1947, 21, fig 1; Gaitzsch 1980, pl. 65.309; Goodman 1964, 121, fig. 127; Liversidge 1950, 27, pl. 2.b; Strong and Brown 1976, 159, fig. 265; Van Buren *AJA* 52 (1948) pl. 50.B)
(fig. 2.3)

Partial depiction of a horizontal lathe operated by a bow, from a Roman-period sarcophagus from Greece. (Kontoleon 1890, 333; Rieth 1940, 99)
(fig. 3.27)

Plane

ACTUAL EXAMPLES

Italy

Pompeii, plane, single (rear) handle, iron with wooden core (House of Marcus Lucretius, IX.3.5), Naples Museum, inv. 71964. Smoothing, 21.3 cm long by 6.2 cm wide by 5 cm high; the blade, 3.5 cm wide, projects 6.5 cm from the top face, cutting angle, 50 degrees. Before A.D. 79 (Ciarello and De Carolis 1999, 127.96; Gaitzsch 1980, 113; Gaitzsch and Matthäus 1981, 231, fig. 39; Goodman 1964, 44, fig. 39; Greber 1956, 70, fig. 37)
(fig. 3.2)

Pompeii, plane, iron and wood (similar in form and dimensions to the previous entry, except that the iron is broken off at the top), Naples Museum. 21 cm long by 6 cm wide by 5 cm high; the blade is 4 cm wide, cutting angle ca. 50 degrees. Before A.D. 79. (Gaitzsch 1980, 113; Goodman 1964, 44, fig. 39; Greber 1956, 70, fig. 37)

Other Mainland Europe

Cologne (Colonia Agrippina), plane, iron sole and side walls, top fretted metalwork, Rheinisches Landesmuseum, Bonn, inv. 36.199. 32.4 cm long by 5.3 cm wide by 4.5 cm high; toothed iron 17.6 cm long by 4 cm wide, rake, 53 degrees. Second century. (Gaitzsch 1980, 379, pl. 60.296; Gaitzsch and Matthäus 1981, 216–17, figs. 14–15; *id.* 1985, pl. 2; Goodman 1964, 50; Greber 1956, 83, fig. 42; Liversidge 1976, 160, fig. 266; Long and Steedman 2002, 18, fig. 3)
(fig. 3.31)

Feldberg, plane, iron sole, four rivets, side plates (no longer attached to the sole), Saalburg-Museum, inv. FMa 165. 36.8 cm long by 4.8 cm wide by 5.5 cm high; iron 3.2 cm wide, rake ca. 65 degrees. (Gaitzsch and Matthäus 1981, 222, figs. 21–24; Goodman 1964, 43; Greber 1956, no. 87, fig. 44.c)

Miltenberg, plane, iron sole with four rivets, Museum Miltenberg. 33 cm long by 4.4 cm wide by 6.3 cm high; rake ca. 62 degrees. (Gaitzsch and Matthäus 1981, 224, figs. 27–28)

Saalburg, plane, iron sole and beechwood stock, front and rear grips cut into the wood (the front grip is intact), Saalburg-Museum, inv. SHa 571. 38 cm long by 4.8 cm wide by 5.5 cm high; iron 2.4 cm wide, 14.4 cm long, rake, 50 degrees. Second–third century. (Gaitzsch and Matthäus 1981, 214–15, figs. 12–13; Goodman 1964, 44, fig. 40; Greber 1956, no. 87, fig. 44.d; Liversidge 1976, 159)

Seltz (Saletio), plane, iron sole with four rivets, Historisches Museum Hagenau. 34 cm long by 4.5 cm wide by 5.5 cm high; iron 3 cm wide, cutting angle not preserved. Third century. (Gaitzsch 1980, 113; Gaitzsch and Matthäus 1981, 225, fig. 29; Schaeffer 1927; Goodman 1964, 46; Greber 1956, no. 92)

Steinkritz, plane, iron sole with four rivets, Saalburg-Museum, inv. S 3171, early publications assign the find to Saalburg. 36.2 cm long by 5.0 cm wide by 4.9 cm high; iron ca. 3.30 cm wide, rake ca. 65 degrees. (Gaitzsch and Matthäus 1981, 223, figs. 25–26; Greber 1956, no. 87, fig. 44.b)

Zugmantel (Kastell), plane, iron sole, rivets, side plates, Saalburg-Museum, inv. ZM 5108. 35.2 cm long by 4.4 cm wide by 5.5 cm high; iron 2.8 cm wide, rake ca. 65 degrees. Second–third century. (Gaitzsch and Matthäus 1981, 218, fig 16–19; Goodman 1964, 43)

United Kingdom

Caerwent (Venta Silurum, Wales), plane, iron sole turned up at both ends, holes for rivets, with blade attached, from House XII, Newport Museum, inv. D2.31. 36.5 cm long by 4.5 cm wide; iron 3 cm wide. (Gaitzsch and Matthäus 1981, 227, figs. 32–33; Goodman 1964, 47, fig. 45)

Caerwent (Venta Silurum, Wales), plane, iron sole similar to previous but badly corroded; from Newport Museum. 32.5 cm long by 5 cm wide; iron 3 cm wide, rake, 66 degrees. (Gaitzsch and Matthäus 1981, 228, figs. 34–35; Goodman 1964, 50, fig. 51)

Goodmanham (East Yorkshire), plane, iron sole turned up at both ends, three rivets, ivory stock, blade attached. 33.0 cm long by 6.0 cm wide by 8.5 cm high; iron 4.0 cm wide by 7.5 cm long, rake 65 degrees. Before late fourth century. (Long and Steedman 2002)

Silchester (Calleva), plane, iron sole, four rivets, side plates, Reading Museum, inv. 07490. 34 cm long by 5.8 cm wide by 6 cm high; iron 3.8 cm wide, rake 65 degrees. Fourth century. (Boon 1974, 282, fig. 41, 10; Evans 1894, 151, fig. 18; Gaitzsch 1980, 113; Gaitzsch and Matthäus 1981, 229, fig. 36; Goodman 1964, 48, figs. 48, 49; Long and Steedman 2002, 18, fig. 4; Mercer 1960, 115, fig. 110; Petrie 1917, 39, pl. 78, M139)
(fig. 3.32)

Verulamium, plane, iron sole turned up at both ends, and four rivets, Verulamium, St. Albans Museum, U.K. 43.9 cm long by 8.4 cm wide by 7 cm high; iron not preserved, probably a jack plane. ca. A.D. 300. (Frere and Manning 1972, 166, fig. 61.14, pl. 50a; Gaitzsch 1980, 378, pl. 59.294; Gaitzsch and Matthäus 1981, 226, figs. 30–31; Goodman 1964, 46, fig. 44; Long and Steedman 2002, 18, fig. 3; McWhirr 1971, 37)
(fig. 3.32)

Egypt

Kom Washim, plane, molding (rounding), Cairo Museum, Egypt. Only the wooden handle survives; ca. 16 cm long by 9.5 cm high by 5 cm thick; Roman period. (Goodman 1964, 41, fig. 38)

PLANE BLADES (IRONS)

Mainland Europe

Augst, 4 plane blades, iron:

a. 15.2 cm long by 2.25 cm wide by 4 mm thick; inv. 64.11459;

b. 17.3 cm long by 1.8 cm wide by 4 mm thick; inv. 64.11458;

c. 18.4 cm long by 3.3 cm wide by 3 mm thick; inv. 64.11457;

d. 23.5 cm long by 2.85 cm wide by 3.5 mm thick; inv. 64.11456.

(Mutz 1980, 125, fig. 3)

(fig. 3.30)

England, Scotland, Wales

Caerleon (Wales), plane blade, iron, Roman Legionary Museum. 16 cm long.

Gorhambury, 2 plane blades, iron. The better preserved of the two is rectangular, one convex face, and pierced by a rectangular hole; 18 cm long and 4 cm wide. Late second–third century. (Neal 1990, 138, fig. 131.367)

Newstead, molding blade, iron, found in a ditch of the early fort, 14.2 cm long, cutting head 2.5 cm wide. (Curle 1911, pl. 59, no. 2)

RELIEF SCULPTURE

Italy

Plane (jack), funerary relief, Museo Nazionale d'Abruzzo, L'Aquila, inv. 888. Front and rear grips cut into the stock, depiction of tool against blank background. Late first or early second century. (Adam 1994, 98, fig. 224; Colini 1958, 619, no. 62; Gaitzsch 1981, 209, fig. 4; Goodman 1964, 45, fig. 42; Greber 1956, fig. 40; Zimmer 1982, 163, no. 85)

(fig. 3.34)

Plane, funerary relief of Eutyches from Priolo, Sicily, Museo Nazionale in Syracuse. Front and rear grips cut into the stock. Late third–early fourth century. (Gaitzsch 1980, 380, no. 302; additional references under *Axe, relief sculpture*)

(fig. 3.15)

Plane, funerary relief, in use by a craftsman standing at a bench, sarcophagus, Vatican, Museo Gregoriano Profano, inv. 3262. Mid-third century. (References under *Adze, relief sculpture*)

(fig. 3.8)

Plane (jack), funerary relief, sandstone, from Kastel-Staadt (Trier-Saarburg), Landesmuseum, Trier, inv. 17237. Front and rear grips cut into the stock are held by a carpenter standing at a bench. (Gaitzsch and Matthäus 1981, 211, fig. 7; Goodman 1964, 45, fig. 41; *Mosel und Saar* 1983, 220, cat. 171)

PAINTING

Man using a long (joining) bench plane, fresco from an *officina lignaria*, Pompeii, VI, 7.8–9, Naples Museum, inv. 8991. First century. (Bianchi Bandinelli 1950, 175, pl. 82, fig. 166; *id.* (1973) 299, fig. 123; Ciarello and De Carolis 1999, 121.74; Felletti Maj 1977, 334, pl. 69, 169a-b; Gaitzsch 1980, pl. 27, fig. 137a; Gaitzsch and Matthäus 1981, 212, fig. 8; *id.* 1985, pl. 10; Greber 1956, 73, fig. 38; Malten 1912, 241, fig. 4)

OTHER RENDITIONS

Long (joining) bench plane used by a shipwright standing at his bench, gold leaf on glass, from the Catacombs of Saturninus, Rome, Museo Biblioteca Apostolica Vaticana, inv. 60788. The craftsman pushes the tool with his right hand and guides it with his left. Early fourth century.

(Full references under *Adze*)

(fig. 3.23)

Planes depicted as mint marks on Roman republican silver (denarius) coinage:

a. flat sole, arched at throat, iron depicted; moneyer L. Roscius Fabatus 64 B.C. (Fava 1969, pl. B, 90–91; Gaitzsch 1980, 384, pl. 74, fig. 321a);

b. wood(?) stock, two handgrips, iron depicted; depicted on the same coin as *a*;

c. wood(?) stock, handgrip at back, iron depicted, sharply sloped front end, moneyer L. Papius, 79 B.C. (Fava 1969, pl. B, 94–95; Gaitzsch 1980, 384, pl. 74, fig. 321c; Gaitzsch and Matthäus 1981, 212, fig. 10) (fig. 3.29)

Saws

ACTUAL EXAMPLES

Handsaws (triangular blade)

Handsaw, iron, with tang and rivet, Verulamium, St. Albans Museum, U.K. inv. 78.535. Triangular, 41.7 cm long, teeth not set, 8 teeth to inch and 2.5 cm. ca. A.D. 160. (Frere and Manning 1972, 166) (fig. 3.40)

Handsaw, iron, Naples Museum, inv. 71719. Forged in one piece with frame of handle, triangular, heavily corroded, blade 13.4 cm long, total length, 24 cm. First century.

Handsaw, iron, Historisches Museum, Rathaus, Lucerne, Switzerland. With tang and two rivets, tapering toward the tip, set teeth, raked toward the front; total length ca. 37 cm; teeth on 21 cm of blade, 7–8 teeth to inch and 2.5 cm. (Goodman 1964, 116, fig. 119)

Handsaw, iron, Great Chesterford, Essex, U.K. With tang and iron nail to secure handle, tapering toward the tip; total length ca. 35.7 x 8.3 cm maximum width, 3.5 teeth to inch and 2.5 cm. (Neville 1856, 10, pl. 2.20)

Backed (hand) saw, iron, Landesmuseum, Zurich. Small with tang for handle (missing). (Goodman 1964, 118, fig. 121) (fig. 3.35)

Bow and frame saws (strap-type thin blade with parallel top and cutting edges, held under tension in a wooden frame)

Blade fragment, iron, Verulamium, St. Albans Museum, U.K. 9.4 cm long, no set to teeth, teeth sloped, 12 teeth to inch and 2.5 cm. A.D. 105–30. (Frere and Manning 1972, 166)

Blade fragment, iron, offset teeth, Landesmuseum, Zurich. (Goodman 1964, 118, fig. 121; Neuberger 1919, 76, fig. 113)

Blade fragment, iron, Pompeii, Naples Museum, inv. 286773. Fragment measures 9.2 cm long by 7.5 cm deep, teeth are set. (Ciarello and De Carolis 1999, 121.74; Gaitzsch 1980, II, no. 309)

Blade fragment, Gorhambury, St. Albans, U.K. Small fragment 4.5 cm long, six teeth to inch and 2.5 cm, teeth not raked. (Neal 1990, 140, fig. 131.375)

Heavy crosscut or band saw (blade with attached handle on either end)

Crosscut saw, blade fragment, iron, Pompeii, Naples Museum, inv. 286772. The fragment measures 12.4 cm long by 11.5 cm high. A handle would have been fitted to each end, according to the most recent analysis. (Ciarello and De Carolis 1999, 123.79)

RELIEF SCULPTURE

Small frame (buck) saws (Gaitzsch Type II, *Rahmensägen*)

Bucksaw, Altar to Minerva, Capitoline Museums, Rome, inv. 1909. Outwardly bowed handles, back of blade convex, toggle stick. Late first century B.C.–early first century (References under *Adze*) (fig. 2.2)

Bucksaw, hanging on the wall of a furniture maker's shop, Capitoline Museums (Montemartini), inv. 2743; plaster cast in Museo della Civiltà Romana, inv. 3408. Outwardly bowed handles terminate in volutes; no toggle stick depicted. Late first century (?) (References under *Lathe*) (fig. 2.3)

Bucksaw operated by two standing men, one in a tunic, the other bare-chested, Isola Sacra, Ostia Museum, inv. 138. The saw depicted is large for its type, nearly shoulder-height. Poorly preserved funerary relief in marble; parts of the frame and the entire blade have been broken off the relief. Handgrips are bowed outward. Late first century. (Calza 1940, 256, fig. 158; Colini 1947, 24, fig. 6; Zimmer 1982, 140, no. 58) (fig. 3.39)

Bucksaws, two variants on a terra-cotta funerary plaque of a *faber ferrarius* (blacksmith) from Isola Sacra, Ostia, tomb no. 29:

a. four sides parallel, double toggle sticks indicated(?);

b. handgrips bowed outward; no toggle shown. Mid–second century. (Calza 1931, 536, fig. 20; Calza 1949, 251, fig. 150; Becatti 1951, 134, fig. 68; Fava 1969, cover; Gaitzsch 1980, pl. 69, no. 314; Gaitzsch 1985, fig. 1a; Meiggs 1973, pl. 27; Squarciapino 1941, 12; Zimmer 1985)

Bucksaw, funerary relief, Autun, Musée Rollin d'Autun. Straight blade, grips bowed outward, tension cord held by voluted ends; toggle obscured by adze in foreground.

(References under *Adze, relief sculpture*)

Large frame saws (Gaitszch Type III, *Klobensägen*)

Frame saw ripping a long board, from a nineteenth-century drawing. (Goodman 1964, 118, fig. 122; Mercer 1960, 152, fig. 142)

Frame saw, hanging on the wall of a furniture maker's shop, Rome, Capitoline Museums (now at Montemartini annex). The tool is attached to a long beam of wood, perhaps to serve for ripping boards or as a hand-operated band saw. Late first century (?) (References under *Bucksaws*)

(fig. 2.3)

Frame saw(?) used in a saw-pit by two sawyers, grave relief from Gaul.

(Meiggs 1982, 348)

Crosscut band saw, for operation by two men (Gaitzsch Type IV, *Bandsäge*). Crosscut saw, Altar to Minerva, Capitoline Museums, Rome. (References under *Adze, relief sculpture*)

(fig. 2.2)

PAINTING

Large frame saw used in a saw-pit by two sawyers, fresco from an *officina lignaria*, Pompeii, VI, 7.8–9, Naples Museum, inv. 8991. (Blümner 1875 (II) 346, fig. 60; Meiggs 1982, 348, fig. 14b; additional references under *Plane, painting*).

(fig. 3.33)

OTHER RENDITIONS

Bucksaw (or large frame saw), gold leaf on glass, from the Catacombs of Saturninus, Rome, Museo Biblioteca Apostolica Vaticana, inv. 60788. A shipwright holds the saw in his right hand while steadying the plank in his left. If the artist intended to depict a frame saw for ripping boards, he is in error, as the blade should be in the center of the frame, not at one side. Early fourth century. (References under *Adze*)

(fig. 3.23)

Bucksaws as mint marks on Roman republican silver coinage:

a. blade side wider than tensioning cord, toggle depicted, moneyer L. Papius, 79 B.C. (Fava 1969, pl. B 94–95; Gaitzsch1980, 384, pl. 74, fig 321c);

b. four sides parallel, corners on tensioning side splayed outward; moneyer L. Papius, 79 B.C.; depicted with the crosscut saw in following entry. (Fava 1969, pl. B, 92–93; Gaitzsch 1980, 384, pl. 74, fig 321e)

(fig. 3.29)

Crosscut two-man saw, mint mark on Roman republican silver coinage, toothed blade slightly arced, handles on both ends depicted; shown with bucksaw (b) of previous entry.

Two miniature model frame saws in bronze, perhaps with sloping teeth, Sussex, U.K., British Museum (Goodman 1964, 24, fig. 13)

Notes

CHAPTER 1. Introduction

Epigraph: Plin. *HN* 12.1: "diu fuere occulta eius beneficia, summumque munus homini datum arbores silvaeque intellegebantur. hinc primum alimenta, harum fronde mollior specus, libro vestis; etiamnunc gentes sic degunt."

CHAPTER 2. The Roman Woodworker

Epigraph: Translation by B. Perrin, *Plutarch: Lives*, vol. 1 (Loeb Classical Library, Cambridge, Mass., 1993).

1. Cic. *Off.* 1.150–51. Agriculture is at the top of Cicero's list, but professions that benefit society, such as medicine and architecture, are rated well above manual labor, retail trade, tax collecting, and money lending. Cicero further differentiates between basic unskilled manual labor and that which involves *ars*.

CHAPTER 3. The Tools of the Trade

1. *Terebra* is also applied to a type of siege machine that was used to punch a hole in enemy fortifications (for example, Vitr. 10. 13.7).
2. See also Grodde 1989, 161.
3. A full discussion of Roman-period turned wooden bowls and pyxides can be found in Pugsley 2003, 66ff.
4. For images of plumb bobs from Pompeii, see Goodman 1964, 200, and Adam 1981, 102, fig. 26.

CHAPTER 5. Foundations

1. For the significance of the text as representation of a Roman building contract, see Anderson 1997, 72. The Latin reads as follows: "villam aedificandam si locabis novam ab solo, faber haec faciat oportet. Parietes omnes, uti iussitur, calce et caementis, pilas ex lapide angulari, tigna omnia, quae opus sunt, limina, postes, iugumenta, asseres, fulmentas.
2. "Rationem pontis hanc instituit. Tigna bina sesquipedalia paulum ab imo praeacuta dimensa ad altitudinem fluminis intervallo pedum duorum inter se iungebat. Haec cum machinationibus immissa in flumen defixerat fistucisque adegerat, non sublicae modo directe ad perpendiculum, sed prone ac fastigate, ut secundam naturam fluminis procumberent, his item contraria duo ad eundem modum iuncta intervallo pedum quadragenum ab inferiore parte contra vim atque impetum fluminis conversa statuebat. Haec utraque insuper bipedalibus trabibus immissis, quantum eorum tignorum iunctura distabat, binis utrimque fibulis ab extrema parte distinebantur; quibus disclusis atque in contrariam partem revinctis tanta erat operis firmitudo atque ea rerum natura, ut, quo maior vis aquae se incitavisset, hoc artius illigata tenerentur. Haec directa materia iniecta contexebantur ac longuriis cratibusque consternebantur; ac nihilo setius sublicae ed ad inferiorem partem fluminis oblique agebantur, quae pro ariete subiectae et cum omni opere coniunctae vim fluminis exciperent, et aliae item supra pontem mediocri spatio, ut, si arborum trunci sive naves deiciendi operis essent a barbaris missae, his defensoribus earum rerum vis minueretur neu ponti nocerent."

CHAPTER 6. Framing and Walls

Epigraph: "arbore exaedificamus tecta." *Tecta* literally means "roofs," but as a general introduction to his topic, Pliny is probably referring to building in general.

1. The term *opus incertum* indicates a facing of small, irregular stones that covers a concrete and rubble fill. Usually the stone facing was covered with stucco and painted.
2. Dimensions from Papaccio 1993. This article also includes a complete list of houses at Herculaneum that employ walls of *opus craticium*; see particularly 611, n. 11.
3. The wall painting is from the so-called Bottega del Fornaio (the shop of the baker) at Pompeii (VIII 3, 30; Naples Museum, inv. 9071). Traditionally considered the scene of a bread seller's shop, it may in fact represent the distribution of free bread (a scene of *largitio*; cf. T. Fröhlich 1991).
4. Capitoline Museum, inv. 16154. Excavations of S. Omobono began in 1938 and continued sporadically until 1959 (Gjerstad 1960, 448).
5. Cf. Mattingly and Sydenham 1926, 2.239, where the image is interpreted as a representation of the Pons Sublicius.
6. Calp. 7. 23–24: "vidimus in caelum trabibus spectacula textis / surgere, Tarpeium prope despectantia culmen." Here *spectacula* refers to a place where games are held: an amphitheater. Additional comment can be found in Coarelli 2001, 46.

CHAPTER 7. Wooden Flooring

1. Refer as well to *contabulatio* in the glossary: a structure made of *tabulae*, or planks.
2. *Opus signinum* is a mixture of cement and small pieces of broken tile or potsherds, used as both a substrate and a functional floor surface. For an ancient description, see Plin. HN 35.165.
3. Personal communication with Prof. Emeritus Norman Doenges of Dartmouth College, Dept. of Classics, Hanover, N.H., and Doenges et al., 2005.
4. Good examples can be seen in the joist holes over the doorway opening off Cardo IV, just opposite the House of the Wooden Partition (Insula IV, 8).
5. Note that there is no second level of sockets into which the lighter set of joists would have been anchored. This suggests that the three main beams would have supported the entire weight of the floor.
6. Dio 55.8.4, 66.24.2. The Diribitorium was originally opened in A.D. 7. Both Pliny (HN 16.201) and Dio (loc. cit.) refer to the use of immense beams (of larch) to span the building.

CHAPTER 8. Roofing and Ceilings

1. See Stamper for a review of scholarship concerning the reconstruction of the Temple of Jupiter Optimus Maximus (2005, 19–33). Anderson (1997, 131) proposes spans of 7.85 m across the facade columns and 10.46 m along the flanks.
2. "sub tectis, si maiora spatia sunt, et transtra et capreoli, si commoda, columen, et cantherii prominentes ad extremam suggrundationem; supra cantherios templa; deinde insuper sub tegulas asseres ita prominentes, uti parietes protecturis eorum tegantur."
3. Izenour (1992, 46) similarly reconstructs a trussed roof over the meeting hall (*ekklesiasterion*) at Priene, ca. 200 B.C., with a clear span of 14.65 m.
4. The broadest clear spans in the Greek world are found in circular buildings. The Arsinoeion in Samothrace is cited by Coulton as the most ambitious, with a span of 16.80 m. Here the rafters of the wooden roof would have been raised with a "wigwam-like cone" of timbers (Coulton 1977, 158).
5. If the rendition is meant to portray a fully timbered arena, as Coarelli suggests (2001, 47), then there must be some explanation for the "masonry" courses carved on the lower half of the building. If not blocks of stone, perhaps they are meant to indicate some kind of screen made of panels placed over a timber framework.
6. For hypothetical reconstructions of the timber roofs built over music halls, see Izenour (1992); note that the specific framing solutions offered are conjectural. Refer as well to table 3 (p. 149); the *odeum* at Augusta Praetoria in northern Italy may have been covered with the greatest clear span ever attempted. Note that this structure was built in the far north of Italy, close to sources of good mountain larch.
7. Wooden flooring and the term *contignatio* are discussed in chapter 7.

8. The English translation of Vitruvius by W. Wilkins (London 1812) also considers the *capreolus* as an interior brace of the truss. Cf. section 2, pl. 2 of this edition.
9. Meiggs (1982, 366) includes discussion of Diocletian's Price Edict and the timber trade. A seventy-five-foot length of fir had a set price of fifty thousand *denarii*. A beam two-thirds of that length cost only twelve thousand. The Latin and Greek text of the Price Edict can be found in Lauffer 1971, 138–39.
10. The Latin text of this passage can be found in the glossary, under "VIII: Roofing and Ceilings." The English translation here is based upon MacDonald 1982, 61–62.
11. Mattingly and Sydenham 1926, 2.205; Mattingly 1930, 81.3; Tamm 1963, 212, fig. 112; MacDonald 1982, pl. 51.
12. The name of the mountain, Abetone, is apparently derived from *abeti* (fir trees); here in Etruscan times there may have been a grove sacred to Silvanus. The fact that the region around Caere was (and is still) heavily forested helps to account for the renditions of timbered ceilings placed in chamber tombs found outside the Etruscan city (Edlund 1987, 53).
13. Wiegand's reconstruction of diagonal braces between the posts and the lower joist system is more speculative; there are no confirming stone cuttings to verify the bracing. If such timbers were used, their effectiveness in keeping the ceiling joists in tension is questionable.
14. Cic. *Tusc.* 5.62. Cicero's depiction of a sword descending from the coffers was imitated by Persius Flaccus (mid–first century A.D.) in *Saturae (Satires)*, book 3, 40: "auratis pendens laquearibus ensis."
15. Suet. *Ner.* 31.2: "cenationes laqueatae." In a similar vein Petronius presents the dining room of Trimalchio, and Seneca (*Ep.* 90.15), who exhibits a particular revulsion for the coffered ceiling in the private setting, says that some wealthy homeowners have dining room ceilings with movable panels that change their patterns as often as the courses (the passage is translated in the glossary, section VIII.2).
16. "reliquum lacunariorum et arcae supra trabes ratio habeatur."
17. The lagging for the centering of a dome would have been applied radially as well as laterally, as has been documented by imprints still observable in concrete domes (Rasch 1991). Ingenious solutions for the centering of large domed spaces such as the Pantheon have been proposed by a number of scholars, most recently by Taylor (2003, 190 ff.)

CHAPTER 9. Interior Woodwork

1. Like other types of luxury doors, those of marble were undoubtedly inspired by Greek models. The surviving building records (409 B.C.) of the Erechtheum in Athens (completed in 405 B.C.) mention the use of interior double-leaved marble doors (Dinsmoor 1975, 1990).
2. The defeat of the Gauls on Parnassus, near the sanctuary of Apollo at Delphi, took place in 279 B.C.; Pausanias provides a dramatic account in his *Description of Greece* 10.23.4 (written in the second century).
3. The discovery in 1998 of sixteen Roman ships embedded in mud in the vicinity of the train station of Pisa holds great promise for future research into Roman woodworking and ship construction. See Bruni (2000).
4. Doors similar to those depicted in *oecus* 6 of the Villa of the Mysteries, including the squamate pattern in the upper *tympana*, include those in *cubiculum* 16, Villa of the Mysteries, alcove *a*, and at the villa found at Boscoreale, *cubiculum m* (now in the Metropolitan Museum of Art in New York [fig. 6.14]). The squamate pattern with a similar color scheme is shown on painted columns in the southeast portico of the Villa at Oplontis and in the dado zone of Room IV in the "House of the Griffins" from the Palatine Hill. See J. Engemann (1967), pl. 21, fig. 1, pl. 31, pl. 37.

CHAPTER 10. Wheels

1. While Granger (Granger 1985, 274) specifically comments that *materia* should *not* be translated as "wood" or "timber" in this passage, the combination of *materia* and *coniunctio* seems to make this meaning clear. Similar to my translation is Rowland (1999, 119): "a continuous piece of joinery."

CHAPTER 11. Furniture and Veneers

1. Excavations at Quinto Fiorentino, near Florence, recovered fragments of an Etruscan-period folding ivory stool; cf. Nicosia 1969, 24.

2. For an actual example of such stretchers on a wooden table, see Mols 1999, 97 (cat. no. 15).
3. For such *lararia* or, as Mols prefers, *aediculae* from Herculaneum, cf. catalog in Mols (1999, nos. 27, 28).
4. The description of the chest from Herculaneum and the dimensions given are from Mols 1999, 217–19.
5. Franchi dell'Orto 1990: Naples, inv. 3277.
6. "media pars arborum crispior, et quo proprior radici minoribus magisque flexilibus maculis. haec prima origo luxuriae arborum, aliam alia integi et vilioris ligni e pretiosiore corticem fieri. Ut una arbor saepius veniret, excogitatae sunt et ligni bratteae."

CHAPTER 12. Classification of Trees and Species of Timber

1. "Illud, quod non ope humana provenit, materiae est magis aptum."
2. *Picea abies* is known by the common names European spruce, Norway spruce, and red fir.
3. In addition to the term *tuber*, the terms *molluscum* and *bruscum* were used to refer to such burls.
4. Not all writing tablets discovered are of boxwood, however. Two well-preserved examples in pine have been reported from the excavations of the Roman fort at Newstead in Scotland (Curle 1911, 308).
5. When the temple was excavated in 1939, a marble image of the god dated stylistically to the second century was discovered. Thus the image of cypress wood reported by Pliny may have been revered for about four hundred years.
6. Plin. HN 16.200: "amplissima arborum ad hoc aevi existimatur Romae visa quam propter miraculum Tiberius Caesar in eodem ponte naumachiario exposuerat advectam cum reliqua materie, duravitque ad Neronis principis amphitheatrum. fuit autem trabs ea e larice, longa pedes CXX, bipedali crassitudine aequalis, quo intellegebatur vix credibilis reliqua altitudo fastigium ad cacumen aestimantibus."
7. Pliny also mentions that larch was considered less desirable for the planking of the hulls of ships, since it was susceptible to boring worms (Plin. HN 16.219); note Vitruvius's contrary opinion (2.9.14).
8. The Italians call *Quercus robur "farnia."* The Roman *aesculus* is probably equivalent to the *Quercus sessiliflora*, the *Quercus petraea*, and the *Quercus frainetto*; all describe essentially the same tree, known, depending on location, as the sessile oak, the durmast oak, the Hungarian oak, and the Italian oak.
9. For the passage of Cato, see *fibula* in the Glossary under "IV.2: Joints."
10. See the previous note.
11. Plin. HN 13.99: "artifices vero frumenti acervis inponunt septenis diebus totidem intermissis, mirumque ponderi quantum ita detrahatur."

CHAPTER 13. The Forests of Italy

Epigraph 1: Strabo, trans. by H. L. Jones (1988).

Epigraph 2: Plin. HN 12.3: "nec magis auro fulgentia atque ebore simulacra quam lucos et in iis silentia ipsa adoramus."

1. "Nondum caesa suis, peregrinum ut viseret orbem / montibus in liquidas pinus descenderat undas."

Bibliography

Abbreviations used for the titles of journals in the following bibliography are those listed in the *American Journal of Archaeology* 104 (2000): 10–24.

Abbate Edlmann, M., et al. 1989. Un relitto romano recuperato nelle acque dell'isola del Giglio: la struttura navale e i legni impiegati per la sua realizzazione. In *Il Restauro del Legno* by G. Tampone.

Adam, J.-P., and T. Adam. 1981. Le techniche costruttive a Pompei. In *Pompei 1748–1980. I Tempi della Documentazione*. Exhibition catalog. Rome.

Adam, J.-P. 1994. *Roman Building*. Bloomington.

Aldred, C. 1956. Furniture: To the End of the Roman Empire. In *History of Technology*, vol. 2.

Allaby, M. 1999. *Temperate Forests*. New York.

Alzinger, W. 1987. Vitruvs Basilika und der archäologische Befund. In *Munus non Ingratum: Proceedings of the International Symposium on Vitruvius' De Architectura and the Hellenistic and Republican Architecture*, edited by Geertman and Jong. Leiden.

Amedick, R. 1993. Stibadia in Herculaneum and Pompeii. In *Ercolano 1738–1988: 250 anni di ricerca archeologica. Atti del Convegno Internazionale Ravello-Ercolano-Napoli-Pompeii 1988*, edited by L. Franchi dell'Orto. Rome.

Amelung, W. 1903. *Die Skulpturen des Vaticanischen Museums* I. Berlin.

Anderson, J. C. 1997. *Roman Architecture and Society*. Baltimore.

Andrén, A. 1959–60. Origine e formazione dell'architettura templare etrusco-italica. *Atti della Pontificia Accademia Romana di Archeologia. Rendiconti* 32:21–59.

———. 1940. *Architectural Terracottas from Etrusco-Italic Temples*. Lund and Leipzig.

Arena M.-Marcello, A. 1957–63. Su di un traffico d'ostriche nell'antica Altino. II Studio dei frammenti de legno e della segature. *Memorie di Biogeografia Adriatica* 5:122–37.

Arnold, B. 1982. The Architectural Woodwork of the Late Bronze Age Village Auvernier-Nord. In *Woodworking Techniques before A.D. 1500*, BAR-IS 129, edited by S. McGrail. Oxford.

Baker, H. 1970. *Plants and Civilization*. London.

Baker, H. G. 1966. *Furniture in the Ancient World*. London.

Barfield, L. 1971. *Northern Italy Before Rome*. London and New York.

Barrett, A. 1989. *Caligula*. New Haven.

Bartoli, A. 1963. *Curia Senatus: Lo scavo e il Restauro*. Rome.

Bartoloni, G., F. Buranelli, V. D'Atri, and A. De Santis. 1987. *Le urne a capanna rinvenute in Italia*. Rome.

Bartoloni, G. 1989. *La Cultura Villanoviana. All'inizio della storia Etrusca* (Studi NIS archeologia No. 9, La Nuova Italia Scientifica). Roma.

Bealer, A. W. 1972. *Old Ways of Working Wood*. Barre, Mass.

Becatti, G. 1951. *Arte e Gusto negli scrittori latini*. Florence.

Behrends, O. 1981. Die Rechtsformen des römischen Handwerks. In *Das Handwerk in vor- und frühgeschichtlicher Zeit* 1, edited by Jankuhn and Jannsen. Göttingen.

Bérenger, Adolfo di. 1859–65/1965. *Studi di Archeologia Forestale*. Reissued by the Italian Academy of Forest Sciences (Accademia Italiana di Scienze Forestali). Florence.

Berti, F. 1990. *Fortuna Maris, la Nave Romana di Comacchio*. Bologna.

Bianchi Bandinelli, R. 1970. *Rome: The Center of Power*. New York.

———. 1950. *Storicità dell'arte classica*. Florence.

Bidwell, P., R. Miket, and B. Ford, eds. 1988. *Portae cum Turribus: Studies of Roman Fort Gates*, BAR-BS 206. Oxford.

Birley, R. 1977. *Vindolanda: A Roman Frontier Post on Hadrian's Wall*. London.

———. 1994. *Vindolanda Research Reports*. New Series, Vol. 1: *The Early Wooden Forts*. Hexham, Northumberland.

Blagg, T. F. C. 2000. The votive model: Etrusco-Italic temples from Nemi. In *Nemi—Status Quo*, edited by J. Brandt, A.-M. Leander Touati, and J. Zahle. Rome.

Blake, M. 1973. *Roman Construction in Italy from Nerva Through the Antonines*. Philadelphia.

Blaser, W. 1985. *Architektur im Möbel: vom Altertum zur Gegenwart = Furniture as Architecture: From Antiquity to the Present*. Zürich.

Bliquez, L. J. 1994. *Roman Surgical Instruments and Other Minor Objects in the National Archaeological Museum of Naples*. Mainz.

Blümner, H. 1885. *Das kunstgewerke im Altertum, II, Mobiliar*. Leipzig.

———. 1875–87. *Technologie und Terminologie der Gewerbe und Künste bei Griechen und Römern* 4 vols. Leipzig.

Boëthius, A. 1978. *Etruscan and Early Roman Architecture*. Middlesex.

———. 1955–56. Vitruvio ed il 'Tempio Tuscanico.' *Studi Etruschi* 24:137–42.

Bonino, M. 1985. Sewn Boats in Italy: *Sutiles Naves* and *Barche Cicite*. In *Sewn Plank Boats: Archaeological and ethnographic papers based on those presented to a conference at Greenwich in November, 1984* (National Maritime Museum, Greenwich, Archaeological Series No. 10, BAR-IS 276), edited by McGrail and Kentley. Oxford.

Boon, G. C. 1974. *Silchester, the Roman Town of Calleva*. London, North Pomfret, Vt., and Vancouver.

Bound, M. 1991. *The Giglio Wreck* (Hellenic Institute of Marine Archaeology). Athens.

Brown, F. E. 1960. *Cosa II: The Temples of the Arx* Pt. I (*Memoirs of the American Academy in Rome* 26). Rome.

Bruni, S., ed. 2000. *Le Nave Antiche di Pisa: The Ancient Ships of Pisa*. Florence.

Budetta, T., and M. Pagano. 1988. *Ercolano; legni e piccoli bronzi, testimonianze dell'arredo e delle suppellettili della casa romana*. Rome.

Bundgård, J. A. 1965. Caesar's Bridges over the Rhine. *ActaArch* 36:87–104.

Burford, A. 1972. *Craftsmen in Greek and Roman Society*. Ithaca.

Cagiano de Azevedo, M. 1957. Qualche osservazione sulla lavorazione del legno nella antichità. *Bollettino dell'Istituto Centrale del Restauro* 29/30:11–29.

Calza, G. 1940. *La necropoli del Porto di Roma nell'isola sacra*. Rome.

Calzoni, U. 1936. Resti di un abitato preistorico a 'Casa Carletti,' sulla montagna di Cetona. *Studi Etruschi* 10:329–39.

Capelle, T. 1981. Zur Producktion hölzerner Gefäße im vor- und frühgeschichtlichen Mittel- un Nordeuropa. In *Das Handwerk in vor- und frühgeschichtlicher Zeit* 1, edited by Jankuhn.

Carettoni, G. 1983. *Das Haus des Augustus auf dem Palatin*. Mainz.

Carver, M. O. H., S. Donaghey, and A. B. Sumpter. 1978. *Riverside Structures and a Well in Skeldergate and Buildings in Bishopshill. The Archaeology of York* 4.1.

Caskey, L. 1910. The Roofed Gallery of the Walls of Athens. *AJA* 14:298–309.

Casson, L. 1991. *The Ancient Mariners: Seafarers and Sea Fighters of the Mediterranean in Ancient Times*. 2d ed. Princeton.

Castelletti, L. 1990. Legni e carboni in archeologia. In *Scienze in Archeologia, II° ciclo di lezioni sulla Ricerca applicata in Archeologia*, edited by T. Mannoni and A. Molinari. Siena.

Cavada, E., ed. 1994. *Archeologia a Mezzocorona: Documenti per la storia del popolamento rustico di eta' romana nell'area atesina*. Trento.

Ceccotti, A., and L. Uzielli. 1990. Sul grado di affidabilita' delle strutture lignee antiche. In *Atti del 2° Congresso Nazionale sul restauro del legno*. Vol. 2, edited by G. Tampone, Nardini ed. Florence.

Champion, B. 1916. Outils en Fer du Musée de Saint-Germain. *Revue archéologique* 211–46.

Chapman, H. 1979. A Roman mitre and trysquare from Canterbury. *Antiquities Journal* 59:403–07.

Childe, V. G. Rotary Motion. Chap. 9 in *History of Technology*, Vol 1.

Choisy, A. 1899. *Histoire de L'Architecture*, 2 vols. Paris.
———. 1873. *L'art de bâtir chez les Romains*. Paris.
Ciarallo, A., and E. De Carolis. 1999. *Homo Faber: natura, scienza e tecnica nell'antica Pompei*. Milan.
Cifani, G. 1995. Aspetti dell'edilizia romana arcaica. *Studi Etruschi* 60:185–226.
Claridge, A. 1998. *Rome: An Oxford Archaeological Guide*. Oxford.
Clarke, J. 1991. *The Houses of Roman Italy 100 B.C.–A.D. 250*. Berkeley.
Coarelli, F. 1974. *Guida archeologica di Roma*. Rome.
———. 2001. Gli anfiteatri a Roma prima del Colosseo. In *Sangue e Arena*, edited by A. La Regina. Milan.
Coles, B. 1986. *Sweet Track to Glastonbury: The Somerset Levels in Prehistory*. New York.
Colini, A. M. 1947. Officina di Fabri Tignarii nei Frammenti di un Ara Monumentale Rinvenuti fra il Campidoglio e il Tevere. *Capitolium* 21–28.
———. 1958. *Museo della Civiltá Romana*. Rome.
Colonna, G., ed. 1985. *Santuari d'Etruria*. Milan.
Colonna, G., E. Paribeni, et al. 1970. *Pyrgi. Scavi del Santuario Etrusco (1959–1967) NSc* suppl. to vol. 24. Rome.
Colonna, G. 1996. *L'altorilievo di Pyrgi: dei ed eroi greci in Etruria*. Rome.
Corona, E., R. D'Alessandro, and M. Follier. 1974. I pali lignei dell'abito neolitico di Fimon Molino Caserotto. *Annali di Botanica* 33:237–49.
Coulton, J. J. 1977. *Ancient Greek Architects at Work*. Ithaca.
Cozzo, G. 1928. *Ingegneria Romana*. Rome.
Cunliffe, B., A. Down, and D. Rudkin. 1996. *Chichester Excavations IX: Excavations at Fishbourne 1969–88*. Chichester.
Cüppers, H. 1969. *Die Trierer Römerbrücken*. Mainz.
Curle, J. 1911. *A Roman Frontier Post and Its People: The Fort of Newstead in the Parish of Melrose*. Glasgow.
Curtius, L. 1929. *Die Wandmalerei Pompejis*. Leipzig.
Déchelette, J. 1924–27. *Manuel d'archéologie préhistorique, celtique et gallo-romaine*. 4 vols. Paris.
De Fine Licht, K. 1968. *The Rotunda in Rome. A Study of Hadrian's Pantheon* (Jutland Archeological Society 8). Copenhagen.
DeLaine, J. 1997. *The Baths of Caracalla: A Study in the Design, Construction, and Economics of Large-Scale Building Projects in Imperial Rome*. Portsmouth, R.I.
———. 1996. The Insula of the Paintings. A Model for the Economics of Construction in Hadrianic Ostia. In *'Roman Ostia' Revisited*, edited by A. G. Zevi and A. Claridge. Rome.
Delbrück, R. 1912. Archäologische Funde im Jahre 1911: Italien. *JdI* 27:270–322.
Della Corte, M. 1965. *Case ed abitanti di Pompei*. 3d ed. Naples.
Desantis, P. 1990. Le suppellettili in legno di uso quotidiano. In *Fortuna Maris, la Nave Romana di Comacchio*, edited by F. Berti. Bologna.
Dessau, H. 1892–1956. *Inscriptiones Latinae Selectae = ILS*. Berlin.
Dillon, J. 1989. A Roman Timber Building from Southwark. *Britannia* 20:229–31.
Dinsmoor, W. B. 1950/1975. *The Architecture of Ancient Greece*. 3d ed. London 1950; reprinted New York 1975. New York.
Disston, Henry, and Sons. 1916. *The Saw in History*. Philadelphia.
Doenges, N. A., M. Orfila, and H. B. Mattingly. 2005. *Pollentia, A Roman Colony on the Island of Mallorca*. BAR-IS 1404. Oxford.
Dwyer, E. 1991. The Pompeian Atrium House. In *Roman Art in the Private Sphere*, edited by E. Gazda. Ann Arbor.
Eckardt, H. 1999. The Colchester 'Child's Grave.' *Britannia* 30:57–89.
Edlund, I. E. 1987. *The Gods and the Place: Location and Function of Sanctuaries in the Countryside of Etruria and Magna Graecia (700–400 B.C.)*. Stockholm.
Edwards, H. J. 1917/1970. Caesar, *The Gallic War*. Loeb Classical Library, Cambridge, Mass., and London.
Elia, O. 1934. Pompei: Relazione sullo scavo dell'Insula X della Regio I. *NSc* ser. 6.10:264–344.
Emy, A. R. 1856. *Tratatto dell'arte del carpentiere*. Venice.
Engemann, J. 1967. *Architekturdarstellungen des frühen zweiten Stils* (Mitteilungen des Deutschen Archaeologischen Instituts, Römische Abteilung 12). Heidelberg.
Ericsson, C. H. 1984. *Navis oneraria: The cargo carrier of late antiquity, studies in ancient ship carpentry*. Åbo.

Espérandieu, E. 1907–81. *Recueil général des bas-reliefs, statues et bustes de la Gaule Romaine*. Vols. 1–16. Paris.

Evans, E. 2000. *The Caerleon Canabae: Excavations in the Civil Settlement 1984–90* (Britannia Monograph Series No. 16). London.

Evans, J. 1894. On Some Iron Tools and Other Articles Formed of Iron Found at Silchester in the Year 1890. *Archaeologia* 54:139–56.

Fava, A. S. 1969. *I simboli nelle monete argentee republicane e la vita dei romani*. Rome.

Fehr, H. 1981. Eine Rheinbrücke zwischen Koblenz und Ehrenbreitstein aus der Regierungszeit des Claudius. *BonnJbb* 181:287–300.

Felletti Maj, B. M. 1977. *La tradizione italica nell'arte romana (Archaeologica 3)*. Rome.

Ferroni, A., and C. Meucci. 1989. Primi osservazioni sulla barca di Ercolano: il recupero e la costruzione navale. In *Il Restauro del Legno*, edited by G. Tampone. Florence.

Fioravanti, M., and R. Caramiello. 1999. Il Legno e la sua Lavorazione. In *Homo Faber*. Exhibition catalog. Milan.

Fitchen, J. 1986. *Building Construction Before Mechanization*. Cambridge.

———. 1961. *The Construction of Gothic Cathedrals: A Study of Medieval Vault Erection*. Oxford.

Fittschen, K. 1974. Zur Herkunft und Entstehung des 2. Stils. In *Hellenismus in Mittelitalien: Kolloquium in Göttingen 1974*.

Fitzpatrick, A. P., and P. R. Scott. 1999. The Roman Bridge at Piercebridge, North Yorkshire-County Durham. *Britannia* 30:111–32.

Follieri, M. Interdisciplinary Studies on Wood and Charcoal in Italy. In *Wood and Archaeology*.

Franchi dell'Orto, L., ed. 1993. *Ercolano 1738–1988: 250 anni di ricerca archeologica. "Atti del Convegno Internazionale Ravello-Ercolano-Napoli-Pompeii 1988*. Rome.

Franchi dell'Orto, L., and A. Varone, eds. 1990. *Rediscovering Pompeii*. Exhibition catalog. Rome.

Frank, T., ed. 1933–40. *An Economic Survey of Ancient Rome*. Vols. 1–5 (esp. R. M. Haywood, vol. 4: *North Africa*; T. Frank, vol. 5: *Rome and Italy of the Empire*). Baltimore.

Fremersdorf, F., and Gutshof Köln-Müngersdorf. 1933. Römisch-Germanische Forschungen 6.

Frere, S., and J. J. Wilkes. 1989. *Strageath: Excavations within the Roman Fort* (Britannia Monograph Series 9). London.

Frere, S. 1972. *Verulamium Excavations*. Vol. 1, Series: Society of Antiquaries of London. Research Committee. Reports, no. 28, cf. especially W. H. Manning, "The Iron Objects," 163–95. London.

———. 1984. *Verulamium*. Vol. 3. Oxford.

Fröhlich, T. 1991. *Lararien- und Fassadenbilder in den Vesuvstädten. Untersuchungen zur 'volkstümlichen' pompejanischen Malerei* (RM Erg. 32). Mainz.

Gabelmann, H. 1973. *Die Werkstattgruppen der oberitalischen Sarkophage* (Beihefte der Bonner Jahrbücher 34). Berlin.

Gaitzsch, W. 1980. *Eiserne römische Werkzeuge. Studien zur römischen Werkzeugkunde in Italien und den nördlichen Provinzen des Imperium Romanum*, 2 vols. BAR-IS 78. Oxford.

———. 1985. Werkzeuge und Geräte in der römischen Kaiserzeit, eine Übersicht. ANRW II.12.3:170–204.

———. 1978. *Römische Werkzeuge*. Stuttgart.

Gaitzsch, W., and H. Matthäus. 1981. Runcinae—römische Hobel. *Bonner Jahrbücher* 181:205–47.

Galliazzo, V. 1994. *I Ponti Romani*. 2 vols. Treviso.

Ganzert, J. 1996. *Der Mars-Ultor-Tempel auf dem Augustusforum in Rom*. Mainz.

Garnsey, P. 1974. Legal Privilege in the Roman Empire. In *Studies in Ancient Society*, edited by M. I. Finley. London and Boston.

Garrucci, R. 1864. *Vetri Ornati di Figure in Oro Trovati nei Cimiteri Cristiani di Roma*. Rome.

Gazzola, P. 1963. *Ponti Romani*. Vol. 2. Florence.

Geertman, H., and J. J. de Jong, eds. 1989. *Munus non Ingratum: Proceedings of the International Symposium on Vitruvius' De Architectura and the Hellenistic and Republican Architecture*. Leiden.

Giachi, G., S. Lazzeri, and S. Paci. 2000. Il Legno Utilizzato per la Costruzione delle Imbarcazioni: Indagini Preliminari. In *Le Nave Antiche di Pisa: The Ancient Ships of Pisa*, edited by S. Bruni. Florence.

Gibson, S., J. DeLaine, and A. Claridge. 1994. The Triclinium of the Domus Flavia: A new Reconstruction. BSR 62:67–97.

Ginouvès, R., and R. Martin, eds. 1985–92. *Dictionnaire méthodique de l'architecture grecque et romaine*. Vols. 1, 2. Rome and Athens.
Giordano, G. 1993. *Tecnica delle costruzioni in legno*. 2d ed. Milan.
Giuliani, C. F. 1977. Domus Flavia: Una Nuova Lettura. RM 84:91–106.
———. 1990. *L'edilizia nell'antichitá*. Rome.
Gjerstad, E. 1960. *Early Rome III*. Lund.
Glasbergen, W., and W. Groenman-van Waateringe. 1974. *The Pre-Flavian Garrisons of Valkenburg*. Amsterdam.
Glendinning, B. D., and A. J. Dunwell. 2000. Excavations of the Gask Frontier Tower and Temporary Camp at Blachhill Wood, Ardoch, Perth and Kinross. *Britannia* 31:255–90.
Göttlicher, A. 1978. *Materialien für ein Corpus der Schiffsmodelle im Altertum*. Mainz.
Goodburn, D. 1991. A Roman timber framed building tradition. *Archaeological Journal* 148:182–204.
———. 1989. Tool marks on the logboat. In "A Late Saxon logboat from Clapton, London Borough of Hackney," edited by P. Marsden. *International Journal of Nautical Archaeology* 18:97–103.
———. 1991. *Wet site excavation and Survey*. Exeter.
Goodman, William Louis. 1964. *The History of Woodworking Tools*. London.
Granger, F. 1932/1998. *Vitruvius, on Architecture*. Loeb Classical Library, Cambridge, Mass., and London.
Greber, J. M. 1956. *Die Geschichte des Hobels*. Zürich.
Greuter (or Gruytere), Janus. 1602–03/1707. *Inscriptiones Antiquae totius orbis Romani*, first published in 2 vols. Heidelberg 1602–03, published in 4 vols. in Amsterdam 1707.
Grodde, B. 1989. *Hölzernes Mobiliar im vor- und frühgeschichtlichen Mittel- und Nordeuropa*. Frankfurt am Main.
Guatoli, M. 1981. Lavinium. *Archeologia Laziale* 4:287–91.
Gummerus, H. 1913. Darstellungen aus dem Handwerk auf römischen Grab- und Votivsteinen in Italien. *JdI* 28:63–125.
Hall, R. A. 1982. 10th Century Woodworking in Coppergate, York. In *Woodworking Techniques before A.D. 1500*, BAR-IS 129, edited by S. McGrail. Oxford.
Hanson, W. S. 1982. Roman Military Timber Buildings: Construction and Reconstruction. In *Woodworking Techniques before A.D. 1500*, BAR-IS 129, edited by S. McGrail. Oxford.
———. 1978. The Organisation of Roman Military Timber Supply. *Britannia* 9:293–305.
Harris, J. 1714. *Lexicon Technicum*. London.
Hawkes, C. F., and M. R. Hull. 1947. *Camulodunum: First Report on the Excavations at Colchester*. Oxford.
Heal, S. 1982. The Wood Age? The Significance of Wood Usage in Pre-Iron Age North-Western Europe. In *Woodworking Techniques before A.D. 1500*, BAR-IS 129, edited by S. McGrail.
Himmelmann, N. 1973. *Typologische Untersuchungen an römischen Sarkophagreliefs des 3. und 4. Jahrhunderts n. Chr.* Mainz.
History of Technology = C. Singer, E. J. Holmyard, A. R. Hall, and T. Williams, eds., *A History of Technology*. Vol. 1: *From Early Times to Fall of Ancient Empires* (Oxford 1954); Vol. 2: *The Mediterranean Civilizations and the Middle Ages* (Oxford 1956)
Hodge, A. T. 1960. *The Woodwork of Greek Roofs*. Cambridge.
———. 1992. *Roman Aqueducts and Water Supply*. London.
Hodges, R., R. Buckley, and A. Sennis. 1994. An early medieval building tradition? A *Pagliaio* at Colli a Volturno. BSR 62:311–20.
Holland, L., and P. Davis. 1934. The Porch Ceiling of the Temple of Apollo on Delos. *AJA* 38:71–80.
Hooper, W. D., and H. B. Ash. 1934/1993. *Marcus Porcius Cato on Agriculture*. Loeb Classical Library, Cambridge, Mass., and London.
Hori, Y. 1993. The Upper Floors in Regio VII Insula 12: The Importance of the Upper Floor in the Study of Town Houses in Pompeii. *Opuscula Pompeiana* 3.
———. 1992. Thresholds in Pompeii. *Opuscula Pompeiana* 2.
Hort, A. 1990. 1916/1990. *Theophrastus: Enquiry into Plants*. 2 vols. Loeb Classical Library, Cambridge, Mass., and London.
Hughes, J. D. 1983. How the Ancients Viewed Deforestation. *JFA* 10:436–45.
Hugot, L. 1982. Ausgraben und Forschungen in Aachen. In *Rheinische Ausgraben*. Cologne.
Izenour, G. C. 1992. *Roofed Theaters of Classical Antiquity*. New Haven.

Jackson, D. A., and T. M. Ambrose. 1976. A Roman Timber Bridge at Aldwincle, Northants. *Britannia* 7:39–72.

Jacobi, G. 1974. *Werkzeug und Gerät aus dem Oppidum von Manching.* Wiesbaden.

Jacobi, H. 1934. Die Ausgrabungen 8: Holz und Leder. *Saalburg Jahrbuch* 6:78–79.

———. 1934. Die Ausgrabungen und Funde der Jahre 1929–33. *Saalburg Jahrbuch* 7:7–28.

Jacobi, L. 1897. *Das Römische Kastell Saalburg.* 2 vols. Saalburg.

Jahn, O. 1870. Über Darstellungen des Handwerks und Handelsverkehrs auf antiken Wandgemälden. *Abhandlungen der königlich sächsischen Gesellschaft der Wissenschaften, Philologisch-Historischen Classe* 5:263–318.

Jankuhn, H., and H. Jannsen. 1981. *Das Handwerk in vor- und frühgeschichtlicher Zeit.* 2 vols. Göttingen.

Jones, H. L. 1927/1988. *The Geography of Strabo.* Loeb Classical Library, Cambridge, Mass., and London.

Jope, Edward Martyn, ed. 1961. *Studies in Building History: Essays in Recognition of the Work of B. H. St. J. O'Neil,* particularly I. Richmond, "Roman Timber Building," 15–26. London.

Jurgeit, F. 1990. Fragmente eines etruskischen Rundthrones in Karlsruhe. RM 97:1–32.

Kampen, N. 1981. *Image and Status: Roman Working Women in Ostia.* Berlin.

Keepax, C., and M. Robson. 1978. Conservation and Associated Examination of a Roman Chest. *The Conservator* 2:35–40.

Kellner, H.-J. 1976. *Die Römer in Bayern.* Munich.

Killen, J. 1980. *Ancient Egyptian Furniture.* Warmington.

Klein, N. L. 1998. Evidence for West Greek Influence on Mainland Greek Roof Construction and the Creation of the Truss in the Archaic Period. *Hesperia* 67:335–74.

Kleiner, D. E. E. 1977. *Roman Group Portraiture: The Funerary Reliefs of the Late Republic and Early Empire.* New York.

———. 1992. *Roman Sculpture.* New Haven.

Kontoleon, A. E. 1890. Epigraphika. *AM* 15:330–37.

La Rocca, E., M. de Vos, and A. de Vos. 1976. *Guida Archeologica di Pompei.* Verona.

Lancaster, L. 2000. Building Trajan's Markets 2: The Construction Process. *AJA* 104:755–85.

Lanciani, R. 1897/1979. *The Ruins and Excavations of Ancient Rome.* New York.

Langsner, D. 1987. *Green Woodworking.* Emmaus, Penn.

Lauffer, S. 1971. *Diokletians Preisedikt.* Berlin.

Lemonnier, P., ed. 1993. *Technological Choices: Arbitraries in Technology from the Neolithic to Modern High Technology.* New York.

Lepper, F., and S. Frere. 1988. *Trajan's Column: A New Edition of the Cichorius Plates.* Gloucester.

Letta, C. 1990. Dormire tra linci e amorini. Scoperta e restauro di un capolavoro dell'artigianato italico: il letto funerario di Amplero. *Archeo* 64:50–57.

Lindsay, W. 1911. *Isidori Hispalensis episcopi Etymologiarvm sive Originvm libri XX.* Oxford.

Ling, R. 1997. *The Insula of the Menander at Pompeii.* Oxford.

Lipke, P. 1984. Retrospective on the Royal Ship of Cheops. In *Sewn Plank Boats: Archaeological and ethnographic papers based on those presented to a conference at Greenwich in November 1984* (National Maritime Museum, Greenwich, Archaeological Series No. 10, BAR-IS 276), edited by S. McGrail and E. Kentley. Oxford.

Littauer, M. A., and J. Crouwel. 1986. The Earliest Known Three-Dimensional Evidence for Spoked Wheels. *American Journal of Archaeology* 90:395–98.

———. The Origin and Diffusion of the Cross-Bar Wheel? *Antiquity* 51:95–105.

Liversidge (1976): see Strong and Brown (1976).

Liversidge, J. 1950. Tables in Roman Britain. *Antiquity* 24:25–29.

———. 1955. *Furniture in Roman Britain.* London.

Long, D. A., K. Steedman, and L. Vere-Stevens. 2002. The Goodmanham Plane: A Unique Roman Plane, of the Fourth Century AD, Discovered in Yorkshire in AD 2000. *Tools and Trades (The Journal of the Tool and Trades History Society)* 13:9–20.

MacDonald, G., and A. Park. 1906. The Roman Forts on the Bar Hill, Dunbartonshire. *Proceedings of the Society of Antiquaries of Scotland* 40:403–546.

MacDonald, W. 1965, 1982. *Architecture of the Roman Empire.* Vols. 1, 2. New Haven.

MacIntosh, J. 1974. Representations of Furniture on the Frieze Plaque from Poggio Civitate. RM 81:15–40.

Maiuri, A. 1958. *Ercolano. I Nuovi Scavi (1927–1958).* Rome.

———. 1933. *La Casa del Menandro e il suo Tesoro di Argenteria.* Rome.

———. 1939. La Croce di Ercolano. *RendPontAcc* 15:193–218.

———. 1931. *La Villa dei Misteri*. Rome.

Malmros, C. Wood Technology in the Viking Age Chamber Grave at Grimstrup, Jutland, Denmark. In *Wood and Archaeology*.

Malten, L. 1912. Hephaistos. *JdI* 27:232–64.

Manning, W. H. 1966. A Group of Bronze Models from Sussex in the British Museum. *Antiquaries Journal* 46:50–59.

———. 1965. A Relief of Two Greek Freedmen. *British Museum Quarterly* 29:25–28.

———. 1975. Roman Military Timber Granaries. *Saalburg Jahrbuch* 32:105–29.

———. 1985. *Catalogue of the Romano-British Iron Tools, Fittings, and Weapons in the British Museum*. London.

Manning, W. H., and I. R. Scott. 1979. Roman Timber Military Gateways in Britain and on the German Frontier. *Britannia* 10:19–62.

Manning, W. H., and C. Saunders. 1972. A Socketed Iron Axe from Maids Moreton, Buckinghamshire, with a Note on the Type. *Antiquaries Journal* 52:276–92.

Mansuelli, G. A. 1967. *Le stele romane del territorio ravennate e del basso Po*. Ravenna.

Marchetti, D. 1891. Di un antico molo per lo sbarco dei marmi reconosciuto sulla riva sinistra del Tevere. *BullCom* ser. 4:45–60.

Marsden, P. 1994. *The Ships of the Port of London: First to Eleventh Centuries* AD (English Heritage Archaeological Report No. 3). London.

Matherat, G. 1936, 1937. Les Ponts-de-Fascines de Jules César à Breuil. *RA* sec. 7 (1936):53–94; 9 (1937):38–67.

Matthäus, H. 1984. Untersuchungen zu Geräte- und Werkzeugformen aus der Umgegung von Pompei: Zu ostmediterran-hellenistischen Traditionen im römischen Handwerk. *BerRGK* 65:73–158.

Mattingly, H., and Edward A. Sydenham. 1926. *Roman Imperial Coinage*. Vol. 2. London.

Mattingly, H. 1923–62. *Coins of the Roman Empire in the British Museum*. 6 vols. London.

Maxwell, G. S. 1976. A Roman Timber Tower at Beattock Summit. *Britannia* 7:33–38.

Mazois, C. F. 1812. *Les ruines de Pompei, dessinées et mesurées par F. Mazois pendant les années 1809, 1810, 1811*. Paris.

McGrail, S., and G. Denford. 1982. Boatbuilding Techniques. In *Woodworking Techniques before A.D. 1500* (BAR-IS 129), edited by S. McGrail. Oxford.

McGrail S., and E. Kentley, eds. 1985. *Sewn Plank Boats: Archaeological and ethnographic papers based on those presented to a conference at Greenwich in November 1984* (National Maritime Museum, Greenwich, Archaeological Series No. 10, BAR-IS 276). Oxford.

McGrail, S., ed. 1982. *Woodworking Techniques before* A.D. *1500: Papers presented to a symposium at Greenwich in September 1980* (BAR-IS 129). Oxford.

McWhirr, A. 1982. *Roman Crafts and Industries* (Shire Archaeology Publication). Aylesbury.

Meiggs, R. 1960. *Roman Ostia*. Oxford.

———. 1980. Sea-Borne Timber Supplies to Rome. *MAAR* 36:185–96.

———. 1982. *Trees and Timber in the Ancient Mediterranean World*. Oxford.

Melville, A. D., trans. 1986. *Ovid: Metamorphoses*. Oxford.

Menotti, F. 2001. *"The Missing Period": Middle Bronze Age Lake-Dwellings in the Alps* (BAR-IS 968). Oxford.

Mensching, E. 1981. Die Koblenzer Rheinbrücke, P. Pomponius Secundus und der Brückenbau an Rhein und Mosel. *Bonner Jahrbücher* 181:325–54.

Mercer, H. C. 1960. *Ancient Carpenters' Tools*. 3d ed. Doyleston.

Middleton, J. H. 1892. *The Remains of Ancient Rome*. London.

Mielke, F. 1977. Treppen in Herculaneum. *AW* 8.3:41–46.

Milne, G. 1982. Recording Timberwork on the London Waterfront. In *Woodworking Techniques before A.D. 1500* (BAR-IS 129), edited by S. McGrail. Oxford.

———. 1985. *The Port of Roman London*. London.

Minto, A. 1953. Decorazione Coroplastica nell'Architettura del Tempio Etrusco. *Studi Etruschi* 22:9–48.

Mollo-Mezzena, R. 1992. Un letto funerario dalla necropoli orientale di Augusta Praetoria (Aosta). In *Bellezza e Lusso: Imagini e Documenti di Piaceri della Vita*, edited by R. Capelli. Rome.

Mols, S. T. 1993. Osservazioni sulla forma e sulla tecnica del mobilio ligneo di Ercolano. In *Ercolano 1738–*

1988: *250 anni di ricerca archeologica.* "*Atti del Convegno Internazionale Ravello-Ercolano-Napoli-Pompeii 1988*, edited by L. Franchi dell'Orto. Rome.
———. 1999. *Wooden Furniture in Herculaneum.* Amsterdam.
Montagni, C. 1993. *Il Legno e il Ferro: Antiche Tecniche Costruttive Liguri.* Genova.
Morey, C. 1959. *The Gold-Glass Collection of the Vatican Library* (Biblioteca Apostolica Vaticano). Vatican City.
Morley, J. 1999. *The History of Furniture.* London.
Morris, P. 1979. *Agricultural Buildings in Roman Britain* (BAR-BS 70). Oxford.
Mosel und Saar = *Die Römer an Mosel und Saar.* 1983. Exhibition catalog. Mainz.
Moss, C. F. Roman Marble Tables. Dissertation, Ann Arbor, 1988.
von Müller, I., ed. 1897. *Atlas zur Archäologie der Kunst.* Munich.
Musty, J., and W. Manning. 1977. A Wooden Chest from the Roman Villa at Bradwell, Milton Keynes, Bucks. *AntJ* 57:330–32.
Mutz, A. 1980. Ein fund von Holzbearbeitungs-Werkzeugen aus Augst Insula 31. In *Jahresberichte aus Augst und Kaiseraugst* (Amt für Museen und Archäologie des Kantons Basel-Landschaft). Liestal.
———. 1976. *Römisches Schmiedehandwerk.* Augst.
Nash-Williams, V. E. 1932. The Roman Legionary Fortress at Caerleon in Monmoutshire: Report of the Excavations Carried out in the Prysg Field 1927–1929 II. *Archaeologia Cambrensis* 87:48–104.
Naso, A. 1996. *Architetture Dipinte: Decorazioni parietali non figurate nelle tombe a camera dell'Etruria meridionale (VII–V sec. a. C.).* Rome.
Neal, D. S., A. Wardle, and J. Hunn. 1990. *Excavation of the Iron Age, Roman and Medieval Settlement at Gorhambury, St. Albans* (English Heritage Archaeological Report 14). London.
Neuburger, A. 1919. Die Bearbeitung des Holzes. In *Die Technik des Altertums.* Leipzig.
Neville, R. C. 1856. Description of a Remarkable Deposit of Roman Antiquities of Iron, Discovered at Great Chesterford, Essex, in 1854. *ArchJ* 13:1–13.
Nicholls, R. V. 1979. A Roman couch in Cambridge. *Archaeologia* 106:1–32.
Nicosia, F. 1969. Selection in *Restauri archeologici: Mostra dei restauri sulle opere d'arte del Museo Archeologico di Firenze danneggiate dall'alluvione del 4 novembre 1966.* Florence.
O'Connor, C. 1993. *Roman Bridges.* Cambridge and New York.
Oleson, J. P. 1978. Technical Aspects of Etruscan Rock-Cut Tomb Architecture. RM 85:283–314.
———, ed. 1989. *The Harbours of Caesarea Maritima: Results of the Caesarea Ancient Harbour Excavation Project, 1980–1985* (vol. 1 by Avner Raban, BAR-IS 491). Oxford.
———. 1982. *The Sources of Innovation in Later Etruscan Tomb Design.* Rome.
———. 1986. *Bronze Age, Greek and Roman Technology. A Select, Annotated Bibliography.* New York.
Orlandos, A. 1966, 1968. *Les matériaux de construction et la technique architecturale des anciens Grecs.* 2 vols. Paris.
———. 1994. *The Wood-Roofed Early Christian Basilica in the Mediterranean Basin* (Library of Athens Archaeological Society, 36). Athens.
Orme, B. 1982. Prehistoric Woodlands and Woodworking in the Somerset Levels. In *Woodworking Techniques before* A.D. *1500* (BAR-IS 129), edited by S. McGrail. Oxford.
Orsi, P. 1891. Priolo. *NSc* 4:359–61.
Painter, K. S. 1961. Roman Iron Implements from London. *British Museum Quarterly* 23:115–18.
Papaccio, V. 1993. Il telaio ligneo (*opus craticium*) ercolanese: Considerazioni e ricerche sui requisiti antisismici. In *Ercolano 1738–1988: 250 anni di ricerca archeologica. Atti del Convegno Internazionale Ravello-Ercolano-Napoli-Pompeii 1988*, edited by L. Franchi dell'Orto. Rome.
Pasquale, G. di. 1994. Studio su un gruppo di compassi romani provenienti da Pompei. *Nuncius. Annali di storia della Scienza* 9:635–44.
Pavolini, C. 1983. *Ostia.* Roma-Bari.
Pearson, C. E., and J. A. Smythe. 1937. Examination of a Roman Chisel from Chesterholm. *Proceedings of the University of Durham Philosophical Society* 9:141–45.
Perring, D., S. Roskams, and P. Allen. 1991. *The Archaeology of Roman London.* Vol. 2 (Council for British Archaeology Research Report 70). London.
Petrie, W. M. F. 1917. *Tools and Weapons Illustrated by the Egyptian Collection in University College, London.* London.

Petrikovitz, H. 1981. Die Spezialisierung des römischen Handwerks. In *Das Handwerk in vor- und frühgeschichtlicher Zeit* 1, edited by H. Jankuhn and H. Jannsen. Göttingen.

Pieresca, G. 1985. *Il Legno e L'Arte di Costruire Mobili e Serramenti.* 5th ed. Milan.

Piggott, S. 1983. *The Earliest Wheeled Transport.* London and New York.

Pigorini, L. 1865. *Le Terramare e le Palafitte del Parmense* (La Civiltá Italiana 10). Rome.

———. 1883. Terramara dell'età del bronzo situata in Castione dei Marchesi (Territorio Parmigiano). *MemLinc* ser. 3, 8:265–318.

Pincelli, R., and C. Morigi Govi. 1975. *La Necropoli Villanoviana di San Vitale* (Istituto per la Storia di Bologna, vol. 1). Bologna.

Pinelli, P., and A. Wasowicz, eds. 1986. *Catalogue des bois et stucs grecs et romains provenant de Kertch.* Paris.

Pitts, L. F., and J. K. St. Joseph. 1985. *Inchtuthil: The Roman Legionary Fortress Excavations 1952–1965* (Britannia Monograph Series No. 6). London.

Pollen, J. H. 1875. *Ancient and Modern Furniture and Woodwork.* London.

Pompei: Abitare sotto il Vesuvio. 1996. Exhibition catalog. Ferrara.

Prayon, F. 1975. *Frühetruskisches Grab- und Hausarchitektur.* Heidelberg.

Puglisi, Salvatore M. 1951. Gli abitatori primitivi del Palatino attraverso le testimonianze archeologiche e le nuove stratigrafiche sul Germalo. *Monumenti Antichi* 41: cols. 1–98.

Pugsley, P. 2003. *Roman Domestic Wood* (BAR-IS 1118). Oxford.

Pulak, C. 1994. 1994 Excavation at Uluburun: The Final Campaign. *INA Quarterly* 21:8–16.

———. 1993. The Shipwreck at Ulu Buran, Turkey: 1992 Excavation Campaign. *INA Quarterly* 19:4–21.

Pyrgi: scavi del santuario etrusco (1959–1967). 1972. Exhibition catalog. Rome.

Quilici, L. 1979. *Roma primitiva e le origini della civiltà Laziale.* Rome.

Rackham, H. 1945/1986. *Pliny: Natural History. Vol 4: Libri XII–XVI.* Loeb Classical Library, Cambridge, Mass., and London.

Rackham, O. 1996. Ecology and Pseudo-Ecology. In *Human Landscapes in Classical Antiquity*, edited by J. Salmon and G. Shipley. London.

Ransom, C. L. 1905. *Studies in Ancient Furniture, Couches and Beds of the Greeks, Etruscans, and Romans.* Chicago.

Ragghianti, C. L. 1963. *Pittori di Pompei.* Milan.

Rasch, J. 1991. Zur Konstruktion spätantiker Kuppeln vom 3. bis 6. Jahrhundert. *JdI* 106:311–83.

Rendeli, M. 1979. *Città aperte: Ambiente e paesaggio rurale organizzato nell'Etruria.* Rome.

Rich, A. 1884. *Dictionary of Roman and Greek Antiquities.* London.

Richardson, L. 1992. *A New Topographical Dictionary of Ancient Rome.* Baltimore.

———. 1988. *Pompeii: An Architectural History.* Baltimore.

Richmond, I. A. 1950. *Archaeology and the After-life in Pagan and Christian Imagery.* London.

———. 1982. *Trajan's Army on Trajan's Column.* London.

Richter, G. M. A. 1926. *Ancient Furniture: A History of Greek, Etruscan and Roman Furniture.* Oxford.

———. 1966. *The Furniture of the Greeks, Etruscans, and Romans.* London.

Rickman, G. 1971. *Roman Granaries and Store Buildings.* Cambridge.

Rieth, A. 1955. Antike Holzgefässe. *AA* 70:1–26.

———. 1940. Das Holzdrechseln. *Ipek: Jahrbuch für prähistorischer und ethnographische Kunst* 13:85–107.

———. 1940. Entwicklung der Dreschseltechnik. *AA* 55:616–34.

———. 1942. *Die Eisentechnik der Hallstattzeit.* Leipzig.

Righini, V. 1965. Forma e struttura delle porte romane: gli esemplari di Sarsina. *Studi Romagnoli* 16:393–418.

Rival, M. 1991. *La charpenterie navale romaine.* Paris.

Robsjohn-Gibbings, T. H. 1963. *Furniture of Classical Greece.* New York.

Rolfe, J. C. 1914/1979. *Suetonius: The Lives of the Caesars.* Loeb Classical Library. Cambridge, Mass., and London.

Rondelet, J. 1814. *Traité théorique et pratique de l'Art de bâtir.* Paris.

Rowland, I. 1999. *Vitruvius. Ten Books on Architecture.* Cambridge.

Rybczynski, W. 2000. *One Good Turn: A Natural History of the Screwdriver and the Screw.* New York.

Ryberg, I. Scott. 1955. *Rites of the State Religion in Roman Art* (MAAR 22). Rome.

Saatmann, K., E. Jüngst, and P. Thielscher. 1938–39. Caesars Rheinbrücke. *BonJbb* 143–44:83–208.

Säflund, G. 1939. *Le Terremare delle Provincie di Modena, Reggio Emilia, Parma, Piacenza* (Acta Instituti Romani Regni Sueciae 7). Lund and Leipzig.

Salmon, J., and G. Shipley. 1996. *Human Landscapes in Classical Antiquity*. London.

Salzman, L. F. 1923. *English Industries of the Middle Ages*. Oxford.

Scarborough, J. 1969. *Roman Medicine*. Ithaca.

Schäfer, T. 1989. *Imperii Insignia: Sella curulis und Fasces* (RM fasc. 29).

Schieferdecker, F.-D. 1981. Brückenbautechnische Überlegungen zur römerzeitlichen Rheinbrücke bei Koblenze. *BonJbb* 181:313–24.

Schindler, R. 1977. *Führer durch das Landesmuseum Trier*. Trier.

Schofield, J., and L. Millar. 1976. New Fresh Wharf: The Roman Waterfront. *London Archaeologist* 1976:390–95.

Schoppa, H. 1951. Römische Holzbrunnen in Köln. *SaalbJb* 10:76–80.

Schweingruber, F. 1982. *Microscopic Wood Anatomy*. Teüfen.

Scrinari, S. M. 1972. *Museo Archeologico di Aquileia: Catalogo delle Sculture Romane*. Rome.

———. 1979. *Le Navi del Porto di Claudio*. Rome.

Sear, F. 1982. *Roman Architecture*. Ithaca.

Selzer, W. 1988. *Römische Steindenkmäler* (Catalog Mainz Museum). Mainz.

Semple, E. C. 1931. *The Geography of the Mediterranean Region: Its Relation to Ancient History*. New York.

Shirley, E. A. M. 2000. *The construction of the Roman legionary fortress at Inchtuthil* (BAR-BS 298). Oxford.

Simpson, E. 1990. Midas' Bed and a Royal Phrygian Burial. *JFA* 17:69–87.

———. 1983. Reconstructing an Ancient Table. *Expedition* 25:11–26.

Simpson, E., and R. Payton. 1986. Royal Wooden Furniture from Gordion. *Archaeology* 39:45–46.

Simpson, E., K. Spirydowicz, and V. Dorge. 1992. *Gordion Wooden Furniture* (Museum of Anatolian Civilization). Ankara.

Smith, C. D. 1996. Where was the wilderness in Roman Times? In *Human Landscapes in Classical Antiquity*, edited by J. Salmon and G. Shipley. London.

Smith, T. 2003. *Vitruvius on Architecture*. New York.

Sokolskii, N. 1969. *Sarcophagi from the northern shore of the Black Sea*. Moscow.

Sommella Mura, A. 1977. L'area sacra di Sant'Omobono: La decorazione architettonica del tempio arcaico. *PP* 32:62–128.

Spinazzola, V. 1953. *Pompei alla luce degli scavi nuovi di Via dell'Abbondanza (1910–1923)*. 2 vols. Rome.

Squarciapino, M. F. 1941. *Artigianato e industria*. Rome.

Stamper, J. W. 2005. *The Architecture of Roman Temples: The Republic to Middle Empire*. Cambridge.

Stefani, E. 1953. Veio. Tempio detto dell'Apollo. Esplorazione e Sistemazione del Santuario. NSc ser. 8, vol. 7:29–112.

St. John Hope, W. H. 1897. Excavations of the Site of the Roman City at Silchester, Hants., in 1896. *Archaeologia* 55:409–30.

Stopponi, S., ed. 1985. *Case e Palazzi d'Etruria*. Milan.

Strack, P. L. 1931. *Untersuchungen zur römischen Reichsprägung des zweiten Jahrhunderts: I: die Reichsprägung zur Zeit des Traian*. Stuttgart.

Stribrny, C. 1987. *Die Herkunft der Römischen Werksteine aus Mainz und Umgebung* (Corpus Signorum Imperii Romani 2.8). Mainz.

Strong, D., and D. Brown, eds. 1976. *Roman Crafts*, especially the chapter by J. Liversidge entitled "Woodwork," 155–66. New York.

Talamo, E. 1993. Ricostruzione di un letto funerario da una tomba dell'Esquilino. *BullCom* 95:285–87.

Tamm, B. 1963. *Auditorium and Palatium: A Study on Assembly-rooms in Roman Palaces during the 1st Century B.C. and the 1st Century A.D.* Stockholm.

Tampone, G. 1989. *Il Restauro del Legno* (Atti del 2° Congresso Nazionale Restauro del Legno). Florence.

———. 1966/2000. *Il Restauro delle Strutture di Legno*. 3d ed. Milan.

Tatton-Brown, T. 1974. Excavations at the Custom House Site, City of London, 1973. *Transactions of the London and Middlesex Archaeological Society* 25:117–219.

Taylor, M. 1992. Flag Fen: The Wood. *Antiquity* 66:476–98.

Taylor, R. 2003. *Roman Builders: A Study in Architectural Process*. Cambridge.
Turcan-Déléani, M. 1958. Les monuments representées sur la Colonne Trajane: schématisme et réalisme. *MÉFR* 70:149–76.
Ucelli, G. 1950. *Le Navi di Nemi*. Rome.
Ulbert, G. 1959. Römische Holzfässer aus Regensburg. *Bayerische Vorgeschictsbilder* 24:6–29.
———. 1959. *Die römische Donau-Kastelle Aislingen und Burghöfe (Limesforschungen 1)*. Berlin.
Ulrich, R. B. 1996. *Contignatio*, Vitruvius, and the Campanian Builder. *AJA* 100:137.
———. 2000. Innovation vs. Tradition: The Roman Vault and the Flat Ceiling. *New England Classical Journal* 27:200–05.
———. 2000. Rediscovering Herculaneum's Wooden Furniture. *JRA* 13:580–83 (review of S. Mols 1999).
Van Giffen. n.d. *De Romeinsche Castella in den Dorpsheuvel Te Valkenburg aan den Rijn-II*. Gronigen.
Vogel, J. C., and H. T. Waterbolk. 1972. Groningen Radiocarbon Dates X. *Radiocarbon* 14:6–110.
Wagner, F. 1973. *Raetia und Noricum* (Corpus Signorum Imperii Romani, Deutschland, 1.1). Bonn.
Walde, A., and J. B. Hofmann. 1982. *Lateinisches Etymologisches Wörterbuch*. Heidelburg.
Ward, J. 1911. *Romano-British Buildings and Earthworks*. London.
Ward-Perkins, J. B. 1976. *Roman Imperial Architecture*. London.
Wasowicz, A., and M. Vaulina. 1974. *Bois grecs et romains de l'Ermitage*. Warsaw.
Watzinger, C. 1905. *Griechische Holzsarkophage aus der Zeit Alexanders des Grossen*. Leipzig.
Weeks, J. 1982. Roman Carpentry Joints: Adoption and Adaptation. In *Woodworking Techniques before A.D. 1500* (BAR-IS 129), edited by S. McGrail. Oxford.
Wiegand, T., ed. 1921–25. *Baalbek: Ergebnisse der Ausgrabungen und Untersuchungen in den Jahren 1898 bis 1905*. 3 vols. Berlin and Leipzig.
———. 1894. Die puteolanische Bauinschrift. *Jahrbücher für classische Philologie. Supplementband* 20:661–775.
Wheeler, R. E. M. 1930. *London in Roman Times* (Museum of London, Catalog no. 13). London.
White, K. D. 1975. *Farm Equipment of the Roman World*. Cambridge.
———. 1984. *Greek and Roman Technology*. Ithaca.
Whitley, J. 2001. *The Archaeology of Ancient Greece*. Cambridge.
Wightman, E. M. 1970. *Roman Trier and the Treveri*. London.
Wilkins, W. 1812. *The Civil Architecture of Vitruvius*. London.
Williams, C. R. 1905. *Studies in ancient furniture: Couches and beds of the Greeks, Etruscans and Romans*. Chicago.
Williams, E., and R. Williams. 1990. *Rough Carpentry Illustrated*. Blue Ridge Summit, Penn.
Wilmott, T. 1982. Excavations at Queen Street, City of London, 1953 and 1960, and Roman Timber-Lined Wells in London. *Transactions of the London and Middlesex Archaeological Society* 33:1–78.
Wood and archaeology = Bois et archéologie: Acts of the European Symposium held at Louvain-la-Neuve, October 1987, edited by T. Hackens, A. V. Munaut, and C. Till. Strasbourg, 1988.
Woytowitsch, E. 1978. *Die Wagen der Bronze- und frühen Eisenzeit in Italien*. Munich.
Young, R. 1974. Phrygian Furniture from Gordion. *Expedition* 16:2–13.
———. 1981. *The Gordion Excavations Final Reports*. Vol. 1: *Three Great Early Tumuli*. Philadelphia.
Zevi, F. 1971. Il tempio del collegio dei 'fabri tignuarii' e una dedica a Pertinace divinizzato. In *Miscellanea ostiense*, III; *RendLinc* 26:472.
Zienkiewicz, J. David. 1986. *The Legionary Fortress Baths at Caerleon*. 2 vols.; vol. 2: *The Finds*. Cardiff.
Zimmer, G. 1982. *Römische Berufsdarstellungen, Archäologische Forschungen* 12. Berlin.
———. 1985. Römische Handwerker. *ANRW* II.12.3:205–28.

Index

www.ingramcontent.com/pod-product-compliance
Lightning Source LLC
Chambersburg PA
CBHW082057090726
47909CB00011B/3067

* 9 7 8 0 3 0 0 1 9 9 8 4 0 *